Zooplankton Diversity and Pelagic Food Webs

Zooplankton Diversity and Pelagic Food Webs: Investigating Present and Past with Different Techniques

Editors

Roberta Piscia
Roberta Bettinetti
Barbara Leoni
Marina Manca

MDPI • Basel • Beijing • Wuhan • Barcelona • Belgrade • Manchester • Tokyo • Cluj • Tianjin

Editors
Roberta Piscia
CNR IRSA
Italy

Roberta Bettinetti
Insubria University
Italy

Barbara Leoni
UNIMIB-DISAT
Italy

Marina Manca
CNR IRSA
Italy

Editorial Office
MDPI
St. Alban-Anlage 66
4052 Basel, Switzerland

This is a reprint of articles from the Special Issue published online in the open access journal *Water* (ISSN 2073-4441) (available at: https://www.mdpi.com/journal/water/special_issues/Zooplankton_Diversity_Pelagic_Food).

For citation purposes, cite each article independently as indicated on the article page online and as indicated below:

LastName, A.A.; LastName, B.B.; LastName, C.C. Article Title. *Journal Name* **Year**, *Article Number*, Page Range.

ISBN 978-3-03943-549-4 (Hbk)
ISBN 978-3-03943-550-0 (PDF)

Cover image courtesy of Andrea Lami.

Contents

About the Editors

Roberta Piscia has a degree in Biological Sciences and PhD in Environmental Sciences and is presently a permanent technician at CNR—Water Research Institute (Verbania), and she has been working at CNR—Institute for Ecosystem Study since 2009. Her fields of expertise include ecology of freshwater zooplankton; analysis of zooplankton resting stages in lacustrine sediments; and analyses of freshwater planktonic food webs and the role of crustacean pelagic zooplankton in the transfer of persistent organic pollutants by means of carbon and nitrogen stable isotope analyses.

Roberta Bettinetti is an ecologist/ecotoxicologist of the water environments. Her interests are mainly based on the interactions between aquatic organisms (both lacustrine and marine) and the environmental risks associated with contaminants.

Barbara Leoni is an Associate Professor in Ecology at the University of Milan Bicocca. Main interests: the ecology of zooplankton, pelagic food-webs, long-term ecological and limnological research on deep lakes, and the impact of anthropogenic pressures (eutrophication, microplastic, climate change) in lentic ecosystems.

Marina Manca is presently Senior Research Associate at CNR IRSA, Verbania, and was previously Senior Researcher at CNR ISE (Institute for Ecosystem Study, formerly "Istituto Italiano di Idrobiologia"). Between 2014 and 2016, she was the director of CNR ISE. Her fields of interest include ecology and population dynamics of freshwater zooplankton; reconstruction of zooplankton communities from fossil remains of lake sediments; functional ecology and long-term changes in freshwater zooplankton communities under human and climate impacts; changes in the diversity of zooplankton communities and the role of lake egg banks; and carbon and nitrogen stable isotope analyses for investigating lacustrine food webs and infra-zooplankton predation.

Preface to "Zooplankton Diversity and Pelagic Food Webs: Investigating Present and Past with Different Techniques"

Zooplankton is of key importance in the structure and functioning of aquatic food webs, providing a large part of the functional and structural biodiversity of predator and prey plankton communities. Promptly responding to long-term and seasonal changes in the physical and chemical environment, zooplankton organisms are sensitive indicators of patterns and mechanisms of impact drivers, both natural and human induced. In this volume, we aim to present evidence for both long-term and seasonal changes in zooplankton community structure and dynamics, investigating different approaches from population dynamics to advanced molecular techniques and reconstructing past communities from subfossil remains in lake sediments.

By applying an innovative statistical technique, Arfè et al. were able to disentangle the long-term impact of changes in trophic status, nutrients supply and climate in a large, deep subalpine lake. Patterns of change in zooplankton and phytoplankton communities at a taxonomic and functional level were addressed, and the role of the two main drivers (climate and trophy) was identified along with different phases of the lake's long-term evolution.

Fundamental ecosystem alteration in high-latitude lakes is of basic importance for holistic understanding of lake responses and resilience to climate warming. In their contribution to this volume, Nevalainen et al. investigated long-term changes in the functioning of a Tundra Lake (Lake Loažžejávri) by means of a paleolimnological approach, based on Cladocera subfossil remain analyses. The study reconstructed taxonomic and functional changes through a sediment core representing 2000-year. Variations in functional diversity were associated with climate oscillations. Higher diversity during warm climate periods suggested the establishment of physical UV refugia for cladocerans by the benthic vegetative substrata under increasing benthic production. A decreasing trend of *Bosmina* sexual reproduction with increasing benthic production also traced a favorable environmental regime since the 17th century.

Changes in planktivory and herbivory regimes over the last 250 years were the object of a paleolimnological study by Carrozzo et al., focused on a shallow lake in Argentina (Lake Blanca Chica). The authors applied Generalized Additive Models (GAM) to time series of fish predation indicators (ephippial abundance and size, micro size, fish scales, and the planktivory index) and pheophorbide α concentration. Profound changes in taxonomic composition and in the size structure of the zooplankton community, and of grazers in particular, resulted in changes in planktivory regime during lake eutrophication. The results of this study provide a reference status for future management strategies of this ecosystem.

High mountain lakes are biodiversity treasures and host endemic taxa that are adapted to live in extreme environments. Among adaptations, production of diapausing eggs allows for harsh conditions to be overcome during the cold season. These sedimentary resting eggs can provide a reservoir of species, thus buffering from extinction risk and biodiversity loss. In the contribution by Piscia et al., the viability of diapausing eggs of Rotifers and Cladocera from sediment cores, representing the last ca. 1100 and 500 years of two lakes in the Khumbu Region of the Himalayas, was tested in the laboratory setting. It was found that only diapausing eggs of the Monogont rotifer *Hexarthra bulgarica nepalensis* were able to hatch, thus suggesting that a permanent egg bank is lacking for the other taxa of the lakes, not least for the two *Daphnia* species described from these sites. The different ability of different taxa to leave viable diapausing eggs in the sediments of high mountain lakes poses serious constraints to the capability of buffering risk of biodiversity loss in these extremely fragile environments.

Body size is a major trait for zooplankton vulnerability to fish predation. The study by Czerniawski and Krepski compared abundance and composition of the zooplankton community in outlet sections of a lake and a waste stabilization pond of a sewage treatment plant. The aim of the study was to determine which zooplankton organisms, among those drifting from the waste stabilization pond or from the lake, were preferentially preyed by roach. Large bodied *Daphnia pulex* was the most abundant in the pond, while small planktonic rotifers prevailed in the lake. The results indicated that fish preferentially fed on zooplankton drifting from the waste stabilization pond than from the lake. The importance of riverine zooplankton in the downstream food web may render these results even more important.

Understanding the impact of environmental drivers on the regional scale on biodiversity is crucial for planning effective biodiversity conservation strategies. Mancinelli et al. used a proxy of phylogenetic diversity of crustacean zooplankton assemblages from 40 ponds and small lakes located in Albania and North Macedonia to assess the level of prediction of changes from waterbodies' landscape characteristics and local bioclimatic conditions. The species–area relationship hypothesis proved inadequate in explaining the diversity of crustacean zooplankton assemblages; in contrast, local biological and climatic factors appeared to influence species richness and composition, but not phylogenetic diversity. The latter is in fact a result of long-term adaptation mechanisms and, as such, is hardly predictable and largely unreplaceable.

Coastal lagoons are very important ecosystems, still poorly studied, particularly with respect to the impact of changes in water quality. In the paper by Oghenekaro and Chigbu, mesozooplankton abundance, diversity and community structure were analyzed in time and space with respect to changes in salinity and temperature. Diversity was higher at sites with high salinity near the Ocean City inlet than at sites near the mouth of tributaries with lower salinity, higher nutrient levels and higher phytoplankton biomass. It was hypothesized that the relatively low salinity and high temperatures in 2013 were responsible for the increase in abundance of hydromedusae, in turn resulting in decreased abundance of bivalve larvae and other taxa, because of increased predation.

Climate change is expected to influence population dynamics of copepods, and effects are likely more pronounced for marine species at high latitudes. The study by Balazy et al. focuses on key zooplankton species of the genus Calanus in the European Arctic. Analyzing changes in body size and phenology of populations from two fjords on the West Spitsbergen Shelf (cold Hornsund vs. warm Kongsfjorden), they found that in Hornsund, the share of *Calanus* in the zooplankton community was greater, and the copepodite structure was progressively older over time, matching the little auks (*Alle alle*; seabird) timing.. The importance of *Calanus* was much lower in Kongsfjorden, and with a prevalence of young copepodites; this result was explained by the Atlantic water advections, which made this area a less favorable feeding ground. Altered age structure towards a domination of young copepodites and the body size reduction of *Calanus* with increasing seawater temperatures was interpreted as resulting in insufficient food availability for these seabirds, a reason for recommending further studies on match/mismatch between *Calanus* and little auks.

Exotic species invasion is recognized as a potential hazard for biodiversity conservation and ecosystem services. Camatti et al. describe the history of the invasion and the present distribution of the copepod *Acartia tonsa* in the Venice lagoon. Temporal trends and changes in coexistence patterns among original *Acartia* species before and after *A. tonsa* settlement and dominance are discussed in their contribution. Disappearance of local species previously dominant and a possible coexistence with the invader are analyzed in detail, highlighting the environmental factors promoting the success of the invasive species. An analysis of spatial distribution of *A. tonsa* revealed significant association with temperature, phytoplankton, particulate organic carbon (POC), chlorophyll α, and counter gradient of salinity. The study overall confirms *A. tonsa* as an opportunistic tolerant species.

Protists, basic components of plankton communities, are the object of the study by Chen et al. Aiming at investigating the impact of sand dust and phosphorus (P) on the microbial food web of the southern Yellow Sea, laboratory experiments investigated changes in the protists community exposed to different concentrations of sand dust and P. The study showed that the effects of the addition of sand dust and phosphorus on size fractions and on autotrophic vs/ mixotrophic protists were similar and dependent on exposure time. Phosphorus addition initially promoted growth of autotrophs, while at the late period of sand dust deposition, small-size heterotrophic protists contributed to the material cycle and food transmission in the ocean. The positive effect of sand dust deposition on autotrophic plankton of the southern Yellow Sea, naturally P deficient, was attributed to a release of P from sand dust. In turn, the release resulted in alteration of the structure of the micro-food web, therefore affecting the ecological function of planktonic protists in the system.

Zooplankton is crucial for the transfer of matter, energy and pollutants through aquatic food webs. Such transfer varies with the season, along with changes in contribution of primary and secondary consumers to zooplankton abundance and standing stock biomass. In the paper by Piscia et al., taxa- and size-specific carbon and nitrogen stable isotope analysis is applied for tracing the path of pollutants through zooplankton along the seasons. Plurennial changes in concentration of polychlorinated biphenyls (PCBs) and dichlorodiphenyltrichloroethane and its relatives (DDTs) were analyzed along with size- and taxa-specific $\delta^{15}N$ signatures in a large deep, subalpine lake. The highly significant correlation between taxa-specific $\delta^{15}N$ and pollutant concentrations, resulting from taxa contribution to standing stock biomass and $\delta^{15}N$ isotopic signatures, proved a basic step in understanding the taxa-specific role in zooplankton enrichment in PCBs and DDTs and the seasonal dynamics of pollutants' transfer to planktivorous fish.

Mining is recognized to deeply influence invertebrate assemblages in aquatic ecosystems. The study by Pociecha et al. used Cladocera fossils in sediment cores to reveal the effects of cessation of mine water discharge on a Cladocera community structure and density. The aquatic system investigated includes the river and associated subsidence ponds in the valley. Some ponds were contaminated during the period of mining, which ceased in 2009, while one of the ponds only appeared after mining had stopped. The concentrations of Zn, Cd, Cu and Pb were much higher in sediments of the ponds formed during peak mining than in the ponds formed after the closure of the mine. Statistical analysis (CCA) showed that *Alonella nana, Alona affinis, Alona* sp. and *Pleuroxus* sp. were strongly correlated with pond age and did not tolerate high concentrations of heavy metals (Cu and Cd). Overall, the study provides evidence that the rate of water exchange by the river flow and the presence of aquatic plants affect species composition more than pond age itself.

The studies provide evidence of the importance of zooplankton in trophic webs of both fresh and salt waters. Providing new useful tools for better manage water ecosystems, these studies also highlight the need to improve our knowledge of zooplankton ecology and factors and mechanisms regulating their relationships with the environment in order to preserve the aquatic ecosystems that are increasingly impacted by human activities in the Anthropocene age we are living in.

Roberta Piscia, Roberta Bettinetti, Barbara Leoni, Marina Manca
Editors

Article

Long-Term Changes in the Zooplankton Community of Lake Maggiore in Response to Multiple Stressors: A Functional Principal Components Analysis

Andrea Arfè [1,*], Piero Quatto [2], Antonella Zambon [3], Hugh J. MacIsaac [4] and Marina Manca [5]

[1] Department of Decision Sciences, Bocconi University, 25, 20100 Milan, Italy
[2] Department of Economics, Quantitative Methods and Business Strategies, University of Milano-Bicocca, 7, 20126 Milan, Italy; piero.quatto@unimib.it
[3] Department of Statistics and Quantitative Methods, University of Milano-Bicocca, 20126 Milan, Italy; antonella.zambon@unimib.it
[4] Great Lakes Institute for Environmental Research, University of Windsor, Windsor, ON N9C 1A2, Canada; hughm@uwindsor.ca
[5] Consiglio Nazionale delle Ricerche, IRSA, 28922 Verbania, Italy; marina.manca@irsa.cnr.it
* Correspondence: andrea.arfe@phd.unibocconi.it

Received: 25 January 2019; Accepted: 25 April 2019; Published: 8 May 2019

Abstract: We describe the long-term (1981–2008) dynamics of several physico-chemical and biological variables and how their changes may have influenced zooplankton structure in Lake Maggiore (Italy). Data was available for the 1981–1992 and 1995–2008 periods. Standardized time-series for temperature and total phosphorus (TP), chlorophyll-α, phytoplankton density (cel m^{-3}), and cell size (μm^3), as well as zooplankton structure (Copepoda, Cladocera, and Rotifera density, ind m^{-3}) were smoothed using penalized B-splines and analyzed using Functional Principal Components (FPCs) to assess their dominant modes of variation. The first four FPCs explained 55% of 1981–1992 and 65% of 1995–2008 overall variation. Results showed that temperature fluctuated during the study period, particularly during 1988–1992 with a general tendency to increase. TP showed a declining trend with some reversions in the pattern observed in the years 1992, 1999, and 2000. Phytoplankton estimators and chlorophyll-α concentration showed a variable trend along the study period. Zooplankton groups also had a variable trend along the study period with a general increase in density of large carnivorous (mainly *Bythotrephes longimanus*) and a decrease of large herbivorous (mainly *Daphnia*), and a similar increase in the ratio of raptorial to microphagous rotifers. Our results suggest that the lake experienced a strong trophic change associated with oligotrophication, followed by pronounced climate-induced changes during the latter period. TP concentration was strongly associated with changes in abundance of some zooplankton taxa.

Keywords: B-Splines smoothing; Functional Data Analysis; limnology; monitoring ecological dynamics; oligotrophication; zooplankton; phytoplankton

---

1. Introduction

Lakes occur at a pivotal position in global landscapes, receiving inputs of both organic and inorganic matter, and generally reflect events occurring in their watersheds. Some of these events stress aquatic life, including soil and water acidification, soil erosion, loss of base cations, release of heavy metals or organic compounds, and application of essential nutrients capable of stimulating primary productivity [1–11]. Superimposed on these changes, climate warming directly impacts lakes via alteration of species' metabolic processes and indirectly by modifying food web interactions [12,13]. The scale of these stressors is influenced by many factors, including loading rate, basin morphometry and depth, and water residence time.

Many stressors interact in a manner that can be difficult to predict [14]. In part, this difficulty stems from the different possible responses by species or entire taxonomic groups to stressors, which may interact additively, synergistically, or antagonistically [2,15]. Mechanistically, negative co-tolerance by a species to interacting stressors yields either additive or synergistic responses with amplified consequences, whereas positive co-tolerance produces antagonistic responses with damped consequences [2]. The onset and termination of stressors often vary temporally, thus it may be possible to use long-term dynamics of lake food webs to discern changes owing to individual stressors or to interactions between them [16]. In a meta-analysis, Jackson et al. [15] determined that warming and nitrification resulted in additive net effects in freshwater systems, while warming and any other stressor was antagonistic.

While excellent long-term analyses exist for individual stressors, such as cultural eutrophication [17], there exists a relative dearth of similar studies to disentangle individual effects in systems subjected to multiple stresses. Long-term studies are essential to fully understand responses by lake ecosystems to perturbations. Schindler [18] warned against extrapolation of short-term or small-scale studies to reveal processes important to cultural eutrophication, noting that only long-term, whole-lake experiments and case histories of recovered lakes can reliably inform abatement policy. Despite this and other calls for long-term studies of lakes [6], there remains a surprising paucity of such studies (but see, for example, References [9–22]).

The objective of this study was to analyze the long-term dynamics of several physico-chemical and biological variables of Lake Maggiore to identify major trends and how this variation impacts zooplankton abundance and community structure. Lake Maggiore is a large, deep lake in the subalpine area of northern Italy, which sustained a wide-ranging series of major perturbations during the 20th century. These changes included cultural eutrophication, organic chemical contamination, commercial fishing, climate warming, and species introductions. To characterize the variation in time of several biological and physico-chemical parameters of Lake Maggiore, we adopted a Functional Data Analysis (FDA) approach [23] based on penalized B-spline smoothing and Functional Principal Components Analysis (FPCA). This approach, seldom considered within the ecological literature, has several features that make it especially suited for the analysis of limnological data.

2. Materials and Methods

2.1. Study Area

Lake Maggiore is an oligomictic, subalpine (194 m above sea level) lake, largely contained within Northern Italy but shared in the Northern end of the basin with Switzerland. The lake has a surface area 212.5 km^2 and a maximum (mean) depth of 370 m (177 m), a product of tectonic-glacial activity. Lake Maggiore is one of the best-studied lakes in the world, with a long-term record of physico-chemical and phytoplankton and zooplankton data dating back to the mid- or even early 1900s for some variables [24]. The drainage basin (including the lake area) covers 6599 km^2, yielding a drainage basin/lake area ratio of 31:1. This high ratio along with unusually steep hillslopes substantially affect the lake's hydrology and many environmental variables. Geological features of the watershed have been influenced by alpine orogenesis and glaciation, and thus are complex. The major topographic feature is a narrow steepsided valley through the Piedmont and Lombardy regions of Italy.

As with many lakes in Europe, P-limited, oligotrophic, Lake Maggiore experienced cultural eutrophication from the late 1960s throughout the 1970s; peak total phosphorus (TP) and chlorophyll-α concentrations were detected at the end of 1970s, with values as high as 40 μg m^{-3} of total phosphorus in-lake concentration at winter mixing (TP$_{mix}$) of 1979, attesting to mesotrophic conditions of the lake [25]. Gradual implementation of sewage treatment plants, and a reduction of the phosphorus content in detergents, led to a gradual decline in TP and eutrophication reversal. A time lag was reported in the response of plankton communities to TP decline: as expected, changes in zooplankton and phytoplankton became evident only after TPmix concentration declined below a threshold of 15 mg m^{-3} in 1988 [26]. According to Sas [27], this is the level at which the total phosphorus becomes limiting to phytoplankton growth. The trophic evolution of the lake is most strictly represented by Cladocera rather than copepods [28,29].

2.2. Data Collection

In this study, we describe dynamics of biological and physico-chemical variables of Lake Maggiore's ecosystem for the periods 1981–1992 and (separately) 1995–2008. Specifically, we considered physical and chemical variables, i.e., water temperature and total phosphorus, as well as biological variables, i.e., phytoplankton and zooplankton population density, biomass, and cell size (measured by cellular volume).

Data for both periods were collected as part of the long-term monitoring of Lake Maggiore, funded since the late 1970s by the International Commission for the Protection of Swiss-Italian Waters (Commissione Internazionale per la Protezione delle Acque Italo-Svizzere, CIPAIS) and published in annual reports (available at www.cipais.org). Phytoplankton and zooplankton measurements were also obtained from the "Plankton and the pelagic food web" research project funded by the National Research Council (CNR). All measurements refer to a single sampling station (45°58′30″ N; 8°39′09″ E) at the lake's maximum depth, which is representative of the pelagic environment [30,31].

In more detail, monthly data on temperature at 0–20 m depth, the water layer representative of the euphotic zone, were based on annual reports of the CNR-ISE meteorological station [32–37]. Total phosphorus concentration was based on a maximum of 13 measures at different depths and was included as indicative of changes in lake trophic status. Data on phytoplankton density and cell size as well as chlorophyll-α were based on integrated phytoplankton samples collected every other week with a 1.5 L van Dorn bottle within the 0–20 m layer. Samples for chlorophyll analysis were filtered through GF/C glass fibre filters (about 1μm pore size). The filters were then stored on silica gel at −20 °C. After about 2–3 weeks the filters were mechanically ground and transferred to acetone 90%. The absorbance of the pigments was measured in a spectrophotometer (Perkin-Elmer Lambda 6); the chlorophyll-α and phaeophytin concentrations were calculated in accordance with Lorenzen [38]. Phytoplankton samples were preserved with acetic Lugol solution and counting was made following the Utermöhl method [39] at 400x in an inverted microscope [40]. Species identification and nomenclature followed the more recent monographs of the series Sußwasserflora von Mitteleuropa, established by A. Pascher (Gustav Fisher Verlag, and Elsevier, Spectrum Akademischer Verlag), specific manuals of the series Das Phytoplankton des Sußwassers, established by G. Huber-Pestalozzi (E. Schweizerbart'sche Verlagsbuchhandlung), and specific papers [41]. Zooplankton population abundance data (*Bythotrephes longimanus*, Leydig 1860; *Eubosmima longispina*, Leydig 1860; cyclopoid copepods, *Daphnia longispina-galeata*, Wagler, 1937; *Diaphanosoma brachyurum*, Liévin 1848; and *Leptodora kindtii*. Focke 1844; and Rotifera) were obtained monthly with a 76-μm nylon net Clarke-Bumpus plankton sampler, towed at a constant speed of ca 3 km h^{-1}, along sinusoidal hauls from the surface to 50 m depth, i.e., in the water layer where zooplankton live [30]. Samples included at least 1000 L of water and were fixed in 4% buffered formaldehyde before counting and identifying the genus or species under a microscope at 6.3x. Taxa identification was based on volumes of the Identification Guides to the Plankton and Benthos of Inland Waters (formerly "Guides to the Identification of the Microinvertebrates of the Continental Waters of the World", H. J. F. Dumont coordinating Editor. Backhuys publisher). Further details on data collection are reported elsewhere [29,31].

2.3. Smoothing Data through Penalized B-Spline Expansions

Although biological and physico-chemical parameters are typically measured at discrete instants, their temporal dynamics can be better represented as a smooth function varying in some continuous time interval $[a, b]$. These smooth representations are of main interest in FDA, where they are typically recovered by penalized splines approximations. Briefly, a *spline basis* is a set of known functions, which can be used to approximate any other function with arbitrary precision. One such set frequently used in FDA consists of smoothly-joined piecewise polynomial functions called *B-splines*. These allow researchers to represent any collection of m time series $s_i(t_{ik})$ (indexed by $i = 1,\dots,m$ and observed at distinct time points $t_{i1},\dots,t_{iK_i}$) with smooth functions $x_i(t)$ of the form

$$x_i(t) = \sum_{j=1}^{B} c_{ij} b_j(t), \tag{1}$$

where t belongs to the time-interval $[a, b]$ and B is the considered number of B-splines $b_j(t)$. The number B is uniquely determined by the polynomial degree and the chosen number of *knots*, i.e. the points where the different polynomial pieces are joined [23]. The coefficients c_{ij}, which provide the best smooth approximation of the observed time-series, are obtained by minimizing the *Penalized Sum-of-Squares Error (PSSE)*

$$PSSE_i = \sum_{k=1}^{K_i} [s_i(t_{ik}) - x_i(t_{ik})]^2 + \lambda_i \int_a^b \left[\frac{d^2 x_i(t)}{dt^2} \right]^2 dt. \tag{2}$$

$PSSE_i$ is the sum of two terms: the first represents the sum of squared approximation errors, which decreases as the B-spline approximation better fits the data; the second is linked to the so-called *"strain energy"* (much like stretched elastic bands, the more the curve $x_i(t)$ is "wiggly", the higher its strain energy). Hence, minimizing $PSSE_i$ results in a B-spline approximation whose fit to the data is a compromise between low approximation error and "wiggliness" of the resulting curve (thus avoiding overfitting). This trade-off is controlled by the *penalization coefficient* $\lambda_i \geq 0$, which is usually chosen by the generalized cross validation method [23].

2.4. Functional Principal Components Analysis (FPCA)

FPCA is a generalization of standard Principal Components Analysis (PCA) to the situation where data is represented by smooth continuous curves (possibly obtained by penalization techniques), as in our setting [23]. Specifically, *Functional Principal Components (FPCs)* are square-integrable orthonormal functions $e_j(t)$ that best represent, in a least-squares sense (see the Appendix A), the m smooth curves $x_i(t)$ by means of expansions such as

$$x_i(t) \cong \overline{x}(t) + \sum_{j=1}^{n} f_{ij} e_j(t), \tag{3}$$

where $\overline{x}(t) = \frac{1}{m} \sum_{i=1}^{m} x_i(t)$ is the mean function, n is the number of FPCs $e_j(t)$, and f_{ij} is the *(FPC) score* for the i-th time series with respect to the j-th FPC [23]. As in standard PCA, FPCs are chosen so as to capture the greatest possible variation in the observed data. More precisely, each FPC $e_j(t)$ (with $j = 1,\dots,n$) is chosen in order to maximize the corresponding explained variation, i.e., the variance σ_j^2 of the m values $f_{1j}, \dots, f_{mj}$, with $\sigma_1^2 \geq \sigma_2^2 \geq \cdots \geq \sigma_n^2$. Thus, for example, the first and the second FPC $e_1(t)$ and $e_2(t)$ represent the mode of variation from the overall mean $\overline{x}(t)$ associated with, respectively, the greatest and the second greatest possible variation between the curves $x_1(t), \dots x_m(t)$. On the other hand, the score f_{ij} measures how well the j-th FPC captures variation over time of the i-th time series (scores of higher absolute magnitudes are associated to the FPCs capturing the greater amount of variation). In more detail, a small f_{ij} score signifies that the j-th FPC does not capture the time variation of the i-th time series. Instead, a high positive (negative) f_{ij} score signifies that the time variation of the i-th time series is captured by the same (opposite) trend than the j-th FPC. For example, if the j-th FPC shows an increasing global trend, then a positive (negative) f_{ij} means that the i-th time series increases (decreases) over time.

When the FPCs have been identified, their interpretation can be aided by plotting the two curves obtained by adding and subtracting from the overall mean $\overline{x}(t)$ the *scaled FPCs*, i.e., the curves $\overline{x}(t) \pm 1.5 \sigma_j e_j(t)$ for each FPC $e_j(t)$. Furthermore, the VARIMAX strategy (commonly used in standard

PCA) can be generalized to the FPCA setting. According to this strategy, the FPCs are rotated so that each component is associated to only a small number of high scores. Rotating the FPCs according to this strategy can therefore aid their interpretation [23].

2.5. Describing Lake Maggiore's Dynamic by Functional Data Analysis

The above-described data analysis methods were implemented to assess long-term temporal dynamics of Lake Maggiore as follows. First, the time-series for the considered variables were smoothed by means of penalized B-splines. Uniformly spaced knots (one about every 30 days) were considered for both the 1981–1992 or 1995–2008 periods, in each case for a total of B = 200 B-splines. This choice ensured that smoothed curves had enough flexibility to represent month-level changes in the underlying variables. Second, FPC analysis was performed on the B-spline smoothed time series of the standardized variables. Specifically, each time series $s_i(t_{ik})$ was standardized as $(s_i(t_{ik}) - \bar{s}_i)/\tau_i$ where $\bar{s}_i$ and τ_i are respectively the mean and standard deviation of the K_i observations $s_i(t_{i1}), \ldots, s_i(t_{iK_i})$. Standardization ensured time-series were expressed on the same dimensionless scale, allowing direct comparisons between them. Smoothed standardized time-series (using B = 200 B-splines) were then used to extract VARIMAX-rotated FPCs. All components that explained at least 10% of total variability of the smoothed standardized time series, i.e., the sum of the σ_j^2 [23], were extracted. All analyses were performed in the statistical software R (version 3.0) using the "fda" library [23].

3. Results

3.1. Long-Term Limnological Change

Thermal conditions changes in Lake Maggiore over the study period. Mean summer temperature across the 0–20 m depth averaged 21.2 °C between 1981 and 1992 and 21.9 °C between 1995 and 2008, as also observed in previous studies [42]. Lake Maggiore underwent a gradual warming of the epilimnion in the decade 1988–1998, according to the trend of increasing heat content of the deep Italian lakes, pointed out by Carrara et al. [43]. The period was characterized by strong fluctuations, in particular between 1988 and 1992. Since 1999, the increase of water temperature was less evident, probably because of the occurrence of particular hydro-meteorological mechanisms reducing the heat content of the water column [43,44]. Lake Maggiore also experienced a dramatic reduction in total phosphorus (TP) concentration over the study period. Mean TP_{mix} concentration declined from 10.9 µg L^{-1} between 1981 and 1992, to 8.4 µg L^{-1} between 1995 and 2008. An exceptionally high value, of 17 µg L^{-1}, was detected in 1991, after the lake's complete winter overturn. Over a general increase, a second peak value was recorded in 1999, after another complete overturn. The overall increase to stable values of 11 µg L^{-1} after 2000 could be mainly explained by meteorological-climatic conditions, rather than by an eutrophication reversal [45]. Eutrophication abatement results were mixed, however, as mean chlorophyll-a concentration fell sharply (from 4.7 to 3.4 µg L^{-1}) between these periods, though results for phytoplankton biomass were less clear (1830 vs. 1181 mm^3 m^{-3}, respectively).

3.2. Description of the Smoothed Time-Series

The observed and smoothed time-series of the considered variables are represented in Figure 2. A visual inspection of the graphs in Figure 2 suggests that the smoothed curves have a good fit to the overall trend of the corresponding variables. Major variations in phosphorus concentration, phytoplankton cell size and chlorophyll-a concentration during 1981–1988 matched an increase in variability and mean density of *Bythotrephes longimanus*, and a decrease in between-year variability and mean annual population densities of *Bosmina* (mainly, *Eubosmina longispina*) and *Daphnia* (*D. longispina-galeata* gr.). The increased within-year variability in *Bythotrephes* observed during the 1990s was maintained across the second time period, whereas *Daphnia* and *Bosmina* variability and population density increased between 2002–2008, along with *Diaphanosoma brachyurum*. During 2002–2008, variability of diaptomid copepods seems to have decreased slightly relative to the periods

1981–1992 and 1995–2000. Rotifer (and to lesser extent) *Leptodora kindtii* density were characterized by increased variability during the last five years of the 1995-2008 period. In addition, rotifer communities changed dramatically through time, with the mean ratio of raptorial to microphagous species abundance decreasing from 1.5 ± 2.0 (mean ± st.dev.) during 1981–1992 to 0.5 ± 0.6 (mean ± st.dev.) during 1995–2002 [42,46].

Phytoplankton density was highly variable during the 1996–1998 interval. Phosphorus concentration, phytoplankton cell size, chlorophyll-*α* concentration, *Bosmina* and *Daphnia* population density all exhibited a more variable trend during 1981–1992 than during 1995–2008 using both observed and smoothed time-series data (Figure 2). Conversely, phytoplankton density, *Bythothrephes*, cyclopoid copepods, and *Diaphanosoma* population density were more variable between 1995–2008.

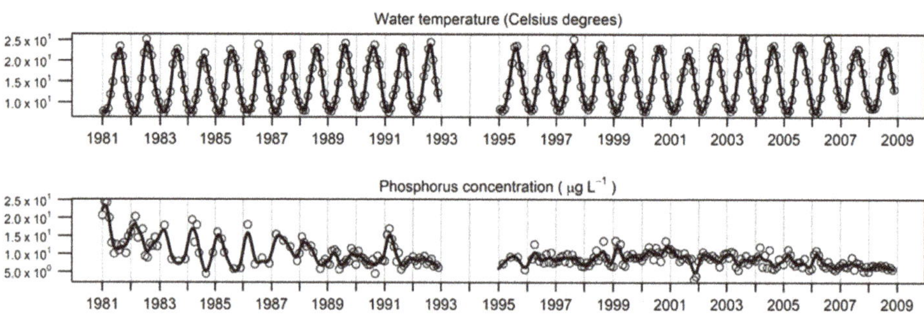

Panel (**A**): Physico-chemical variables.

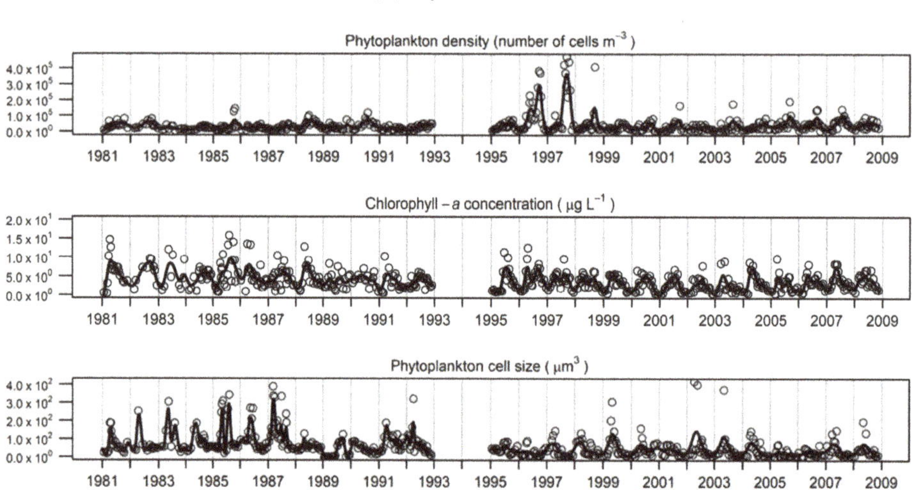

Panel (**B**): Phytoplankton-related variables.

Figure 1. *Cont.*

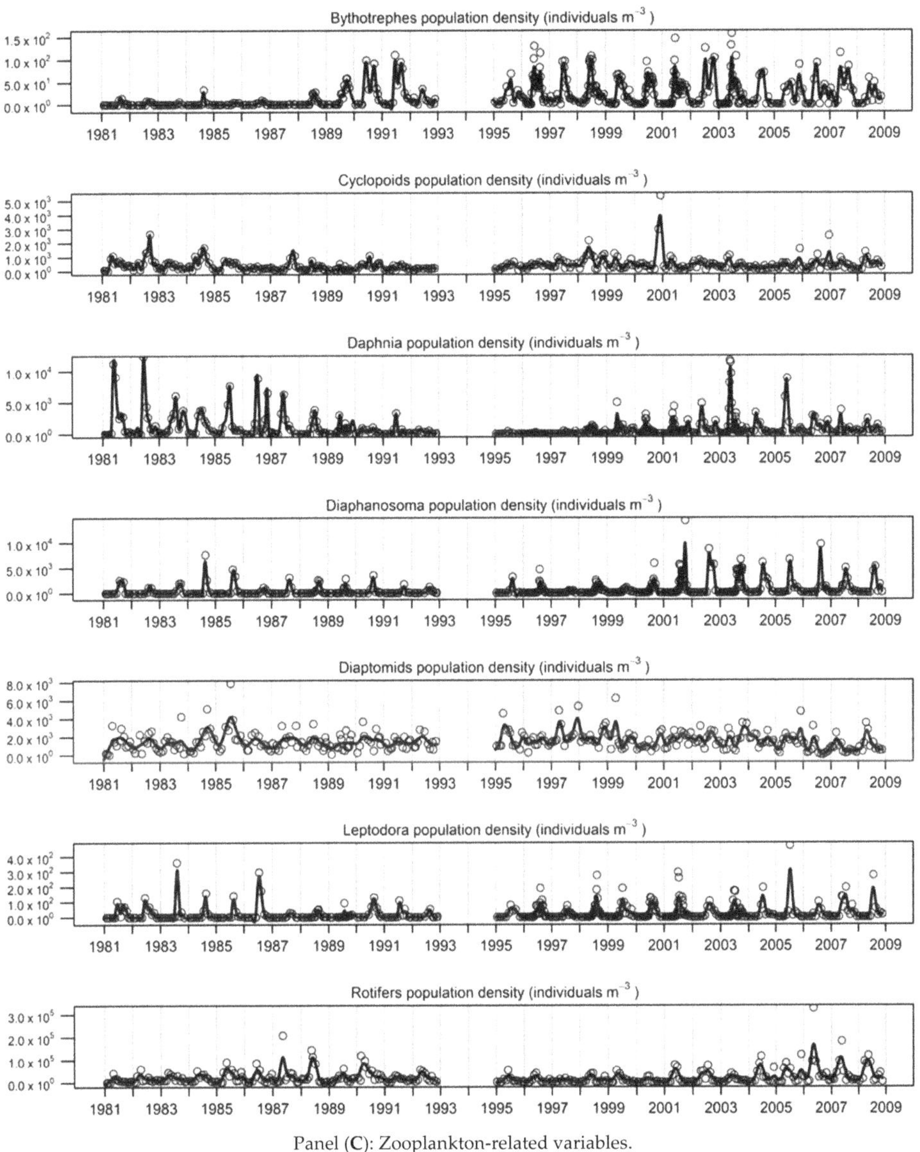

Panel (**C**): Zooplankton-related variables.

Figure 1. Observed and smoothed time series for all physico-chemical and biological variables. Continuous line = smoothed curve; circles = original data points. Panel (**A**): Physico-chemical variables. Panel (**B**): Phytoplankton-related variables. Panel (**C**): Zooplankton-related variables. Data were not available for 1993-1995. Points are monthly means, lines are the smoothed value.

3.3. Extracted Functional Principal Components

For both periods, the first four rotated FPCs were retained, and explained 55% and 65% of the overall variation of smoothed standardized variables for 1981–1992 and 1995–2008 periods, respectively. The effect of each FPC (denoted FPC1-4 in decreasing order of explained variation for both periods) on the trend of considered variables is represented in Figure 2. This shows, for the two periods, the overall mean (continuous line) of the smoothed standardized variables, along with curves representing the effect of adding (dashed line) or subtracting (dotted line) each scaled FPC on the mean. The FPC1 corresponds to an overall increasing trend in both the 1981–1992 and 1995–2008 periods, as in each case the dotted (increasing) and dashed (decreasing) lines cross the graph of the overall mean as time increases (Figure 2a). In addition, the FPC1 for the 1981–1992 period also corresponds to more extreme of mid-year values during the 1990s. Similarly, the FPC1 for 1995–2008 corresponds to more extreme mid-year values during the second half of the 2000s, as well as very extreme values at the end of 2000. Fluctuations around the overall mean for FPC1 mirror the high variability of phosphorus concentration and of *Bythotrephes* density during 1981–1993.

Dynamics of the other principal components (FPC2-4) are highlighted in Figure 2b,c. Briefly, FPC2 corresponds to more extreme values during the 1980s (especially in 1981, 1982, 1983 and 1986) and during the 2000s. Strong deviations observed during the 1980s can be attributed to fluctuations of two parameters with the highest scores on this component: The density of *Bosmina* and the concentration of chlorophyll-α. FPC3 gives more weight to 1982 and 1984, as well as 1996, 1997, 2003, and 2005. The high variability, in some of these years, of chlorophyll concentration and *Daphnia* density explains much of the observed fluctuations around the overall mean. Lastly, FPC4 gives more weight to 1985 and the 2000s, especially 2005. The observed perturbations from the mean during the 2000s can be explained by variation in *Leptodora* population density.

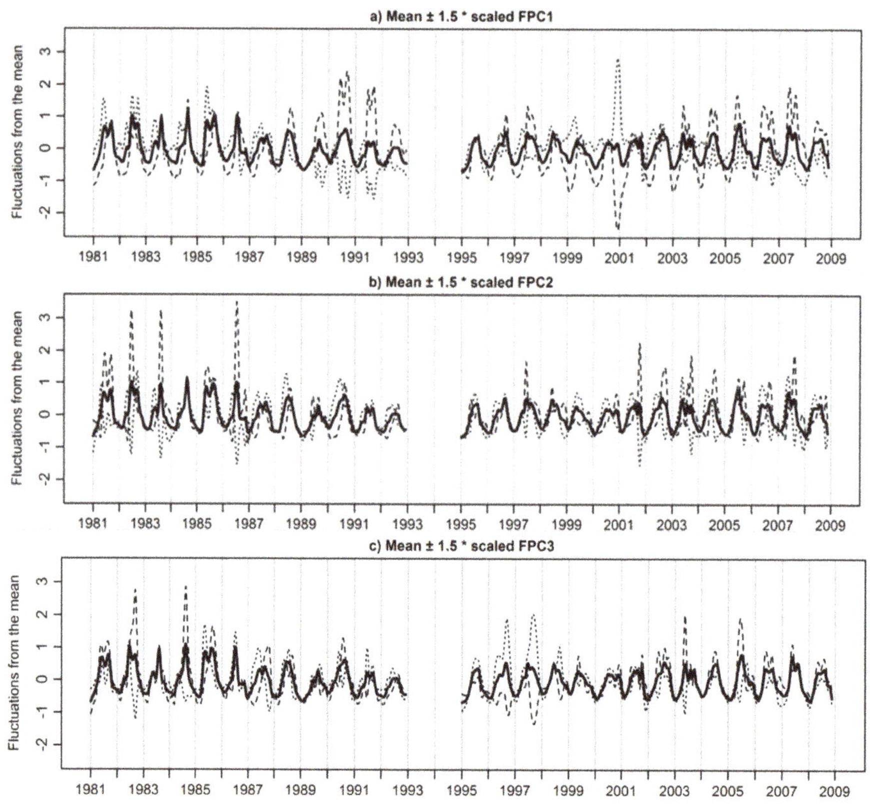

Figure 2. *Cont.*

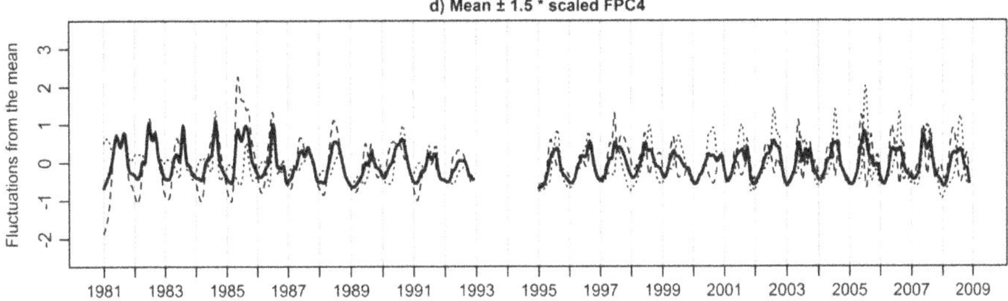

Figure 2. Temporal fluctuations from the mean of all standardized functional variables (solid line) obtained by adding (dashed line) or subtracting (dotted line) the extracted FPCs, each scaled by 1.5 times the square root of its explained variance (panels: **a**, FPC1; **b**, FPC2; **c**, FPC3; and **d**, FPC4).

3.4. Functional Principal Components Scores

The scores for the FPC1 show how, during 1981–1992, *Bythotrephes* population density (score 48.2) increased dramatically during the 1990s, while phosphorus concentration decreased over time (score −37.8) (Table 1). Phosphorus concentration maintained a similar trend during 1995–2008 (score −45.2), while water temperature increased during both considered periods (score of 30.5 in 1981–1992 and of 30.2 in 1995–2008), fitting FPC1 better than *Bythotrephes* density (score 19.8) (Table 1).

Table 1. Functional Principal Components (FPCs) scores for three different groups (in italics) of considered variables. The most extreme positive and negative scores are highlighted (italics) for each FPC. Pop. dens. = population density.

	Functional Principal Components (FPCs) Scores							
Calendar period	1981–1992				1995–2008			
Variables	FPC1	FPC2	FPC3	FPC4	FPC1	FPC2	FPC3	FPC4
Physico-Chemical								
Water temperature	30.5	11.9	16.7	13.1	*30.2*	10.9	−12.6	−25.7
Phosphorus concentration	−37.8	−16.6	−25.9	−46.9	−45.2	−15.2	−14.2	−2.0
Phytoplankton-related								
Phytoplankton pop. dens.	13.7	−10.6	12.6	−3.8	8.6	1.1	−39.6	5.5
Chlorophyll-α concentration	−11.2	−9.7	3.8	8.6	12.6	−29.1	−16.5	1.9
Phytoplankton cell size	−20.8	−18.7	−21.0	3.9	−17.0	−19.5	12.1	10.4
Zooplankton-related								
Bosmina pop. dens.	−13.0	*37.5*	−17.7	26.0	20.6	*46.1*	1.4	20.0
Bythotrephes pop. dens.	*48.2*	−17.2	−14.7	−17.1	19.8	1.0	−1.8	−16.6
Cyclopoid pop. dens.	−16.0	−10.6	*41.8*	0.1	−38.5	1.4	7.6	15.9
Daphnia pop. dens.	−14.0	34.6	−4.6	7.9	9.2	−15.0	33.8	17.9
Diaphanosoma pop. dens.	12.4	−2.1	24.8	−9.7	4.3	34.0	10.2	−32.9
Diaptomid pop. dens.	−1.3	−9.1	2.9	13.8	−28.1	5.4	−3.0	24.7
Leptodora pop. dens.	13.3	31.8	4.6	−12.2	11.4	0.7	4.7	−33.5
Rotifer pop. dens.	−4.1	−21.1	−23.2	16.3	12.0	−21.8	18.0	14.4

Scores for FPC2 during 1981–1992 show how this principal component had the most effect on *Bosmina* density (score 37.5), which reached higher values at the start of 1980 (Figure 2c). By contrast, rotifer density (score −21.1) was stable at the start of the 1980s and more variable at the end of the decade (Table 1). During 1995-2008, FPC2 was affected in a similar way by both *Bosmina* density (score 46.1) and rotifer density (score −21.8). Chlorophyll-α concentration (score −29.1), which reaches its lowest values in 2001 (Figure 2b), also becomes important for FPC2 in this period.

The scores for FPC3 during 1981–1992 show that this FPC were most affected by cyclopoid copepod density (score 41.8), which reached high values in both 1982 and 1984 (Figure 2c), and phosphorus (score −25.9) which reached a minimum concentration in 1984 as oligotrophication proceeded (Figure 2a). Between 1995–2008, FPC3 was most affected by *Daphnia* density (score 33.8), which reached high values in 2003 and 2005 (Figure 2b), and phytoplankton density (score −39.6), which had very high values in 1996–1997 (Figure 2b).

Lastly, in 1981–1992, FPC4 was most affected by phosphorus concentration (score −46.9), which had a minimum in 1985 (Figure 2a), and *Bosmina* density (score 26.0), which reached one of its highest values in the same year (Figure 2c). During 1995–2008, FPC4 was most affected by diaptomid copepod density (score 24.7), whose within-year dynamic changed substantially during the 2000s (Figure 2c), and *Leptodora* density (score −33.5), which reached high values in the summer 2005 (Figure 2c).

4. Discussion

We used Functional Data Analysis to describe long-term abiotic and biotic dynamics of Lake Maggiore. FDA has features that made it ideal for long-term studies like ours. FDA provides a useful framework to analyzing short and long-term dynamics of a lake ecosystem, its data smoothing respects the continuous nature of the underlying biological and physico-chemical processes while accounting for potential irregularities during measurement periods, and it allows analysis of dominant modes of variation in trajectories of measured parameters in a way that respects the time ordering of observations.

FPCA proved useful in identifying Lake Maggiore's responses to both trophic and climatic variability. Among the extracted FPCs, FPC1 illustrated the increasing trend of *Bythotrephes* population density and decreasing trend of phosphorus concentration at the beginning of oligotrophication, between 1981–1992. In 1996 and 1997, large blooms of ultraplankton—including small cyanobacteria—occurred in the lake, coincident with very low mean densities of *Daphnia*. However, as lake productivity declined owing to phosphorus abatement [25–27,42] below 15 mg m^{-3}, the level at which TP mix becomes limiting for phytoplankton growth [27], *Bythotrephes* increased in density, while *Daphnia* declined [47,48]. A previous study has demonstrated that *Bythotrephes* occurs primarily in European lakes with low summer chlorophyll concentration [49], consistent with its resurgence during oligotrophication in Lake Maggiore. The decline in *Daphnia* (see Table 1) could be due to either reduced food supply (bottom-up response) or to increased predation by *Bythotrephes* (top-down response), or both. Available evidence suggests that food reduction may have caused the decline between 1983 and 1987, as declining phytoplankton stocks may have caused food limitation. However, beyond this point, predation by *Bythotrephes* likely accelerated the decline in *Daphnia* density, as the latter's death rate increased when *Bythotrephes* populations surged beyond 20 ind·m^{-3} beginning in 1987 [47]. Larger and seemingly more potent *Bythotrephes* sharply curtail crustacean abundance when present at levels beyond ~2 ind·m^{-3} in Ontario lakes [50]. Interestingly, *Leptodora* and *Bythotrephes* have inverse distributions in Ontario lakes with the former often replaced by the latter after it invades new systems [51]. These species show no such pattern in Lake Maggiore, where both species are native and presumably co-adapted to one another's presence. In addition, body size differences between the species are much smaller than in Ontario lakes, suggesting less physical dominance by *Bythotrephes*.

In addition to dramatic changes in population density, *Bythotrephes* has also experienced large shifts in seasonality, with peak density shifting from August to May (see Figure 2c) as the lake warmed [31,36]. This, in turn, may have increased predation pressure on *Daphnia* [47,48], which previously typically achieved population maxima in July but now peaks months earlier [52]. The second phase of *Daphnia* increase, during full and stable oligotrophy, resulted both from *Daphnia's* ability to cope with increased invertebrate predation by means of changes in population phenology and by a release from fish predation consequent to a decrease in coregonid (i.e., whitefish) abundance, which also favors consumption of *Bythotrephes* over *Daphnia* [53,54]. Previous research has demonstrated strong preference of coregonid fishes for *Bythotrephes* prey [55].

Oligotrophication was associated with a decline in phytoplankton density between 1995–2008, as clearly shown by both FPC2 and FPC3. However, this reduction in plankton food resources does not seem to have affected most cladocerans as FPC2 highlighted an increase of *Bosmina* whereas FPC3 captured the same trend for *Daphnia*. We hypothesize that the long-term increase in water temperature stimulated the growth of *Daphnia*—as demonstrated by peak population density achieved during the heat wave of summer 2003—possibly owing to reduced development time [56,57].

FPC analyses are compatible with a response by planktonic communities to both temperature increase and total phosphorus concentration decline. The former variable enhances primary production with its consequent indirect, positive effect on herbivores and negative effect on *Bythotrephes*, whereas the latter appeared to favor *Bythotrephes* and its strong top-down effects.

Raptorial dominance declined after 1988 concomitant with changes in cladocerans: *Bythotrephes longimanus* increased by an order of magnitude, *Daphnia* gr. *hyalina-galeata* showed a sharp decline and small cladocerans such as *Eubosmina longispina* and *Diaphanosoma brachyurum* increased [26,29]. An increase in *B. longimanus* releases rotifers, especially large *Conochilus hippocrepis* (or *C. volvox ehrbg*, Schrank, 1803) colonies, from competition with cladocerans [58]. The decrease in abundance and mean size of phytoplankton cells with lake restoration might have enhanced microphagous species able to ingest multiple food items of ~15–20 μm, while raptorials prefer larger particles [46,59–61]. Decreased abundance of competitors and decreased food size, therefore, might enhance microphagous species, while raptorial species were able to coexist with their competitors [46,59,62]. Large (≥1 mm diameter) *Conochilus* colonies have feeding rates comparable to *Daphnia* and lead to a seasonal increase in water clarity like that caused by *Daphnia* grazing [58].

The response of rotifers to changes in the macrozooplankton assemblage is probably more complex than previously hypothesized. Manca [63] reported a strong increase of *Conochilus hippocrepis* in Lake Maggiore when *Bythotrephes* became increasingly abundant in the late 1980s. However, the FPCA highlighted that the increase in rotifer density preceded that of *Bythotrephes* and the two populations seemed to follow divergent development trajectories during the periods examined. Therefore, the behavior of rotifers during oligotrophication of Lake Maggiore is different from that observed in Ontario lakes, where they benefited numerically from competitive release associated with reduced density of *Daphnia* competitors when *Bythotrephes* was abundant [64]. Once established, *C. hippocrepis* populations would be relatively immune to growing *Bythotrephes* populations owing to its gelatinous colonial matrix.

Exceptional meteoclimatic events can be detected over the long term [65]. As an example, the almost complete vertical mixing of February 1999 and the record flood of 2000 were also reported to impact zooplankton seasonal dynamics. As reported by Manca, Cavicchioni and Morabito [66], nutrient replenishment of surface waters from deeper waters consequent to vertical mixing triggered a cascading effect, involving all planktonic communities. Larger input of new nutrients stimulated phytoplankton productivity and higher food availability for herbivorous zooplankton, resulting in an increased density of rotifers, copepods, and cladocerans, as well as of *Daphnia* clutch size. Finally, an increase of predatory cladocerans (*Leptodora*) was recorded in July, five months after the complete mixing episode. Temperature can control nutrient supply, by affecting the extent of the mixing layer at the time of spring overturn: This mechanism is particularly effective in deep Italian lakes [67,68], where a strong link between winter temperature, mixing depth and was established. Similarly, the record flood of 2000 is well identified in the time series analyzed. Impact on zooplankton included detection in the pelagic of littoral taxa such as *Bosmina longirostris* and altered the seasonality of *Bythotrephes longimanus* [52].

5. Conclusions

Long-term data allow for disentangling the contribution of different stressors to changes in response variables. In Lake Maggiore, many stressors altered the system during the 20th century, prominent among them cultural eutrophication in the 1960s and 1970s. Phosphorus abatement began in 1977, with strong observed TP reductions observed thereafter. This change had pronounced effects on numerous biological response variables, notably chlorophyll-a concentration and phytoplankton community size composition, and *Daphnia* population density. We also observed increased variability and enhanced population density of the invertebrate predator *Bythotrephes longimanus* during the initial study period, and by small-bodied cladocerans *Bosmina* and *Diaphanosoma* during the second period. The increase in population density and inter-annual variability of small cladocerans might be viewed as a response to decreasing *Daphnia* population density. *Bythotrephes* population densities increased during the second study period, resulting in enhanced predation pressure on *Daphnia* prey. Impact on *Daphnia* during this period was reduced by a temperature increase - which served to reduce its development time and increase its birth rate - and by earlier phenology such that it exhibited less temporal overlap with *Bythotrephes*. As previously observed, over the long term, both cladocera and rotifers responded more clearly to eutrophication reversal and to warming, than copepods. The latter seem to be more influenced by exceptional meteoclimatic events, such as mixing depth. This result agrees well with analyses of ecological roles of Lake Maggiore freshwater zooplankton by carbon and nitrogen stable isotope analyses. Copepods, particularly cyclopoids, occupy a distinct functional group from the other zooplankton secondary consumers, which is an observation also made in previous studies [54,69,70].

Author Contributions: Authors contributed as follows: conceptualization, M.M., H.J.M., A.Z.; methodology, A.A., A.Z., P.Q., M.M.; software, A.A.; validation, A.A.; formal analysis, A.A., A.Z., P.Q.; resources, M.M.; investigation, A.Z., P.Q., H.J.M.; resources, M.M.; Writing-Original Draft Preparation, A.A., A.Z., P.Q., H.J.M.; Writing-Review & Editing, A.A., A.Z., P.Q., H.J.M., M.M.; visualization, A.A., supervision, A.Z., P.Q., M.M.; project administration, A.Z., A.A.; funding acquisition, M.M.

Funding: This research was funded to M.M. by the "Long-term limnological research in Lago Maggiore of the International Commission for the Protection of Swiss-Italian Waters" (CIPAIS), in the frame of a cooperation agreement between Swiss and Italian Governments. H.J.M. was supported by the CNR Short Term Mobility Program year 2013, by a Canada Research Chair. A.A. was supported by a CNR cooperation contract on "Management of biological time series with advanced statistical methods.

Acknowledgments: We dedicate this paper in memory of Giuseppe Morabito, who studied and loved Lake Maggiore. We are grateful to three anonymous reviewers for very helpful comments.

Conflicts of Interest: The authors have no conflicts of interest to declare.

Appendix A. Technical Details

In FDA, data is assumed to be represented by a collection of smooth curves $x_1(t), \ldots, x_m(t)$ defined on some interval $[a, b]$, possibly obtained by smoothing some time series as described in the main text. These curves are then thought as single points $x_1, \ldots, x_m$ belonging to the Hilbert space $L^2[a, b]$ of square-integrable functions.

In order to clarify how FPCs provide the best representation of the variation in the data $x_1, \ldots, x_m$ in a least-squares sense, we introduce the following definitions.

First, the inner product $\langle \cdot, \cdot \rangle$ and norm $\| \cdot \|$ in the space $L^2[a, b]$ are, respectively, defined as

$$\langle f, g \rangle = \int_a^b f(t)g(t)dt \text{ and } \|f\| = \sqrt{\langle f, f \rangle} \tag{A1}$$

for all functions f, $g \in L^2[a, b]$.

Second, a set of functions $f_1, \ldots, f_n$ is defined to be *orthonormal* if $\langle f_i, f_i \rangle = 1$ and $\langle f_i, f_j \rangle = 0$ whenever $i \neq j$ for all i and j between 1 and n.

Third, the functions $\widetilde{x}_1, \ldots, \widetilde{x}_m$ in $L^2[a, b]$ are defined as $\widetilde{x}_i = x_i - \overline{x}$ for each $i = 1, \ldots, m$ (where $\overline{x} = m^{-1} \sum_{i=1}^{m} x_i$), so that $\widetilde{x}_1, \ldots, \widetilde{x}_m$ represent the centered data.

In FPCA, we seek orthonormal elements $e_1, \ldots, e_n$ of $L^2[a,b]$ and scores f_{ij} such that the approximations

$$\widetilde{x}_i \cong \sum_{j=1}^{n} f_{ij} e_j \tag{A2}$$

yield the lowest possible *Sum-of-Squares Error (SSE)*

$$SSE = \sum_{i=1}^{m} \|\widetilde{x}_i - \sum_{j=1}^{n} f_{ij} e_j\|^2. \tag{A3}$$

Chosen any specific orthonormal set $e_1, \ldots, e_n$, the *SSE* can be written as

$$\begin{aligned}
&\sum_{i=1}^{m} \|\widetilde{x}_i\|^2 - 2\sum_{i=1}^{m}\sum_{j=1}^{n} f_{ij}\langle \widetilde{x}_i, e_j \rangle + \sum_{i=1}^{m}\sum_{j=1}^{n} f_{ij}^2 \\
&= \sum_{i=1}^{m} \|\widetilde{x}_i\|^2 - \sum_{i=1}^{m}\sum_{j=1}^{n} \langle \widetilde{x}_i, e_j \rangle^2 + \sum_{i=1}^{m}\sum_{j=1}^{n} \left(\langle \widetilde{x}_i, e_j \rangle - f_{ij} \right)^2
\end{aligned} \tag{A4}$$

which achieves its minimum when $f_{ij} = \langle \widetilde{x}_i, e_j \rangle$.

With this choice of coefficients and since $f_{ij} = \langle x_i - \overline{x}, e_j \rangle = \langle x_i, e_j \rangle - \langle \overline{x}, e_j \rangle$, the *SSE* becomes

$$\sum_{i=1}^{m} \|x_i - \overline{x}\|^2 - \sum_{i=1}^{m}\sum_{j=1}^{n} \left(\langle x_i, e_j \rangle - \langle \overline{x}, e_j \rangle \right)^2, \tag{A5}$$

which attains its minimum value when each quantity $\sigma_j^2 = \frac{1}{n}\sum_{j=1}^{n}\left(\langle x_i, e_j \rangle - \langle \overline{x}, e_j \rangle \right)^2$, corresponding to the variance of the scores $f_{1j}, \ldots, f_{mj}$, is maximized. Since FPCs are selected in order to maximize these variances, FPCs provide the minimum *SSE* among all possible orthonormal set $e_1, \ldots, e_n$.

References

1. Binelli, A.; Provini, A. DDT is still a problem in developed countries: The heavy pollution of Lake Maggiore. *Chemosphere* **2003**, *52*, 717–723. [CrossRef]
2. Vinebrooke, R.D.; Cottingham, K.L.; Norberg, J.; Scheffer, M.; Dodson, S.I.; Maberly, S.C.; Sommer, U. Impacts of multiple stressors on biodiversity and ecosystem functioning: The role of species co-tolerance. *Oikos* **2004**, *104*, 451–457. [CrossRef]
3. Yan, N.D.; Girard, R.; Heneberry, J.H.; Keller, W.B.; Gunn, J.M.; Dillon, P.J. Recovery of copepod, but not cladoceran, zooplankton from severe and chronic effects of multiple stressors. *Ecol. Lett.* **2004**, *7*, 452–460. [CrossRef]
4. Schindler, D.W. Recent advances in the understanding and management of eutrophication. *Limnol. Oceanogr.* **2006**, *51*, 356–363. [CrossRef]
5. Jeziorski, A.; Yan, N.D.; Paterson, A.M.; DeSellas, A.M.; Turner, M.A.; Jeffries, D.S.; Keller, B.; Weeber, R.C.; McNicol, D.K.; Palmer, M.E.; et al. The widespread threat of calcium decline in fresh waters. *Science* **2008**, *322*, 1374–1377. [CrossRef] [PubMed]
6. Ormerod, S.J.; Dobson, M.; Hildrew, A.G.; Townsend, C.R. Multiple stressors in freshwater ecosystems. *Freshw. Biol.* **2010**, *55* (Suppl. 1), 1–4. [CrossRef]
7. Strayer, D.L. Alien species in fresh waters: Ecological effects, interactions with other stressors, and prospects for the future. *Freshw. Biol.* **2010**, *55* (Suppl. 1), 152–174. [CrossRef]
8. Maberly, S.C.; Elliott, J.A. Insights from long-term studies in the Windermere catchment: External stressors, internal interactions and the structure and function of lake ecosystems. *Freshw. Biol.* **2012**, *57*, 233–243. [CrossRef]
9. Guilizzoni, P.; Levine, S.N.; Manca, M.; Marchetto, A.; Lami, A.; Ambrosetti, W.; Brauer, A.; Gerli, S.; Carrara, E.A.; Rolla, A.; et al. Ecological effects of multiple stressors on a deep lake (Lago Maggiore, Italy) integrating neo and paleolimnological approaches. *J. Limnol.* **2012**, *71*, 1–22. [CrossRef]

10. Allan, A.J.D.; McIntyre, P.B.; Smitha, S.D.P.; Halpern, B.S.; Boyer, G.L.; Buchsbaum, A.; Burton, G.A., Jr.; Campbell, L.M.; Chadderton, W.L.; Ciborowski, J.J.H.; et al. Joint analysis of stressors and ecosystem services to enhance restoration effectiveness. *Proc. Nat. Acad. Sci. USA* **2013**, *110*, 372–377. [CrossRef]

11. Verdonschot, P.F.M.; Spears, B.M.; Feld, C.K.; Brucet, S.; Keizer-Vlek, H.; Borja, A.; Elliott, M.; Kernan, M.; Johnson, R.F. A comparative review of recovery processes in rivers, lakes, estuarine and coastal waters. *Hydrobiologia* **2013**, *704*, 453–474. [CrossRef]

12. Smol, J.P.; Wolfe, A.P.; Birks, H.J.B.; Douglas, M.S.V.; Jones, V.J.; Korhola, A.; Pienitz, R.; Rühland, K.; Sorvari, S.; Antoniades, D.; et al. Climate-driven regime shifts in the biological communities of arctic lakes. *Proc. Nat. Acad. Sci. USA* **2005**, *102*, 4397–4402. [CrossRef]

13. Moss, B. Climate change, nutrient pollution and the bargain of Dr Faustus. *Freshw. Biol.* **2010**, *55* (Suppl. 1), 175–187. [CrossRef]

14. Thompson, P.L.; Shurin, J.B. Regional zooplankton biodiversity provides limited buffering of pond ecosystems against climate change. *J. Anim. Ecol.* **2012**, *81*, 251–259. [CrossRef] [PubMed]

15. Jackson, M.C.; Loewen, C.J.G.; Vinebrooke, R.D.; Chimimba, C.T. Net effects of multiple stressors in freshwater ecosystems: A meta-analysis. *Glob. Chang. Biol.* **2016**, *22*, 180–189. [CrossRef] [PubMed]

16. Hering, D.; Borja, A.; Carvalho, L.; Feld, C.K. Assessment and recovery of European water bodies: Key messages from the WISER project. *Hydrobiologia* **2013**, *704*, 1–9. [CrossRef]

17. Jeppesen, E.; Sondergaard, M.; Jensen, J.P.; Havens, K.E.; Anneville, O.; Carvalho, L.; Coveney, M.F.; Deneke, R.; Dokulil, M.; Foy, B.; et al. Lake responses to reduced nutrient loading—An analysis of contemporary long-term data from 35 case studies. *Freshw. Biol.* **2005**, *50*, 1747–1771. [CrossRef]

18. Schindler, D.W. The dilemma of controlling eutrophication of lakes. *Proc. R. Soc. B* **2012**, *279*, 4322–4333. [CrossRef]

19. Yan, N.D.; Pawson, T.W. Changes in the crustacean zooplankton community of Harp Lake, Canada, following invasion by *Bythotrephes cederstrœmi*. *Freshw. Biol.* **1997**, *37*, 409–425. [CrossRef]

20. Straile, D.; Geller, W. The response of *Daphnia* to changes in trophic status and weather patterns: A case study from Lake Constance. *ICES J. Mar. Sci.* **1998**, *55*, 775–782. [CrossRef]

21. May, L.; Spears, B.M. A history of scientific research at Loch Leven, Kinross, Scotland. *Hydrobiologia* **2012**, *681*, 3–9. [CrossRef]

22. Lottig, N.R.; Carpenter, S.R. Interpolating and forecasting lake characteristics using long-term monitoring data. *Limnol. Oceanogr.* **2012**, *57*, 1113–1125. [CrossRef]

23. Ramsay, J.; Silverman, B.W. *Functional Data Analysis*, 2nd ed.; Springer: New York, NY, USA, 2005.

24. de Bernardi, R.; Giussani, G.; Manca, M.; Ruggiu, D. Long-term dynamics of plankton communities in Lago Maggiore (N. Italy). *Internationale Vereinigung für Theoretische und Angewandte Limnologie: Verhandlungen* **1988**, *23*, 729–733. [CrossRef]

25. de Bernardi, R.; Giussani, G.; Manca, M.; Ruggiu, D. Trophic status and the pelagic system in Lake Maggiore. *Hydrobiologia* **1990**, *191*, 1–8. [CrossRef]

26. Manca, M.; Ruggiu, D. Consequences of pelagic food-web changes during a long-term lake oligotrophication process. *Limnol. Oceanogr.* **1998**, *43*, 1368–1373. [CrossRef]

27. Sas, H. *Lake Restoration by Reduction of Nutrient Loading: Expectations, Experiences, Extrapolation*; Academic Verlag: St. Augustin, FL, USA, 1989.

28. de Bernardi, R.; Canale, C. Ricerche pluriennali (1948–1992) sull'ecologia dello zooplancton del Lago Maggiore. *Doc Ist. Ital. Hydrobiol.* **1995**, *55*, 1–68.

29. Manca, M.; Calderoni, A.; Mosello, R. Limnological research in Lago Maggiore: Studies on hydrochemistry and plankton. *Memorie Istituto Italiano di Idrobiologia* **1992**, *50*, 171–200.

30. Tonolli, L. Holomixy and oligomixy in Lake Maggiore: Inference on the vertical distribution of zooplankton. *Verh. Int. Ver. Limnol.* **1969**, *17*, 231–236. [CrossRef]

31. Manca, M.; DeMott, W.R. Response of the invertebrate predator *Bythotrephes* to a climate-linked increase in the duration of a refuge from fish predation. *Limnol. Oceanogr.* **2009**, *54* Pt 2, 2506–2512. [CrossRef]

32. Arca, G.; Barbanti, L. *Annuario dell'Osservatorio Meteorologico di Pallanza*; Documenta Ist. Ital. Idrobiol; Consiglio Nazionale delle Ricerche, Istituto Italiano di Idrobiologia: Pallanza, Italy, 1991; pp. 1–83. (In Italian)

33. Barbanti, L. *Annuario dell'Osservatorio Meteorologico di Pallanza*; Documenta Ist. Ital. Idrobiol; Consiglio Nazionale delle Ricerche, Istituto Italiano di Idrobiologia: Pallanza, Italy, 1997; pp. 1–81. (In Italian)

34. Ambrosetti, W.; Barbanti, L.; Rolla, A. *Annuario dell'Osservatorio Meteorologico di Pallanza. 1997–2003*; CNR Istituto per lo Studio degli Ecosistemi: Verbania Pallanza, Italy, 2006; pp. 1–38.

35. Ambrosetti, W.; Barbanti, L.; Rolla, A. *Annuario dell'Osservatorio Meteorologico di Pallanza. 2004–2005*; CNR Istituto per lo Studio degli Ecosistemi: Verbania Pallanza, Italy, 2006; pp. 1–70.

36. Ambrosetti, W.; Barbanti, L.; Rolla, A. *Annuario dell'Osservatorio Meteorologico di Pallanza. 2006*; CNR Istituto per lo Studio degli Ecosistemi: Verbania Pallanza, Italy, 2007; pp. 1–51.

37. Ambrosetti, W.; Barbanti, L.; Rolla, A. *Annuario dell'Osservatorio Meteorologico di Pallanza. 2007*; CNR Istituto per lo Studio degli Ecosistemi: Verbania Pallanza, Italy, 2008; pp. 1–56.

38. Lorenzen, C.J. Determination of chlorophyll and pheo-pigments: Spectrophotometric equations. *Limnol. Oceanogr.* **1967**, *12*, 343–346. [CrossRef]

39. Rott, E. Some results from phytoplankton counting intercalibrations. *Schweiz. Z. Hydrol.* **1981**, *43*, 34–63. [CrossRef]

40. Lund, J.W.G.; Kipling, C.; Le Cren, E.D. The inverted microscope method of estimating algal numbers and the statistical basis of estimations by counting. *Hydrobiologia* **1958**, *11*, 143–170. [CrossRef]

41. Wacklin, P.; Hoffmann, L.; Komárek, J. Nomenclatural validation of the genetically revised cyanobacterial genus. *Fottea* **2009**, *9*, 59–64. [CrossRef]

42. Morabito, G.; Manca, M. Eutrophication and recovery of the large and deep subalpine Lake Maggiore: Patterns, trends and interactions of planktonic organisms between trophic and climatic forcings. In *Chapter 6 in Eutrophication*; Lambert, A., Roux, C., Eds.; Nova Science Publishers: New York, NY, USA, 2013; ISBN 978-1-62808-498-6.

43. Carrara, E.; Ambrosetti, W.; Barbanti, L. The management of Lake Maggiore water levels: A study of low water episodes. In Proceedings of the Symposium on the Role of Hydrology in Water Resources Management, Capri, Italy, 29–31 October 2008; Volume 327, pp. 114–123.

44. Ambrosetti, W.; Barbanti, L.; Carrara, E. Mechanisms of hypolimnion erosion in a deep lake (Lago Maggiore, N. Italy). *J. Limnol.* **2010**, *69*, 3–14. [CrossRef]

45. Ruggiu, D.; Morabito, G.; Panzani, P.; Pugnetti, A. Trends and relations among basic phytoplankton characteristics in the course of the long-term oligotrophication of Lago Maggiore (Italy). *Hydrobiologia* **1998**, *369–370*, 243–257. [CrossRef]

46. Obertegger, U.; Manca, M. Response of rotifer functional groups to changing trophic state and crustacean community. *J. Limnol.* **2011**, *70*, 231–238. [CrossRef]

47. Manca, M.; Ramoni, C.; Comoli, P. The decline of *Daphnia hyalina galeata* in Lago Maggiore: A comparison of the population dynamics before and after oligotrophication. *Aquat. Sci.* **2000**, *62*, 142–153. [CrossRef]

48. Manca, M.; Portogallo, M.; Brown, M. Shifts in phenology of *Bythotrephes longimanus* and its modern success in Lake Maggiore as a result of changes in climate and trophy. *J. Plankton Res.* **2007**, *29*, 515–525. [CrossRef]

49. MacIsaac, H.J.; Ketelaars, H.A.M.; Girgorovich, I.A.; Ramcharan, C.W.; Yan, N.D. Modeling *Bythotrephes longimanus* invasions in the Great Lakes basin based on its European distribution. *Arch. Hydrobiol.* **2000**, *141*, 1–21. [CrossRef]

50. Boudreau, S.A.; Yan, N.D. The differing crustacean zooplankton communities of Canadian Shield lakes with and without the nonindigenous zooplanktivore *Bythotrephes longimanus*. *Can. J. Fish. Aquat. Sci.* **2003**, *60*, 1307–1313. [CrossRef]

51. Weisz, E.J.; Yan, N.D. Shifting invertebrate zooplanktivores: Watershed-level replacement of the native *Leptodora* by the non-indigenous *Bythotrephes* in Canadian Shield lakes. *Biol. Invasions* **2011**, *13*, 115–123. [CrossRef]

52. Manca, M. Considerazioni generali sull'evoluzione a lungo termine dei popolamenti planctonici. In *Ecologia delle Acque Interne. Ricerche sull'evoluzione del Lago Maggiore. Aspetti limnologici. Programma quinquennale*; Campagna 2002; Commissione Internazionale per la protezione delle acque italo-svizzere, sezione di Idrobiologia, C.N.R.-I.S.E.: Verbania, Italy, 2014; pp. 134–136. (In Italian)

53. de Bernardi, R.; Giussani, G.; Manca, M. *Cladocera: Predators and Prey*; Springer: Dordrecht, The Netherlands, 1987; pp. 225–243.

54. Visconti, A.; Volta, P.; Fadda, A.; Di Guardo, A.; Manca, M. Seasonality, littoral versus pelagic carbon sources, and stepwise δ^{15}N-enrichment of pelagic food web in a deep subalpine lake: The role of planktivorous fish. *Can. J. Fish. Aquat. Sci.* **2013**, *71*, 436–446. [CrossRef]

55. Coulas, R.A.; MacIsaac, H.J.; Dunlop, W. Selective predation by lake herring (*Coregonus artedi*) on exotic zooplankton (*Bythotrephes cederstroemi*) in Harp Lake, Ontario. *Freshw. Biol.* **1998**, *40*, 343–356. [CrossRef]

56. Visconti, A.; Manca, M.; De Bernardi, R. Eutrophication-like response to climate warming: an analysis of Lago Maggiore (N. Italy) zooplankton in contrasting years. *J. Limnol.* **2008**, *67*, 87–92. [CrossRef]

57. Gillooly, J.F.; Charnov, E.L.; West, G.B.; Savage, V.M.; Brown, J. Effects of size and temperature on developmental time. *Nature* **2003**, *417*, 70–73. [CrossRef]

58. Manca, M.; Sonvico, D. Seasonal variations in population density and size structure of *Conochilus* in Lago Maggiore: A biannual study. *Mem. Ist. Ital. Idrobiol.* **1996**, *54*, 97–108.

59. Obertegger, U.; Smith, H.A.; Flaim, G.; Wallace, R.L. Using the guild ratio to characterize pelagic rotifer communities. *Hydrobiologia* **2011**, *662*, 157–162. [CrossRef]

60. Pourriot, R. Food and feeding habits of Rotifera. *Arch. Hydrobiol. Beih. Ergebn. Limnol.* **1977**, *8*, 243–260.

61. Wallace, R.L.; Snell, T.W.; Ricci C Nogrady, T. Rotifera Volume 1: Biology, Ecology and Systematics. In *Guides to the Identification of the Microinvertebrates of the Continental Waters of the World*, 2nd ed.; Dumont, H.J.F., Ed.; Kenobi Productions: Ghent, Belgium; Backhuys Publishers: Leiden, The Netherlands, 2006; Volume 23.

62. temberger, R.S.; Gilbert, J.J. Rotifer threshold food concentrations and the size-efficiency hypothesis. *Ecology* **1987**, *68*, 181–187. [CrossRef]

63. Manca, M. Invasions and re-emergences: An analysis of the success of *Bythotrephes* in Lago Maggiore (Italy). *J. Limnol.* **2011**, *70*, 76–82. [CrossRef]

64. Hovius, J.T.; Beisner, B.E.; McCann, K.S.; Yan, N.D. Indirect food web effects of *Bythotrephes* invasion: Responses by the rotifer *Conochilus* in Harp Lake, Canada. *Biol. Invasions* **2007**, *9*, 233–243. [CrossRef]

65. Manca, M.; Torretta, B.; Comoli, P.; Amsinck, S.L.; Jeppesen, E. Major changes in trophic dynamics in large, deep sub-alpine Lake Maggiore from 1940s to 2002: A high resolution comparative paleo-neolimnological study. *Freshw. Biol.* **2007**, *52*, 2256–2269. [CrossRef]

66. Manca, M.; Cavicchioni, N.; Morabito, G. First observations on the effect of a complete, exceptional overturn of Lake Maggiore on plankton and primary productivity. *Int. Rev. Hydrobiol.* **2000**, *85*, 209–222. [CrossRef]

67. Salmaso, N.; Buzzi, F.; Garibaldi, L.; Morabito, G.; Simona, M. Effects of nutrient availability and temperature on phytoplankton development: A case study from large lakes south of the Alps. *Aquat. Sci.* **2012**, *74*, 555–570. [CrossRef]

68. Manca, M.; Rogora, M.; Salmaso, N. Inter-annual climate variability and zooplankton: Applying teleconnection indices to two deep subalpine lakes in Italy. *J. Limnol.* **2014**, *74*. [CrossRef]

69. Visconti, A.; Caroni, R.; Rawcliffe, R.; Fadda, A.; Piscia, R.; Manca, M. Defining Seasonal Functional Traits of a Freshwater Zooplankton Community Using δ^{13}C and δ^{15}N Stable Isotope Analysis. *Water* **2018**, *10*, 108. [CrossRef]

70. Visconti, A.; Manca, M. Seasonal changes in the δ^{13}C and δ^{15}N signatures of the Lago Maggiore pelagic food web. *J. Limnol.* **2011**, *70*, 263–271. [CrossRef]

Article

Cladoceran (Crustacea) Niches, Sex, and Sun Bathing—A Long-Term Record of Tundra Lake (Lapland) Functioning and Paleo-Optics

Liisa Nevalainen [1,*], E. Henriikka Kivilä [1], Marttiina V. Rantala [1,2] and Tomi P. Luoto [1]

[1]   Faculty of Biological and Environmental Sciences, Ecosystems and Environment Research Programme, University of Helsinki, Niemenkatu 73, 15140 Lahti, Finland; henriikka.kivila@helsinki.fi (E.H.K.); marttiina.rantala@helsinki.fi (M.V.R.); tomi.luoto@helsinki.fi (T.P.L.)

[2]   CNR-Water Research Institute (IRSA), Unit Verbania, VialeTonolli 50, 28922 Verbania, Italy

*   Correspondence: liisa.nevalainen@helsinki.fi

Received: 2 September 2019; Accepted: 24 September 2019; Published: 27 September 2019

Abstract: Under fundamental ecosystem changes in high latitude lakes, a functional paleolimnological approach may increase holistic understanding of lake responses and resilience to climate warming. A ~2000-year sediment record from Lake Loažžejávri in the tundra of northern Finnish Lapland was examined for fossil Cladocera assemblages to examine long-term environmental controls on aquatic communities. In addition, cladoceran functional attributes, including functional diversity (FD), UV absorbance (ABS_{UV}) of *Alona* carapaces, and sexual reproduction (ephippia) in *Bosmina* and Chydoridae were analyzed. Cladoceran communities responded to a major change in benthic habitat quality, reflected as elevated (increasingly benthic) sediment organic matter $\delta^{13}C$ signal since the 17th century. FD fluctuations showed association with climate oscillation, FD being generally higher during warm climate periods. These ecological changes were likely attributable to diversification of littoral-benthic consumer habitat space. ABS_{UV}, irrespective of increases during the Little Ice Age (LIA) due to higher UV transparency of lake water, was lower under increasing autochthony (benthic production) suggesting establishment of physical UV refugia by the benthic vegetative substrata. *Bosmina* ephippia exhibited a decreasing trend associated with increasing benthic production, indicating favorable environmental regime, and, together with chydorid ephippia, transient increases during the climate cooling of the LIA driven by shorter open-water season.

Keywords: autochthony; cladocera; functional ecology; organic carbon; paleolimnology; tundra lakes; UV radiation

---

1. Introduction

Impacts of recent climate warming are emphasized in small and shallow high latitude lakes being driven by higher air and water temperature and longer open-water season [1]. In a long-term perspective, these major physical drivers shift arctic and subarctic aquatic ecosystems toward an unprecedented ecological status [2,3]. Identification and characterization of the new high latitude lake trajectories is significant since they connect lake food webs, by aquatic-terrestrial coupling, to global-scale biogeochemical processes [4]. Northern ecotonal tree line lakes and their sedimentary environmental archives act as sentinels for estimating current ecosystem functioning and organization with respect to natural variability over the course of the Holocene [5].

The balance between autochthonous (in-lake produced) and allochthonous (catchment-originated) organic matter in aquatic systems is of high importance to the global carbon cycle, since lakes store and transfer carbon and act across the atmospheric-terrestrial-aquatic boundaries [4]. A common character of high latitude lakes is high water transparency, i.e., low amount of terrestrially derived dissolved organic matter (usually estimated as dissolved organic carbon (DOC)) due to lack or scarcity of catchment vegetation [6]. While typically systems with DOC concentrations over 5 µg L^{-1} are estimated to be heterotrophic [7], the high latitude lake food webs with low (approximately 1–3 µg L^{-1}) DOC are principally dependent on autochthonous production. High water transparency induces low attenuation of sunlight and UV radiation and organisms must therefore cope with high UV exposure [8]. Furthermore, food webs and secondary production may be supported primarily by benthic primary production that is fueled by the ample light [9]. However, under climate warming, i.e., due to the advancing tree line, expanding tundra vegetation or thawing permafrost, previously autochthonous and transparent lakes may experience increases in input of allochthonous organic matter from the catchment causing fundamental changes in light climate and energy pathways [5,10,11]. Alternatively, climate warming may enhance autochthony through lengthening of the open-water season and increasing habitat availability for benthic primary producers [12,13].

Knowledge of ecological and biogeochemical functions of aquatic organisms, e.g., feeding guilds, habitat preferences, and photoprotective mechanisms, is important when examining lake food webs and the organic matter cycle. Paleolimnological research focusing on long-term distribution of microscopic aquatic fossils (e.g., crustacean cladocerans or diatom algae) has slowly shifted toward biodiversity sciences [14–16] but an actual functional approach, where functional rather than taxonomic classification of fauna and flora is assessed in relation to natural and anthropogenic ecosystem variability, still remains rather rare. The functional approach may allow a holistic understanding of changes and drivers occurring in lakes and their surroundings since ecosystem functions are not dependent on taxonomic identity [17,18]. In addition to the use of functional diversity (FD) as an index for assessing functional trait distribution in fossil biological assemblages, functionality may be estimated, for example, by patterns in feeding, habitats, reproduction, and morphology [19–22].

In the current study, our focus was on Cladocera (Crustacea) communities and their functioning over the past millennia in a subarctic tundra lake. More specifically, we analyzed fossil cladoceran communities and functional attributes including functional diversity (FD index), reproduction patterns (Chydoridae and *Bosmina* fossil ephippia), and photoprotection (melanization based on carapace UV absorbance) in a 2000-yr sediment core and compared these data to previously available paleoclimate and biogeochemical proxies. The objectives were to track long-term changes in cladoceran community functioning and lake water bio-optics, and to identify interconnections of functional ecology and past climate-driven limnological changes.

2. Materials and Methods

Lake Loažžejávri is located in the subarctic (mean July temperature +12.3 °C, mean annual temperature −2 °C) shrub tundra of Finnish Lapland (69°53′ N, 26°55′ E) at an altitude of 255 m a.s.l. (Figure 1). The lake has an area of ~3 ha with a catchment size of ~200 ha and it is shallow (water depth 1.2 m), oligotrophic (total phosphorus (TP) 5.9 µg L^{-1}, chlorophyll-a (chl-a) 1.6 µg L^{-1}) and transparent (dissolved organic carbon (DOC) 3.4 mg L^{-1}, UV attenuation coefficient at 305 nm (K_{dUV}) 11.1 m^{-1}). A 38-cm sediment sequence was cored from the basin in late summer 2014 with a Limnos gravity corer and subsampled at 1-cm intervals and dated with the ^{14}C dating method. The core covers approximately the past two millennia. The lake has previously been investigated as a part of regional limnological survey [23], and its sediments for late Holocene paleoclimate [24] and carbon utilization of aquatic macroinvertebrate communities [25], from where detailed information on limnological characteristics, catchment and sedimentology, chronology, and biostratigraphy can be found. The age-depth model for the core was based on two plant macrofossils at sediment depths 20 (1390–1440 C.E.) and 25 cm (970–1025 C.E.) analyzed for AMS (accelerator mass spectrometry) ^{14}C dates. Additionally, ^{137}Cs and ^{210}Pb were analyzed but their concentrations remained low and therefore insufficient to add further temporal resolution to the age-depth model [24].

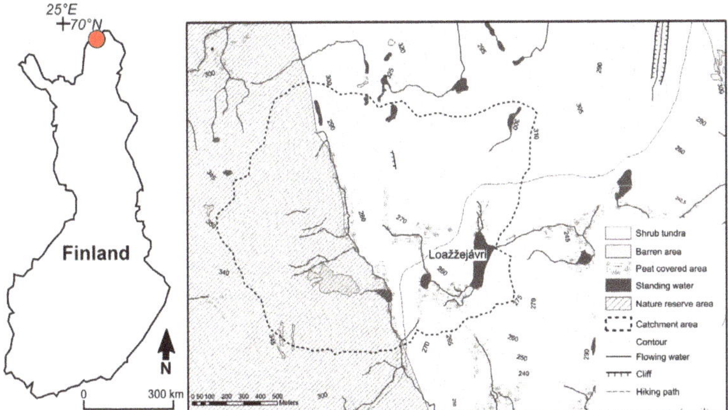

Figure 1. Location of Lake Loažžejávri in northern Finnish Lapland (**red dot in left column**) and its catchment characteristics (**right column**).

Fossil Cladocera were analyzed from the sediment subsamples (1-cm resolution) to examine past community changes. The samples were prepared by heating and stirring ~2 cm^3 of fresh sediment in 10% potassium hydroxide (KOH on a hot plate following the standard protocol [26]. The samples were then sieved through a 51-μm mesh and the residues were centrifuged for 10 min at 4000 rpm. Permanent microscope samples were mounted in glycerol gelatin stained with safranine on a hot plate. The micropscope slides were examined for fossil cladoceran remains (carapaces, head shields, postabdomens, claws, ephippia) with a light microscope at 100–400× magnifications and a minimum of 70 individuals were identified [27] from each sample (53 individuals from sample at 3 cm due to low amount of remains).

In addition to standard community analysis described above, functional attributes of sexual reproduction, functional diversity, and melanization were analyzed. Total bosminid (Bosminidae, *Bosmina*) and chydorid (Chydoridae) ephippia (chitinous envelops for diapausal resting eggs, indicative of sexual reproduction) were enumerated during the standard counting and were used for estimates of sexual reproduction (ratio of sexual to asexual reproduction) by enumerating *Bosmina* and chydorid ephippia (indicative of sexual reproduction) and carapaces (indicative of asexual reproduction) [28]. Functional diversity of the cladoceran community was expressed as Rao's FD index, i.e., Rao's quadratic entropy [29]. For this index, each fossil cladoceran taxa was assigned with selected qualitative functional characters (traits, Table 1) including body size (small <500 μm, intermediate 500–1000 μm, large >1000 μm), body shape (elongated, oval, globular), feeding type (filterer, scraper-detritivore, predator) and habitat (pelagic, benthic, attached to vegetation) following [21]. To examine melanin in cladoceran remains as an index for UV exposure and photoprotection [30], fossil cladoceran carapaces from a large chydorid species *Alona affinis* (Leydig) were extracted from sediment subsamples with fine forceps under a binocular microscope. The carapaces were then measured under UV wavelengths 340 and 305 nm for their UV absorbance (ABS$_{UV}$) to indicate the degree of melanization. The absorbance measurements were performed with a UV/VIS spectrophotometer (UV-1800, Shimadzu Corporation, Kyoto, Japan). Seven carapaces were measured for a mean UV absorbance value per sample omitting highest and lowest absorbance values. UV absorbance measurements are expressed as anomalies from the mean absorbance value of the data series.

The data were further analyzed with numerical methods to elucidate relationships between cladoceran community and functional attributes and paleoenvironmental variation. Redundancy analysis (RDA) was utilized to examine impacts of sediment biogeochemistry (indicative of habitat quality), i.e., elemental and isotopic composition (C%, N%, C/N ratio, δ^{13}C, and δ^{15}N) and amount of organic matter (OM%, data from [25]), on cladoceran community succession though time. In addition, external atmospheric forcers, i.e., sun spot numbers (SSN) and summer (June–August) air temperature (T-JJA) as 10-year averages [following 31, 32, respectively] were used in the RDA as environmental factors. SSN reconstruction is based on dendrochronologically dated radiocarbon concentrations and physics-based models [31] and T-JJA reconstruction is based on maximum latewood density tree-ring chronologies [32]. Species data were square-root transformed prior to data analyses. Paleoenvironmental parameters with high inflation factors were omitted to include a set of variables with inflation factors <10 (C/N, δ^{13}C, δ^{15}N, OM%, SSN, T-JJA). Forward selection of environmental parameters was performed with 999 permutations and significant parameters were assigned as $p < 0.05$. Selected bivariate environmental correlations for functional attributes were examined with Pearson's correlation coefficient and linear regression. RDA was performed with Canoco 5 software [33] and linear regressions with PAST3 software [34]. In addition to OM%, organic matter δ^{13}C, and SSN, mean carbon source contributions in the benthic food web (based on fossil Chironomidae δ^{13}C), modeled with a three-source (benthic, planktonic, and terrestrial) Bayesian mixing model previously applied in [22], were utilized as paleoenvironmental reference data. Here, the modeled planktonic and benthic carbon contributions (to Chironomidae diet) were used in a planktonic to benthic (P/B) ratio. Further, autochthonous (planktonic + benthic) and allochthonous (terrestrial) carbon sources were used in autochthonous to allochthonous ratio. These ratios were calculated based on the Chironomidae δ^{13}C based source data [25].

Table 1. Functional traits of cladoceran taxa encountered from Lake Loažžejávri sediment core. Characterization is based on body size (S = small, M = intermediate, L = large), body shape (G = globular, O = oval, E = elongated), feeding (F = filterer, S-D = scraper-detritivore, P = predator, including parasitism), and habitat (P = pelagial, S = sediment, V = vegetation).

	Body Size			Body Shape			Feeding			Habitat		
	S	M	L	G	O	E	F	S-D	P	P	S	V
Bosmina longispina		*		*			*			*		
Daphnia spp.			*		*		*			*		
Ceriodaphnia spp.		*			*		*			*		*
Polyphemus pediculus			*			*			*	*		
Bythotrephes longimanus			*			*			*	*		
Simocephalus spp.			*		*		*					*
Ophryoxus gracilis			*		*		*	*			*	*
Eyrycercus spp.			*		*			*			*	*
Camptocercus rectirostris			*		*			*			*	*
Acroperus harpae		*			*			*			*	*
Alonopsis elongata		*			*			*			*	
Graptoleberis testudinaria		*			*			*				*
Alona affinis		*			*			*			*	*
Alona quadrangularis		*			*			*			*	
Alona guttata	*				*			*			*	*
Alona guttata f. tuberculata	*				*			*			*	*
Alona rustica		*			*			*			*	*
Alona intermedia	*				*			*			*	*
Alona werestschagini		*			*			*			*	
Alonella excisa	*			*				*			*	*
Alonella nana	*			*				*			*	*
Chydorus sphaericus-type	*			*				*		*		*
Paralona pigra	*			*				*			*	
Rhynchotalona falcata	*				*			*			*	
Drepanothrix dentata		*			*			*			*	
Iliocryptus spp.			*		*			*			*	
Anchistropus emarginatus		*		*					*		*	*

3. Results

A total of 27 cladoceran taxa were identified from the sediment subsamples. The most frequent taxa (occurring in every subsample) included *Alonella nana* (Hill's N2 36.5, mean relative abundance 42.5%), *Bosmina longispina* (36.4, 32.5%), *Alona affinis* (34.5, 7.9%), *Chydorus sphaericus*-type (24.3, 6.1%), and *Paralona pigra* (25.1, 1.5%). In the cladoceran stratigraphy (Figure 2), *B. longispina* and *A. affinis* dominated with ~20%–60% abundances throughout the sediment sequence, *B. longispina* peaking between 1400 and 1500 C.E. and A. nana between 1600 and 1800 C.E. Many littoral-benthic taxa, including *C. sphaericus*-type, *Acroperus harpae*, *Alonella excisa*, and *Eurycercus* spp. increased slightly during 600–1200 C.E. and others, including *Alona quadrangularis* and *A. guttata*, increased or emerged at the top sequence after 1600 C.E.

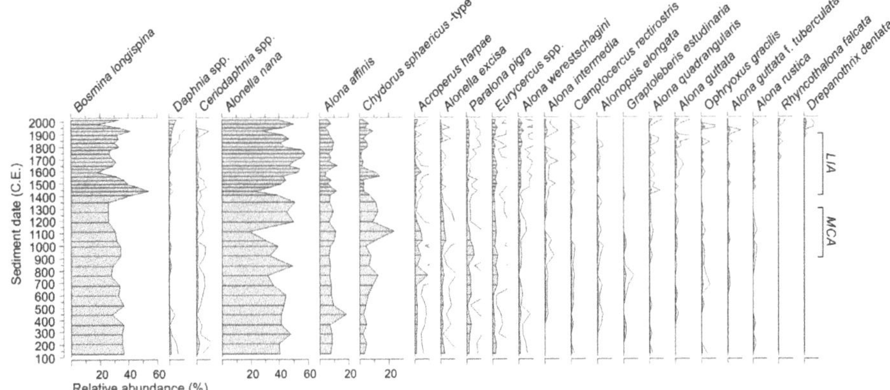

Figure 2. Relative abundance of most common (Hill's N2 > 3.5) cladoceran taxa in the Loažžejávri sediment core indicated with a gray silhouette, where black horizontal lines represent sediment sub-samples. Five-fold exaggeration curves of less abundant taxa are shown with gray lines. Temporal extensions of the cold Little Ice Age (LIA) and warm Medieval Climate Anomaly (MCA) are approximated.

RDA and forward selection of environmental variables resulted in eigenvalues of 0.14 and 0.07 for axes 1 and 2, respectively, explaining 35.2% of variation in species data (Figure 3). It identified δ^{13}C (41.7%, F = 6.2, p < 0.001), OM% (20.1%, F = 3.2, p < 0.001), and SSN (11.5%, F = 1.9, p = 0.035) as the most significant environmental parameters explaining variation in cladoceran species data, together accounting for 73.3% of the explained variation. Most recent samples (12–0 cm, ~1650 C.E.—present) had positive axis 1 scores in relation to increasing δ^{13}C, OM% and SSN. *Alonella nana, Alona werestschagini, Alona quadrangularis*, and *Alona guttata* were associated with increasing δ^{13}C. *Rhynchotalona falcata, Drepanothrix dentata, Alona intermedia*, and *Daphnia* spp. were associated with increasing OM% and SSN having more positive scores for RDA axis 2. Older than ~1650 C.E. samples had negative axis 1 scores and the abundant taxa, such as *Bosmina longispina* and *Chydorus sphaericus*-type, and vegetation-associated taxa *Alonella excisa, Acroperus harpae*, and *Eurycercus* spp. were related to the negative end of RDA axis 1 (i.e., low δ^{13}C).

Of the cladoceran-based functional indices (Figure 4), chydorid and *Bosmina* sexual reproduction both increased around 700–800 C.E. and peaked later on during 1400, 1800, and 1900 (chydorids) and 1300 (*Bosmina*) C.E., after which *Bosmina* sexual reproduction declined. Minimum values occurred around 1700 and 1900 C.E. in chydorids and *Bosmina*, respectively. Mean abundance of chydorid and *Bosmina* ephippia encountered in the samples was high, 33 (min. 6, max. 65) and 18 (min. 1, max. 50) ephippia, respectively. Cladoceran FD was highly variable, but exhibited increases during 600–1100 C.E. and after 1700 C.E. until the present. Mean ABS$_{UV}$ for the sediment core was 1.2 AU (absorbance units). ABS$_{UV}$ exhibited several peaks, indicative of more intensive melanization (i.e., higher UV exposure) during the early core at ~300, 600, and 1000 C.E. and in the younger layers ~1500 and 1700–1800 C.E.

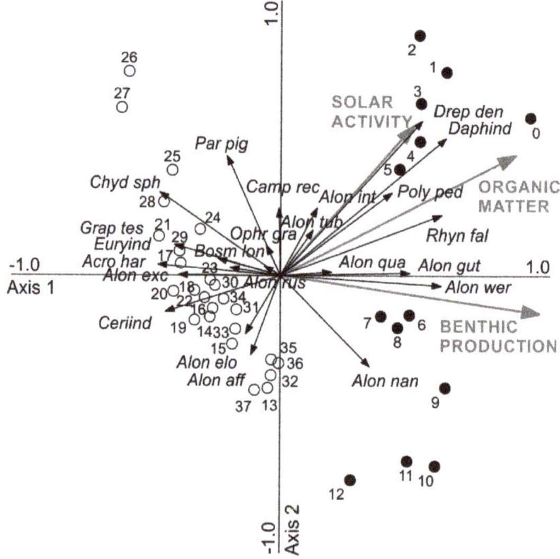

Figure 3. Redundancy analysis ordination diagram for the cladoceran taxa and samples in relation to significant environmental variables δ^{13}C of sediment organic matter (benthic production), organic matter content (as loss-on-ignition), and solar activity (as reconstructed sun spot numbers). Pre ~1650 C.E. samples (13–37 cm) are indicated by white dots and the recent samples (0–12 cm) by black dots. Data for sediment geochemistry are from [25] and for solar activity from [30]. Species abbreviations include four letters of the genus and three from the species name.

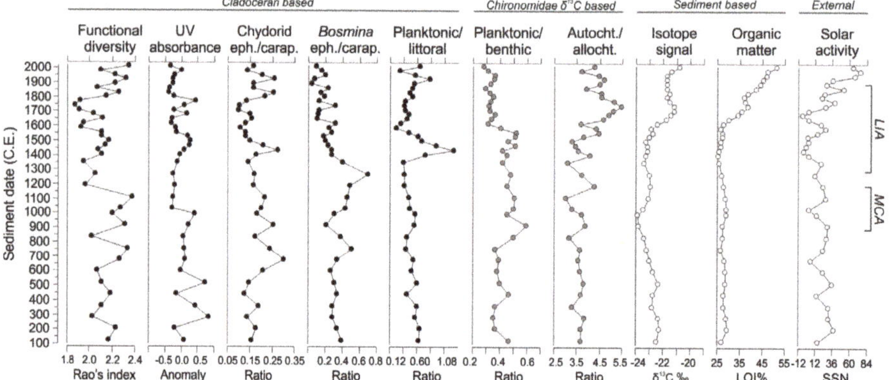

Figure 4. Cladoceran based functional attributes; functional diversity, UV absorbance of *Alona* carapaces, ratio of ephippia (sexual reproduction) to carapaces (asexual reproduction) of Chydoridae and *Bosmina*, and cladoceran planktonic to littoral ratio. In addition, ratios of mean source contributions (planktonic to benthic and autochthonous to allochthonous) in the benthic food web (based on Chironomidae δ^{13}C and Bayesian mixing modeling), sediment organic matter δ^{13}C, organic matter content (as loss-on-ignition), and solar intensity (as reconstructed sun spot numbers) are shown. Data for sediment geochemistry and source contributions are after [25] and for solar intensity after [31]. Temporal extensions of the cold Little Ice Age (LIA) and warm Medieval Climate Anomaly (MCA) are approximated.

Significant Pearson's correlations were found between T-JJA and FD (r = 0.48, *p* = 0.002; Figure 5a), sediment C/N ratio and carapace UV absorbance (r = 0.46, *p* = 0.004; Figure 5b), and sediment carbon content and *Bosmina* ephippia/carapaces ratio (r = −0.60, *p* = < 0.001; Figure 5c). In addition, ABS_{UV} had significant (*p* = <0.05) correlations with C% (r = −0.37) and N% (r = −0.33) and *Bosmina* ephippia with N% (r = −0.56), C/N (r = 0.32), $\delta^{13}C$ (r = −0.59), δ15N (r = 0.64), and OM% (r = −0.59).

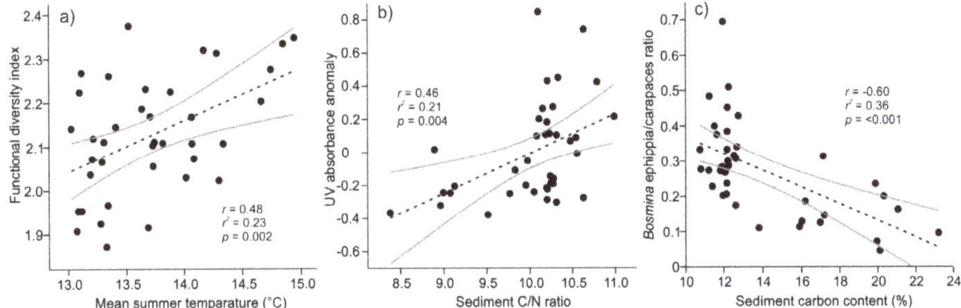

Figure 5. Bivariate relationships (*r* = Pearson's correlation, *r²* = coefficient of determination in linear regressions, *p* = significance) of (**a**) cladoceran functional diversity and reconstructed summer (June–August) mean temperature, (**b**) carapace UV absorbance anomaly and sediment carbon to nitrogen ratio (C/N), and (**c**) ratio of ephippia (sexual reproduction) to carapaces (asexual reproduction) in *Bosmina* and sediment carbon content. Black dots represent sediment core subsamples, black dashed lines indicate linear regression models and gray lines 95% confidence intervals. Sediment geochemical data originates from [25] and summer mean temperature reconstruction from [32].

4. Discussion

4.1. Lake Functioning

While recent research has highlighted the role of northern waters in the cycling of terrestrial organic matter, there also exists contrasting evidence suggesting insignificant contribution of terrestrial inputs to aquatic systems [35–37]. For example, a trend of increasing autochthony rather than allochthony has been observed in small lakes from northern Finnish Lapland [13,25]. Benthic autochthonous production prevailed in Lake Loažžejávri over the past two millennia supporting aquatic consumers such as macrobenthic Chironomidae (chironomid) larvae [24,25]. Of the autochthonous organic material consumed, planktonic carbon component, inferred and modeled from $\delta^{13}C$ of chironomid head capsules, was slightly elevated (increase in P/B, Figure 4) during the warm Medieval Climate Anomaly (MCA) at around 900–1300 C.E. This increase is likely indicative of higher algal production and its more efficient utilization by consumers [25]. Benthic production and benthic carbon component in chironomid diet (inferred from $\delta^{13}C$ of chironomid head capsules) clearly dominated in the lake after ~1600 C.E. and may have been a reflection of reduced ice-cover period after the LIA climax creating wider benthic niche space for autotrophic organisms generally increasing the relative importance of the benthic habitat over the pelagic one [38].

The production in the lake being mostly benthic, the main driver for long-term cladoceran community shifts in Lake Loažžejávri sediment profile was related to the benthic habitat quality. According to the RDA, $\delta^{13}C$ signature indicating the origin of organic matter, i.e., autochthonous (planktonic or benthic) or allochthonous, was the most significant environmental parameter explaining cladoceran assemblages (Figure 3). The isotopic carbon signature separated the sediment core subsamples along the RDA axis 1 to pre- and post-1650 C.E. sample clusters (Figure 3). Organic matter $\delta^{13}C$ varied relatively little during the first 1500 years of the sequence but started to increase at ~1600 C.E. indicative of more pronounced benthic primary production [25]. In concert, the cladoceran community diversified with the inclusion and increase of several benthic taxa, e.g., *Alona werestschagini*, *A. quadrangularis*, *A. guttata*, *R. falcata*, and *Drepanothrix dentata* until the 21st century (Figure 2). Chydorid benthos has been shown to utilize unselectively periphyton and detritus for food in similar lake environments in the region [13] and the assemblage change was thus likely driven by a diversification of microhabitats among the benthic vegetative substrata [39]. Alternatively, the diversification of the benthic habitats may have allowed species with more specialized feeding strategies to increase in the lake following the benthic succession. In common, consistent increases of specialized cladoceran taxa in relation to climate warming have been reported from arctic Canada [40].

Despite its shallowness, the lake supported a planktonic food web, as evidenced by the high abundance of euplanktonic *Bosmina longispina*, presence of *Ceriodaphnia*, and scattered occurrence of *Daphnia* throughout the core (Figure 2) as well as presence of some free living Chironomidae (*Ablabesmya* and *Procladius*, [24]). In the current results, *Bosmina* was associated with more negative $\delta^{13}C$ values of organic matter (Figure 3), indicative of an association with phytoplankton production. *Bosmina* is known to feed selectively on phytoplankton in small and shallow lakes in northern Finland [13,41]. In contrast to *Bosmina*, some planktonic cladocerans, e.g., *Daphnia* and *Ceriodaphnia* may assimilate benthic food particles grazing directly over the sediment microbial active layer [9]. In the RDA, *Daphnia* was associated with elevated $\delta^{13}C$ of organic matter suggesting a connection with the benthic production. Related to the shift in autochthonous production (i.e., increase in benthic production), *Bosmina* decreased slightly but still remained as the dominant planktonic taxon, while *Daphnia* increased at the top of the sediment core (Figure 2). Accordingly, it is possible that the establishment and increase of *Daphnia* ~1850 C.E. onward was related to its ability to utilize benthic resources [42]. In all, planktonic cladocerans, *B. longispina* as the governing taxon, showed rather constant presence in the sediment sequence aside from a transient increase at ~1400 C.E. (Figure 2) This abrupt event was likely associated with a short-term (~100 years) hydroclimatic event of the early Little Ice Age (LIA) causing higher lake level and relative increase of the planktonic habitat.

Despite the shift toward benthic production, *Bosmina* showed high resilience [41] as it remained abundant up to the surface (Figure 2). Though, a change in its reproduction occurred with a decreasing trend in sexual reproduction after a peak around 1300 C.E. suggesting less environmental stress (lower sexual reproduction) during the recent centuries. Sexual reproduction in Cladocera is evidenced in the fossil record through their ephippia and has previously been investigated in relation to climate oscillation, i.e., open-water season length, which dictates the relative importance of asexual (parthenogenetic) vs. sexual (gamogenetic) reproduction [28,43]. Accordingly, it has been suggested that sexual reproduction is less significant (lower sexual reproduction) during warm climate conditions as asexual reproduction prevails under long open-water season. As cold climate of the LIA started to prevail in northern Finland already during the 15th century, the long-term trend in *Bosmina* reproduction was not apparently strictly dictated by open-water season length but also other environmental stimuli counteracted in the reproduction patterns. *Bosmina* sex ratios had a strong negative relationship with sediment carbon content (and organic matter content, Figure 5c) possibly indicative of the past low productive conditions (i.e., low sediment carbon content) being more unfavorable to *Bosmina* than the new, i.e., post-1650 C.E., regime with increasing benthic autotochthonous production. Chydorid sexual reproduction increased during the 15th century, and later peaks ~1800 and 1900, that were likely attributable to reduced length of the open-water season driven by cold climate events of the LIA [44,45].

FD combines the observations of biodiversity and ecological functions to detect trends and patterns in ecosystem functions [17,46]. In the current study, cladoceran FD was generally lower during the early core and between 1200 and 1800 C.E. and higher between ~900 and 1200 C.E. and in the top sequence, from ~1850 onward (Figure 4). The pristine Lake Loažžejávri is mostly governed by natural forcers such as climate due to its remote location in the tundra [25] and, in agreement, there existed a positive relationship between FD and reconstructed T-JJA (Figure 5a) as FD was higher during the MCA and during the post-LIA climate warming (Figure 4). Based on previous paleoecological records [21,47], it has been suggested that long-term development of cladoceran FD is connected with lake productivity, although dependent on the size and eutrophication history of the system in question. We did not find any statistical relationships between FD and biogeochemical proxies related to productivity but the established relationship between FD and T-JJA likely reflects the ultimate climatic (i.e., temperature) control on habitat development and productivity in this cold tundra environment through dictating the length of ice-cover period [48].

4.2. Paleo-Optics

Natural variability in light and UV attenuation in forest and tundra lakes may be strongly related to coupling processes with temperature and precipitation via surrounding vegetation and permafrost [5,30]. However, in aquatic systems located on barren catchments terrestrial inputs of UV screening compounds are low and autochthonous organic matter, e.g., algal biomass acts as an important component in UV attenuation [49]. The tundra catchment of Lake Loažžejávri is currently covered with only ~5% of wetlands [23] but the lake shore is directly connected with paludified shorelines and water channels up in the nested drainage basin (Figure 1). Even though Loažžejávri features low lake water DOC (3.4 mg L^{-1}) and chl-a (1.6 µg L^{-1}), it has been estimated that only a few percent of UV penetrates below 0.5 m water depth [50]. However, the shallow littoral and benthic habitats of the lake are exposed to UV and aquatic biota are thus likely to respond to UV. The ABS$_{UV}$ record of Loažžejávri, indicating melanization and past UV exposure of the benthic *Alona*, was highly variable in the past suggesting large natural variations in underwater UV regimes (Figure 4).

In arctic and subarctic lakes, long-term changes in UV exposure have been mostly governed directly, or indirectly through photodegradation of OM, by solar radiation intensity [51–53]. Although solar forcing interacted partly in cladoceran assemblage succession, SSN reconstruction explaining ~12% of the species variation (Figure 3), it did not have substantial association with the current ABS$_{UV}$ record (Figure 4). In an adjacent lake, which differed from Loažžejávri by its location in the mountain birch woodland at a lower altitude, millennial variation in ABS$_{UV}$ was in clear connection with solar intensity [13,53]. In Loažžejávri, highest carapace ABS$_{UV}$ was recorded in the early core (~300 and 500 C.E.) and later on during 1400–1500 and 1700–1800 C.E. The high variability of ABS$_{UV}$ in the early core does not seem to clearly correlate with changes in SSN reconstruction, sediment geochemistry or cladoceran communities (Figures 2 and 4). However, the latter ABS$_{UV}$ peaks were likely related to a more transparent water column during the LIA, related to constrained organic matter flux from the catchment and reduced in-lake production driven by longer ice-cover period. A similar pattern of increased aquatic UV exposure during the LIA has occurred in lakes of northern and eastern Finland and in the Alps [30,53]. The pre MCA period at the middle of the first millennium of the Common Era has been characterized to be hydroclimatically dry in Fennoscandia. This period overlaps with the Dark Ages Cold Period characterized by noticeable climatic fluctuations [54,55], that may have been driving the early fluctuations in ABS$_{UV}$ (higher peaks ~300 and 500 C.E.) through climate-driven catchment coupling.

There existed a conspicuous long-term relationship between sediment C/N ratio and ABS$_{UV}$ indicating lower UV exposure under increasing autochthony (Figure 5b). This result contradicts a previous investigation from a regional lake set across the subarctic tree line suggesting an intrinsic control of terrestrial DOM (wetland origin) on underwater UV exposure and carapace UV absorbance [50]. As said, Lake Loažžejávri is located on a barren tundra catchment with little terrestrial organic matter contribution throughout its examined history, although connected with some wetland impact [18]. Since ABS$_{UV}$ was lower under prevalently autochthonous conditions, it is possible that UV screening properties of phytoplankton, or specifically those of phytobenthos in this benthic dominated system, have impacted carapace melanization of the benthic *Alona*. In subarctic and arctic lakes, benthic phototrophic communities, including cyanobacteria, algae, and hetero- and chemoautotrophic micro-organisms grow to form thick and stratified biofilms or mats [56], which support benthic and pelagic secondary production and bacterial planktonic production [57,58]. As such, these massive biofilms may act as physical UV refugia for benthic invertebrates allowing them to crawl deeper into the shelter of the mat, since top layers of the benthic mats are abundant in algal pigments screening UV [59,60]. At the same time as acting as a UV screen, the benthic mats likely provided fertile grazing surface for the microbenthos. It has been suggested previously that UV responses of shallow water benthic communities are principally driven by their access to physical UV refugia [61]. Accordingly, we propose that the generally lower carapace UV absorbance values in the top of the core, with the exception of the two high peaks apparently related to the LIA, were caused by increased benthic autochthonous production providing a physical UV shelter for benthic invertebrates.

5. Conclusions

Climatic fluctuations of the cold Little Ice Age and after were seen in the studied tundra lake as increases in benthic autochthonous production due to reduction in the length of ice-cover period. This centennial ecosystem scale succession toward dominance of benthic production has altered community structure, induced higher functional diversity, promoted relative importance of asexual reproduction, and reduced UV exposure in cladocerans. Main ecological mechanisms were related to diversification of benthic niche space, likely as a development of benthic microbial mats.

Fossil cladoceran communities in the studied tundra lake, consisting of planktonic and benthic species, seem to have been relatively resilient to climatic fluctuations until their habitat structure was disturbed. Periodically highly abundant sexual reproduction as a way for dormancy or better fitness (genotypic variation) has likely contributed to cladoceran community resilience. The diversification of the benthic niches induced functionally richer cladoceran communities including the keystone planktonic grazer *Daphnia* and specialized benthic species suggesting that functional diversity is coupled with lake productivity. In addition, the establishment of benthic mats likely provided physical UV refugia for benthic cladocerans emphasizing the fundamental importance of habitat quality for these microinvertebrates.

Author Contributions: Conceptualization, L.N. and E.H.K.; methodology, L.N., E.H.K., M.V.R. and T.P.L.; software, L.N.; validation, L.N., E.H.K., M.V.R. and T.P.L.; formal analysis, L.N.; investigation, L.N.; resources, L.N.; data curation, L.N.; writing—original draft preparation, L.N.; writing—review and editing, E.H.K., M.V.R. and T.P.L.; visualization, L.N. and E.H.K.; supervision, L.N.; project administration, L.N., E.H.K. and M.V.R.; funding acquisition, L.N.

Funding: This research was funded by Academy of Finland, grant number 287547 (VIOLET-project) and 308954 + 314107 (SCUM-project).

Acknowledgments: Annukka Galkin and staff of the Kevo Subarctic Research Station are thanked for assistance with the field work. We thank two anonymous reviewers for their constructive comments.

Conflicts of Interest: The authors declare no conflict of interest.

References

1. Smol, J.P. Arctic and Sub-Arctic shallow lakes in a multiple-stressor world: A paleoecological perspective. *Hydrobiologia* **2016**, *778*, 253–272. [CrossRef]
2. Smol, J.P.; Wolfe, A.P.; Birks, H.J.B.; Douglas, M.S.; Jones, V.J.; Korhola, A.; Pienitz, R.; Rühland, K.; Sorvari, S.; Antoniades, D.; et al. Climate-driven regime shifts in the biological communities of arctic lakes. *Proc. Nat. Acad. Sci. USA* **2005**, *102*, 4397–4402. [CrossRef] [PubMed]
3. Taranu, Z.E.; Gregory-Eaves, I.; Leavitt, P.R.; Bunting, L.; Buchaca, T.; Catalan, J.; Domaizon, I.; Guilizzoni, P.; Lami, A.; McGowan, S.; et al. Acceleration of cyanobacterial dominance in north temperate-subarctic lakes during the Anthropocene. *Ecol. Lett.* **2015**, *18*, 375–384. [CrossRef] [PubMed]
4. Cole, J.J.; Prairie, Y.T.; Caraco, N.F.; McDowell, W.H.; Tranvik, L.J.; Striegl, R.G.; Duarte, C.M.; Kortelainen, P.; Downing, J.A.; Middelburg, J.J.; et al. Plumbing the global carbon cycle: Integrating inland waters into the terrestrial carbon budget. *Ecosystems* **2007**, *10*, 172–185. [CrossRef]
5. Pienitz, R.; Vincent, W.F. Effect of climate change relative to ozone depletion on UV exposure in subarctic lakes. *Nature* **2000**, *404*, 484–487. [CrossRef] [PubMed]
6. Sobek, S.; Tranvik, L.J.; Prairie, Y.T.; Kortelainen, P.; Cole, J.J. Patterns and regulation of dissolved organic carbon: An analysis of 7500 widely distributed lakes. *Limnol. Oceanogr.* **2007**, *52*, 1208–1219. [CrossRef]

7. Jansson, M.; Bergström, A.; Blomqvist, P.; Drakare, S. Allochthonous organic carbon and phytoplankton/bacterioplankton production relationships in lakes. *Ecology* **2000**, *81*, 3250–3255. [CrossRef]

8. Rautio, M.; Bonilla, S.; Vincent, W.F. UV photoprotectants in arctic zooplankton. *Aquat. Biol.* **2009**, *7*, 93–105. [CrossRef]

9. Rautio, M.; Vincent, W.F. Benthic and pelagic food resources for zooplankton in shallow high-latitude lakes and ponds. *Freshw. Biol.* **2006**, *51*, 1038–1052. [CrossRef]

10. Frey, K.E.; Smith, L.C. Amplified carbon release from vast west Siberian peatlands by 2100. *Geophys. Res. Lett.* **2005**, *32*, L09401. [CrossRef]

11. Vonk, J.E.; Tank, S.E.; Bowden, W.B.; Laurion, I.; Vincent, W.F.; Alekseychik, P.; Amyot, M.; Billet, M.F.; Canário, J.; Cory, R.M.; et al. Reviews and syntheses: Effects of permafrost thaw on Arctic aquatic ecosystems. *Biogeosciences* **2015**, *12*, 7129–7167. [CrossRef]

12. Michelutti, N.; Wolfe, A.P.; Vinebrooke, R.D.; Rivard, B.; Briner, J.P. Recent primary production increases in arctic lakes. *Geophys. Res. Lett.* **2005**, *32*, L19715. [CrossRef]

13. Rantala, M.V.; Luoto, T.P.; Nevalainen, L. Temperature controls organic carbon sequestration in a subarctic lake. *Sci. Rep.* **2016**, *6*, 34780. [CrossRef] [PubMed]

14. Gregory-Eaves, I.; Beisner, B.E. Palaeolimnological insights for biodiversity science: An emerging field. *Freshw. Biol.* **2011**, *56*, 2653–2661. [CrossRef]

15. Froyd, C.A.; Willis, K.A. Emerging issues in biodiversity & conservation management: The need for a palaeoecological perspective. *Quat. Sci. Rev.* **2008**, *27*, 1723–1732.

16. Helmens, K.F.; Katrantsiotis, C.; Salonen, J.S.; Shala, S.; Bos, J.A.A.; Engels, S.; Kuosmanen, N.; Luoto, T.P.; Väliranta, M.; Luoto, M.; et al. Warm summers and rich biotic communities during N-Hemisphere deglaciation. *Glob. Planet. Chang.* **2018**, *167*, 61–73. [CrossRef]

17. Barnett, A.J.; Finlay, K.; Beisner, B.E. Functional diversity of crustacean zooplankton communities: Towards a trait-based classification. *Freshw. Biol.* **2007**, *52*, 796–813. [CrossRef]

18. Kivilä, E.H. Functional Paleoecology and Allochthonous Inputs in High Latitude Lake Food Webs. Ph.D. Thesis, University of Jyväskylä, Jyväskylä, Finland, 2019. Available online: https://jyx.jyu.fi/handle/123456789/63530 (accessed on 27 September 2019).

19. Luoto, T.P.; Nevalainen, L. Climate-forced patterns in midge feeding guilds. *Hydrobiologia* **2015**, *742*, 141–152. [CrossRef]

20. Nevalainen, L.; Kivilä, E.H.; Luoto, T.P. Biogeochemical shifts in hydrologically divergent taiga lakes in response to late Holocene climate fluctuations. *Biogeochemistry* **2016**, *128*, 201–215. [CrossRef]

21. Nevalainen, L.; Luoto, T.P. Relationship between cladoceran (Crustacea) functional diversity and lake trophic gradients. *Funct. Ecol.* **2017**, *31*, 488–498. [CrossRef]

22. Kivilä, E.H.; Luoto, T.P.; Rantala, M.V.; Kiljunen, M.; Rautio, M.; Nevalainen, L. Environmental controls on benthic food web functions and carbon resource use in subarctic lakes. *Freshw. Biol.* **2019**, *64*, 643–658. [CrossRef]

23. Rantala, M.V.; Nevalainen, L.; Rautio, M.; Galkin, A.; Luoto, T.P. Sources and controls of organic carbon in lakes across the subarctic treeline. *Biogeochemistry* **2016**, *129*, 235–253. [CrossRef]

24. Luoto, T.P.; Kivilä, E.H.; Rantala, M.V.; Nevalainen, L. Characterization of the Medieval Climate Anomaly, Little Ice Age and recent warming in northern Lapland. *Int. J. Climatol.* **2017**, *37*, 1257–1266. [CrossRef]

25. Kivilä, E.H.; Luoto, T.P.; Rantala, M.V.; Nevalainen, L. Long-term variability in chironomid functional assemblages and carbon utilization in a tundra lake food web. *Hydrobiologia* **2019**, submitted.

26. Szeroczyńska, K.; Sarmaja-Korjonen, K. *Atlas of Subfossil Cladocera from Central and Northern Europe*; Friends of the Lower Vistula Society: Swiecie, Poland, 2007.

27. Kurek, J.; Korosi, J.B.; Jeziorski, A.; Smol, J.P. Establishing reliable minimum count sizes for cladoceran subfossils sampled from lake sediments. *J. Paleolimnol.* **2010**, *44*, 603–612. [CrossRef]

28. Sarmaja-Korjonen, K. Chydorid ephippia as indicators of past environmental changes—A new method. *Hydrobiologia* **2004**, *526*, 129–136. [CrossRef]

29. Rao, C.R. Diversity and dissimilarity coefficients: a unified approach. *Theor. Popul. Biol.* **1982**, *21*, 24–43. [CrossRef]

30. Nevalainen, L.; Rautio, M. Spectral absorbance of benthic cladoceran carapaces as a new method for inferring past UV exposure of aquatic biota. *Quat. Sci. Rev.* **2014**, *84*, 109–115. [CrossRef]

31. Solanki, S.K.; Usoskin, I.G.; Kromer, B.; Schüssler, M.; Beer, J. An unusually active Sun during recent decades compared to the previous 11,000 years. *Nature* **2004**, *431*, 1084–1087. [CrossRef]

32. Matskovsky, V.V.; Helama, S. Testing long-term summer temperature reconstruction based on maximum density chronologies obtained by reanalysis of tree-ring data sets from northernmost Sweden and Finland. *Clim. Past* **2014**, *10*, 1473–1487. [CrossRef]

33. Šmilauer, P.; Leps, J. *Multivariate Analysis of Ecological Data Using Canoco 5*; Cambridge University Press: Cambridge, UK, 2014.

34. Hammer, Ø.; Harper, D.A.T.; Ryan, P.D. Past: Paleontological statistics software package for education and data analysis. *Palaeontol. Electron.* **2001**, *4*, 9.

35. Finstad, A.G.; Andersen, T.; Larsen, S.; Tominaga, K.; Blumentrath, S.; De Wit, H.A.; Tømmervik, H.; Hessen, D.O. From greening to browning: Catchment vegetation development and reduced S-deposition promote organic carbon load on decadal time scales in Nordic lakes. *Sci. Rep.* **2016**, *6*, 31944. [CrossRef] [PubMed]

36. Wauthy, M.; Rautio, M.; Christoffersen, K.S.; Forsström, L.; Laurion, I.; Mariash, H.L.; Peura, S.; Vincent, W.F. Increasing dominance of terrigenous organic matter in circumpolar freshwaters due to permafrost thaw. *Limnol. Oceanogr. Lett.* **2018**, *3*, 186–198. [CrossRef]

37. Bogard, M.J.; Kuhn, C.D.; Johnston, S.E.; Striegl, R.G.; Holtgrieve, G.W.; Dornblaser, M.M.; Spencer, R.G.M.; Wickland, K.P.; Butman, D.E. Negligible cycling of terrestrial carbon in many lakes of the arid circumpolar landscape. *Nat. Geosci.* **2019**, *12*, 180. [CrossRef]

38. Schindler, D.E.; Scheuerell, M.D. Habitat coupling in lake ecosystems. *Oikos* **2002**, *98*, 177–189. [CrossRef]

39. Griffiths, K.; Michelutti, N.; Sugar, M.; Douglas, M.S.V.; Smol, J.P. Ice-cover is the principal driver of ecological change in High Arctic lakes and ponds. *PLoS ONE* **2017**, *12*, e0172989. [CrossRef] [PubMed]

40. Thienpont, J.R.; Korosi, J.B.; Cheng, E.S.; Deasley, K.; Pisaric, M.F.; Smol, J.P. Recent climate warming favours more specialized cladoceran taxa in western Canadian Arctic lakes. *J. Biogeogr.* **2015**, *42*, 1553–1565. [CrossRef]

41. Rantala, M.V.; Luoto, T.P.; Weckström, J.; Perga, M.-E.; Rautio, M.; Nevalainen, L. Climate controls on the Holocene development of a subarctic lake in northern Fennoscandia. *Quat. Sci. Rev.* **2015**, *126*, 175–185. [CrossRef]

42. Mariash, H.L.; Devlin, S.P.; Forsström, L.; Jones, R.I.; Rautio, M. Benthic mats offer a potential subsidy to pelagic consumers in tundra pond food webs. *Limnol. Oceanogr.* **2014**, *59*, 733–744. [CrossRef]

43. Kultti, S.; Nevalainen, L.; Luoto, T.P.; Sarmaja-Korjonen, K. Subfossil chydorid (Cladocera, Chydoridae) ephippia as paleoenvironmental proxies: Evidence from boreal and subarctic lakes in Finland. *Hydrobiologia* **2011**, *676*, 23–37. [CrossRef]

44. Luoto, T.P.; Nevalainen, L.; Sarmaja-Korjonen, K. Multi-proxy evidence for the 'Little Ice Age' from Lake Hampträsk Southern Finland. *J. Paleolimnol.* **2008**, *40*, 1097–1113. [CrossRef]

45. Nevalainen, L.; Helama, S.; Luoto, T.P. Hydroclimatic variations over the last millennium in eastern Finland disentangled by fossil Cladocera. *Palaeogeogr. Palaeoclimatol. Palaeoecol.* **2013**, *378*, 13–21. [CrossRef]

46. Hooper, D.U.; Solan, M.; Symstad, A.; Dıaz, S.; Gessner, M.O.; Buchmann, N.; Degrange, P.; Grime, P.; Hulot, F.; Mermillod-Blondin, F.; et al. Species diversity, functional diversity, and ecosystem functioning. In *Biodiversity and Ecosystem Functioning. Synthesis and Perspectives*; Loreau, M., Naeem, S., Inchausti, P., Eds.; Oxford University Press: Oxford, UK, 2002; pp. 95–281.

47. Nevalainen, L.; Brown, M.; Manca, M. Sedimentary record of cladoceran functionality under eutrophication and reoligotrophication in Lake Maggiore, northern Italy. *Water* **2018**, *10*, 86. [CrossRef]

48. Clark, G.F.; Stark, J.S.; Johnston, E.L.; Runcie, J.W.; Goldsworthy, P.M.; Raymond, B.; Riddle, M.J. Light-driven tipping points in polar ecosystems. *Glob. Chang. Biol.* **2013**, *19*, 3749–3761. [CrossRef] [PubMed]

49. Laurion, I.; Ventura, M.; Catalan, J.; Psenner, R.; Sommaruga, R. Attenuation of ultraviolet radiation in mountain lakes: Factors controlling the among-and within-lake variability. *Limnol. Oceanogr.* **2000**, *45*, 1274–1288. [CrossRef]

50. Nevalainen, L.; Luoto, T.P.; Rantala, M.V.; Galkin, A.; Rautio, M. Role of terrestrial carbon in aquatic UV exposure and photoprotective pigmentation of meiofauna in subarctic lakes. *Freshw. Biol.* **2015**, *60*, 2435–2444. [CrossRef]

51. Nevalainen, L.; Rantala, M.V.; Luoto, T.P.; Rautio, M.; Ojala, A.E.K. Ultraviolet radiation exposure of a high arctic lake in Svalbard during the Holocene. *Boreas* **2015**, *44*, 401–412. [CrossRef]

52. Nevalainen, L.; Rantala, M.V.; Luoto, T.P.; Ojala, A.E.K.; Rautio, M. Long-term changes in pigmentation of arctic *Daphnia* provide potential for reconstructing aquatic UV exposure. *Quat. Sci. Rev.* **2016**, *144*, 44–50. [CrossRef]

53. Nevalainen, L.; Rantala, M.V.; Rautio, M.; Luoto, T.P. Spatio-temporal cladoceran (Branchiopoda) responses to climate change and solar radiation in subarctic ecotonal lakes. *J. Biogeogr.* **2018**, *45*, 1954–1965. [CrossRef]

54. Helama, S.; Jones, P.D.; Briffa, K.R. Dark Ages Cold Period: A literature review and directions for future research. *Holocene* **2017**, *27*, 1600–1606. [CrossRef]

55. Linderholm, H.W.; Nicolle, M.; Francus, P.; Gajewski, K.; Helama, S.; Korhola, A.; Solomina, O.; Yu, Z.; Zhang, P.; D'Andrea, W.J.; et al. Arctic hydroclimate variability during the last 2000 years-current understanding and research challenges. *Clim. Past* **2018**, *14*, 473–514. [CrossRef]

56. Quesada, A.; Fernandez-Valiente, E.; Hawes, I.; Howard-Williams, C. Benthic primary production in polar lakes and rivers. In *Polar Lakes and Rivers. Limnology of Arctic and Antarctic Aquatic Ecosystems*; Vincent, V.F., Laybourn-Parry, J., Eds.; Oxford University Press: Oxford, UK, 2008; pp. 179–196.

57. Rodríguez, P.; Ask, J.; Hein, C.L.; Jansson, M.; Karlsson, J. Benthic organic carbon release stimulates bacterioplankton production in a clear-water subarctic lake. *Freshw. Sci.* **2013**, *32*, 176–182. [CrossRef]

58. Sierszen, M.E.; McDonald, M.E.; Jensen, D.A. Benthos as the basis for arctic lake food webs. *Aquat. Ecol.* **2003**, *37*, 437–445. [CrossRef]

59. Bonilla, S.; Villeneuve, V.; Vincent, W.F. Benthic and planktonic algal communities in a high arctic lake: Pigment structure and contrasting responses to nutrient enrichment. *J. Phycol.* **2005**, *41*, 1120–1130. [CrossRef]

60. Lionard, M.; Péquin, B.; Lovejoy, C.; Vincent, W.F. Benthic cyanobacterial mats in the high arctic: Multi-layer structure and fluorescence responses to osmotic stress. *Front. Microbiol.* **2012**, *3*, 140. [CrossRef] [PubMed]

61. Vinebrooke, R.D.; Leavitt, P.R. Differential responses of littoral communities to ultraviolet radiation in an alpine lake. *Ecology* **1999**, *80*, 223–237. [CrossRef]

Article

Changes in Planktivory and Herbivory Regimes in a Shallow South American Lake (Lake Blanca Chica, Argentina) Over the Last 250 Years

David Carrozzo [1], Simona Musazzi [2], Andrea Lami [2], Francisco E. Córdoba [3] and María de los Ángeles González Sagrario [4,*]

[1] Departamento de Biología, Universidad Nacional de Mar del Plata, J. B. Justo 2550,
     Mar del Plata 7600, Argentina; davidrcarrozzo@gmail.com
[2] National Research Council (CNR),WaterResearchInstitute (IRSA), Largo Tonolli50, 28922 Verbania, Italia;
     simona.musazzi@irsa.cnr.it (S.M.); andrea.lami@cnr.it (A.L.)
[3] Instituto de Ecorregiones Andinas (INECOA, CONICET-UNJu), Instituto de Geología y Minería,
     Universidad Nacional de Jujuy, Av. Bolivia 1661, San Salvador de Jujuy Y4600GNE, Argentina;
     francisco.e.cordoba@gmail.com
[4] Instituto de Investigaciones Marinas y Costeras (IIMYC), Facultad de Ciencias Exactas y Naturales,
     Universidad Nacional de Mar del Plata, CONICET, J. B. Justo 2550, Mar del Plata 7600, Argentina
*   Correspondence: gonsagra@gmail.com or gonsagra@mdp.edu.ar

Received: 24 December 2019; Accepted: 18 February 2020; Published: 22 February 2020

Abstract: Shallow lakes are vulnerable ecosystems impacted by human activities and climate change. The Cladocera occupy a central role in food webs and are an excellent paleoecological indicator of food web structure and trophic status. We conducted a paleolimnological study in Lake Blanca Chica (Argentina) to detect changes on the planktivory and herbivory regimes over the last 250 years. Generalized additive models were fitted to the time series of fish predation indicators (ephippial abundance and size, mucrone size, fish scales, and the planktivory index) and pheophorbide *a* concentration. The cladoceran assemblage changed from littoral-benthic to pelagic species dominance and zooplankton switched from large-bodied (*Daphnia*) to small-bodied grazers (*Bosmina*) ca. 1900 due to increased predation. The shift in planktivory regime (ca. 1920–1930), indicated by fish scales and the planktivory index, as well as herbivory (ca. 1920–1950), was triggered by eutrophication. Changes in planktivory affected the size structure of *Bosmina*, reducing its body size. This study describes the baseline for the lake as well as the profound changes in the composition and size structure of the zooplankton community due to increased predation and the shift in the planktivory regime. These findings will provide a reference status for future management strategies of this ecosystem.

Keywords: *Daphnia*; *Bosmina*; pheophorbide *a*; fish predation; grazing; ephippia; cladocera sub-fossil remains

1. Introduction

Zooplankton play a pivotal role in aquatic ecosystems and global biogeochemical cycles. In fact, zooplankton act as a hinge in the aquatic food web because they exert a control role on phytoplankton through grazing (performed by herbivorous filter organisms such as cladocerans and rotifers) and are the food resource of higher trophic levels [1,2]. Zooplankton are highly sensitive to changes in aquatic ecosystems [3]. The effects of environmental disturbances can be detected through changes in species composition, abundance, and body size distribution. Grazing capacity on algae is directly related to the body size of organisms—the larger size, the greater grazing capacity [4]. Zooplankton are sensitive to the predation pressure exerted by fish (decreasing their body size) and macroinvertebrates (increasing their body size [5]). Jeppesen et al. [6] showed that the top-down control in shallow lakes is stronger due to the lower abundance of piscivorous relative to planktivorous and omnivorous fish. Cladocerans are an important component of the zooplankton and the chitinous structures of their exoskeleton (postabdomen, jaws, claws, antenna segments) and resistance eggs (ephippia) are an important, well-preserved, and taxonomically well-known autochthonous element of lake sediment. Due to their potential role as indicators of various environmental conditions, the study of subfossil remains of Cladocera has been very useful in identifying and inferring changes in trophic status, predation by fish and invertebrates, macrophyte coverage, as well as chemical and physical properties of lake water [7–9].

Shallow lakes are considered as an ecosystem model in which regime shift, that is, rapid and abrupt transition from one persistent regime to a different regime, can be studied [10]. Driving forces may be both stochastic events (for example, violent storms, massive fish death, and herbicide use) and gradual processes that slowly degrade system resilience (eutrophication or global warming) [10]. The fact that a shallow lake is in a clear or murky regime implies that it presents distinctive feedback mechanisms as well as community structure and functioning. Thus, lakes in a clear water regime colonized by submerged macrophytes with a moderate–low plant coverage, have good light penetration, low pelagic primary production, zooplankton of large body size, high diversity of invertebrates and fish, and a high ratio of piscivorous to planktivorous fish [11]. In these systems, the control of zooplankton over phytoplankton (top-downregulation) is favoured because the piscivorous fish reduce the pressure of predation on the Cladocera community, made up of species of large body size and greater capacity for grazing, and also macrophytes might offer refuge to large cladocerans against planktivores [12]. In contrast, in systems with a turbid regime, phytoplankton become dominant because these lakes have a high level of nutrients and there is weak regulation from above as zooplankton are composed of species with less grazing capacity. Furthermore, the piscivorous to planktivorous fish ratio is low [6]. The shift from a clear macrophyte-dominated regime to a turbid one implies a drastic change on the balance and the interaction between the benthic-littoral and pelagic environments, with impacts on the fluxes of organic matter and nutrients, as well as the composition of the different assemblages and communities [6]. For example, the loss of macrophytes implies a change in the composition of the Cladocera assemblage with a decrease in the proportion of littoral and benthic representatives.

The stability of the zooplankton community in shallow lakes is strongly linked to small changes in the surface sediment condition (e.g., bank of seeds and eggs, physical changes of sediment). These changes cannot be detected by short-term studies because they are visible over long-time scales (e.g., decades or centuries) but can potentially be revealed by high resolution paleolimnological studies [9].

Lake sediments are rich sources of paleoecological information derived from the functioning of the lake ecosystem as well as the surrounding watershed. Sedimentary archives have been used to address several issues such as acidification, eutrophication, and lake ontogeny [13,14]. Among the proxies preserved in the sediment, algal pigments have been shown to have the same periodicity as fish fluctuations, and the deposition rates of pigments derived from edible algae are increased by the grazing activity of large-bodied zooplankton [15]. Further, herbivory creates novel derivatives such as pheophorbide *a* [16], and can increase the overall concentration of pigments in the remaining undigested material [16]. Thus, both accumulation rates and concentrations of sedimentary pigments are connected to the dynamic of zooplankton populations [17].

Size-selective fish predation has an impact on the zooplankton community structure, leading to the dominance of small-bodied species. As zooplankton remains are deposited in lake sediments in a direct and frequent way in proportion to zooplankton abundance, microfossils can be used to infer planktivorous predation pressure or even reconstruct fish stocks [18,19]. Several indicators of fish predation have been recognized as changes in ephippia size; the length of cladoceran mandibles, carapaces, or mucrones; or the planktivory index((*Daphnia*)/(*Daphnia* + *Bosmina*)) [11,20,21]. For example, the size of *Daphnia* ephippium has been demonstrated to be directly related to the size of female-bearing eggs [22], whereas the length of the mucrone with the total length of the individual is the case for *Bosmina huaronensis* Delauchaux 1978 [23].

The main objective of this paper was to establish the occurrence of changes in the size of subfossil Cladocera remains and in the concentration of pheopigments in order to infer changes in the planktivory and herbivory pressure during the last 250 years in a shallow lake located on the Pampa Plain(Argentina).In this study, we identified changes in a herbivory biomarker on the size of *B. huaronensis* (mucrone) and the replacement of *Daphnia* species by a small cladoceran (*Bosmina*), indicating a change in two main ecological processes, grazing and planktivory over the last 250 years. The change in predation and grazing was mainly triggered by eutrophication. To evaluate changes in the planktivory and herbivory regimes, we determined the abundance of *Daphnia* and *Bosmina* species, the size of *Daphnia* species ephippia and *Bosmina* mucrone, the abundance of fish scales, and the concentration of pheophorbide *a*.

2. Materials and Methods

2.1. Study Area

Lake Blanca Chica (36°50′00.9″ S; 60°28′00.9″ W) (Figure 1) is located in the Pampa Plain area (Central Argentina, South America), which constitutes one of the largest areas of wetlands of South America [24]. The lake is a shallow (1–2 m), turbid (Secchi Disc depth: 0.2–0.3 m, chlorophyll *a* concentration: 90–500 mg m^{-3}), warm temperature, and a polymictic lake of alkaline water (pH: 8–9.8). Currently, it is in eutrophic status (Total Phosphorus: 0.3–1.2 ppm), lacking submerged vegetation. The fish community is dominated by the planktivorous *Odontesthes bonariensis* (Valenciennes, 1835) in open waters and by the small fish *Cheirodon interruptus* (Jenyns, 1842) in the littoral zone. The zooplankton are composed of small–medium sized Cladocera, Rotifera, and Copepoda species [25].

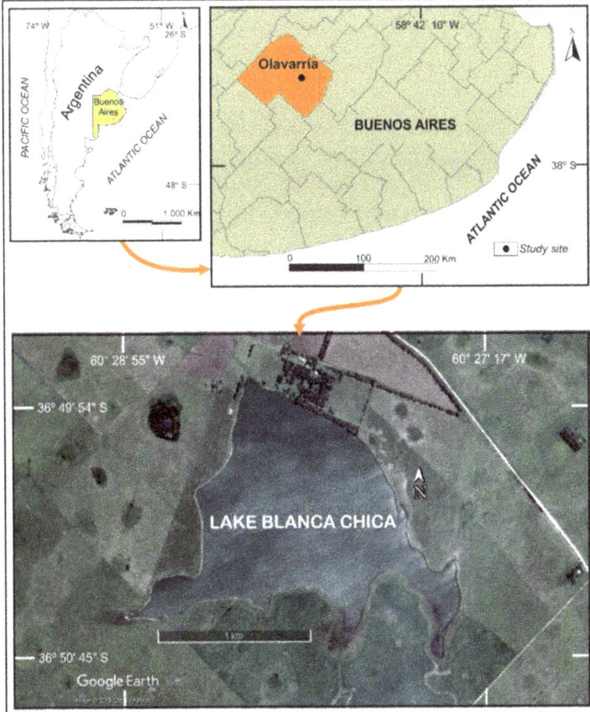

Figure 1. Location of Lake Blanca Chica in the the Argentinean Pampa Plain, South America.

2.2. Core Chronology and Paleolimnological Analyses

A core of 60 cm was collected at the deepest part of the lake using a vibracorer on October 2015. Sub-sampling (every 1cm interval) was carried out on the entire core. Results described in this study correspond to the top 30 cm of the core. Below this level, no pigments or ephippia were detected. The age–depth model was developed applying the constant flux/constant sedimentation (CFCS) model to the radionuclide activities of ^{137}Cs, ^{210}Pb, and ^{226}Ra (detected at the Radiochronology Laboratory of the Laval University, Canada). Three known chronomarkers were used to constrain the ^{210}Pb model: the^{137}Cs maximum peak at 6–7 cm as a time marker of A.D. 1964/1965; the first occurrence of *Eucalyptus* pollen (exotic tree)in the sedimentary record as a result of an extensive forestation since 1880 in Argentina; and a lithology change corresponding to an extreme regional dry pulse registered across the Pampean Plain at the end of the Little Ice Age (estimated to be A.D. 1775 ± 10 years) and recognized in other lakes, for instance, Mar Chiquita, Melincué [26,27]. For further details on the age–depth model, see [28].

Herbivory was inferred in determining the concentration of the pheophorbide *a* as it represents a grazer biomarker of invertebrate herbivory [20]. This pigment was extracted with 5 ml of a 90:10 acetone/MilliQwater solution overnight in the dark at 4 °C, after flushing with nitrogen from around 1 g of wet sediment. Then, sediments were centrifuged at 3000 rpm for 10 minutes before detection using high pressure liquid chromatography with an Ultimate 3000 system (Thermo Scientific, Waltham). The elution program and the methodology for pigment identification and quantification followed previous protocols [29].

Changes in the planktivory regime were inferred considering several indicators of fish predation: the abundance and size of *Daphnia* ephippia, the relative abundance and mucrone size of *Bosmina*, and the planktivory index((*Daphnia*)/(*Daphnia* + *Bosmina*)) [11,21]. This index was calculated considering the relative contribution of ephippia from *Daphnia spinulata* Birabén 1917, *Daphnia obtusa* Kurz 1875, and *B. huaronensis*, as chitinous remains from *Daphnia* species were very scarce or not found in the sedimentary record. Values of the planktivory index closeto 1 indicate a low planktivorous pressure, but when the ratio decreases it represents a higher fish predation pressure. Chitinous remains, ephippia, and fish scales were retained through washing 2–8 cm^3 of sediment gently through a 50 μm sieve. Ephippia and fish scales were identified and were counted under a stereomicroscope at 10–40× magnification, whereas *Bosmina* remains were enumerated at 100–200xmagnification using a Zeiss Primo Star microscope. *Bosmina* remains were analysed on approximately a quarter of the sample after heating (70–80 °C) in 10% KOH, a deflocculating agent, for 45 min [30]. The length of the mucrone was considered as the distance from its base to its extreme, and the length of ephippia as the dorsal length, excluding anterior and posterior appendages. A total of 576 ephippia from *D. spinulata*, 324 from *D. obtusa*, 754 from *Moina* sp., and 1078 mucrones of *B. Huaronensis* were measured. At least 10 ephippia of each *Daphnia* species, 30 of *Moina*, and 30–50 mucrones of *Bosmina* were measured to calculate the mean size per stratigraphic level. The contribution of cladoceran species to the ephippia assemblage was expressed as weighted relative abundance [19], that is, ephippial relative abundance was weighted by that sample's total ephippial abundance. Several keys were used for taxonomic identification [23,31–33].

2.3. Data Analyses

The time series of the different biological indicators was modelled using generalized additive models (GAMs) [34]. In all the cases, fitted GAMs were estimated using maximum likelihood-based smoothness selection procedures, in particular the restricted maximum likelihood (REML). A continuous time first-order autoregressive process (CAR(1)) was chosen to account for the correlation between residuals [35]. To identify periods of transition, we estimated simultaneous confidence intervals from the posterior distribution of the model (under an empirical Bayesian formulation), and the first derivative of the fitted trend [35]. Periods of significant change are identified as those time points where the simultaneous confidence interval on the first derivative bounded away from zero [35].

The estimation of the models and derivatives were performed using the *mgcv* and *gratia* packages. In the case of the size of ephippia or mucrones, GAMs were fitted to the time series of the mean size for each indicator. Fish scales abundance and chitinous remains of *B. huaronensis* data were provided by González Sagrarioand co-workers [28].

To summarize changes in the ephippial assemblage, correspondence analysis was performed (package *vegan*) [36] and the number of axes that best explained data variance were selected according to the Kaiser–Guttman criterion and the broken stick model [37]. For further details, see Figure S1.

All analyses were performed using R version 3.5.1 [38].

3. Results

3.1. Herbivory Biomarker

The fitted GAM to the time series of pheophorbide *a* concentration explained a high percentage of data deviance and showed an increasing trend since ca. 1930. In particular, two increasing transitional periods occurred: ca. 1931–1949 and ca. 1991–2009 (Figure 2; Table 1), which indicate an increase in the total herbivory in the system.

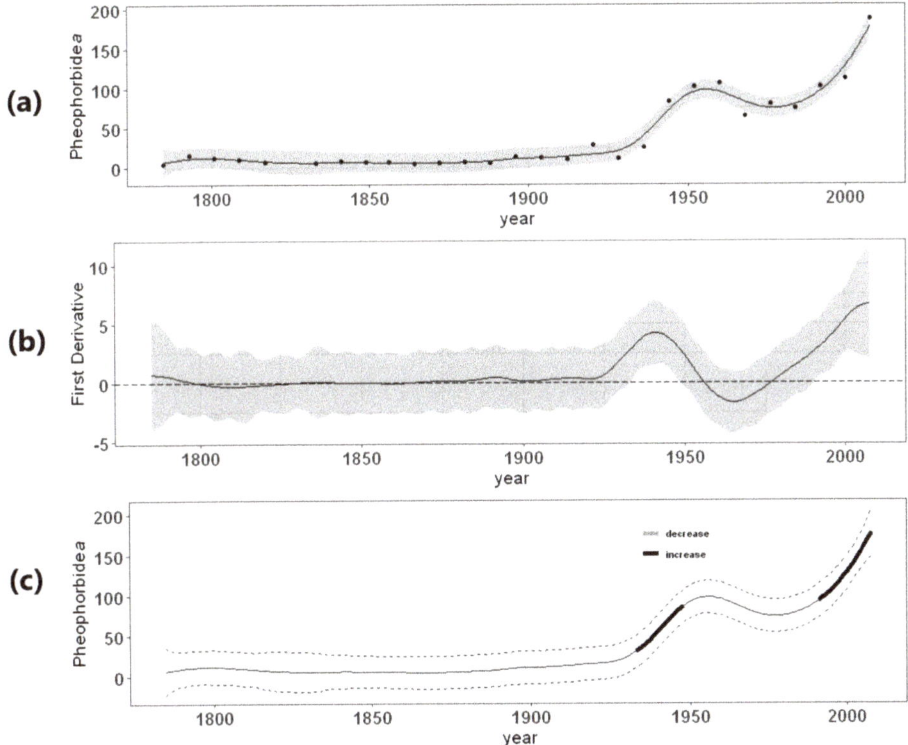

Figure 2. Generalized additive model (GAM) fitted to the time series of pheophorbide *a* (expressed as η Molesg OM^{-1})in the sedimentary archive of Lake Blanca Chica. (**a**) Observed values, GAM-based trend fitted and its simultaneous interval; (**b**) estimated first derivative of the GAM-fitted trend and the 95% simultaneous interval; (**c**) period of transition.

3.2. Ephippial Assemblage

This study represents the first record of the ephippia of *Daphnia spinulata*, *Daphnia obtusa*, *Bosmina Leiderobosmina huaronensis*, *Moina* sp. Baird 1850, *Ceriodaphnia* sp. Dana 1853, *Leydigia louisi* Jenkin 1934cf, *Pleuroxus* sp. Baird1843, and *Chydorus sphaericus* (O. F: Müller 1776) group for the shallow lakes from the Pampa Plain. The weighted relative abundance of each species and the trend detected along the sedimentary record of Lake Blanca Chica can be found in Figures S1 and S2. Additionally, a decrease in the total sum of ephippia occurred after ca. 1915–1920, mostly represented by the loss of littoral species (Figures S1 and S3).

3.3. Fish Predation Indicators

The ephippia relative abundance and mean size of *Daphnia* species showed contrasting patterns. The effect of the smooth term for ephippia abundance and mean size was significant in all cases, and the fitted GAMs explained 25–38%of time series deviance (Table 1). *D. spinulata* showed an increasing trend in the mean size of the ephippium along the entire time series. In contrast, the relative abundance of ephippia in the sedimentary record showed an increasing trend since ca. 1785–1815 until ca. 1900, and after that a decreasing trend, diminishing its relative contribution to the lake zooplankton towards the present time (Figures 3 and 4). Similarly, *D. obtusa* showed an increase in ephippia mean size ca.1950–1995 and its contribution to the ephippial assemblage decreased since ca. 1900 (Figures 3 and 5). No trend for *Moina* ephippia contribution or mean size was detected (Table 1). Indeed, both variables showed a high inter-decadal/inter-annual variation (Figure 3 and Figure S1).

The fitted GAM to *B. huaronensis* relative abundance (chitinous remains) explained a high percent of the data deviance of the time series (81.7%) (Table 1). The first derivative of the fitted trend and its simultaneous interval showed an increase in *Bosmina* contribution ca. 1850–1880; after which no further change occurred (Figure 6). In contrast, a decrease in its size occurred ca. 1915–1935, stabilizing in a smaller size after 1940 (Figures 3 and 6; Table 1). For descriptive statistics on mucrone size, see Table S1.

Table 1. Results from generalized additive models (GAMs) fitted to the temporal series of the different proxies from the sedimentary record of Lake Blanca Chica. The estimated degrees of freedom (*edf*), *F*-statistic (Gaussian distributions) or *chi*-statistic (scaled or gamma distributions, values denoted by *), and *p*-values of the smooth term and the percent (%) of the deviance explained by the fitted model are shown. The significance level was set at $p < 0.05$; ns: non-significant results. All the fitted models were estimated using continuous time first-order autoregressive process (CAR(1)) and restricted maximum likelihood (REML)smoothness selection.

Indicator	Species	%	r^2	*p*-Value	*Edf*	*F or Chi* *
Ephippia weighted	*D. spinulata*	25.2	0.205	0.01	0.870	3.362
Abundance (%)	*D. obtusa*	25.8	0.213	0.00918	1.532	7.294 *
	Moina sp.	9.63	0.0645	Ns	0.924	0.207
Ephippia mean	*D. spinulata*	35.4	0.329	0.000789	1	14.25
Size	*D. obtusa*	37.1	0.464	$<1 \times 10^{-6}$	2.714	32.01 *
	Moina sp.	8.01	0.0447	Ns	1	2.262
Abundance (%)	*B. huaronensis*	81.7	0.778	9.12×10^{-11}	4.796	3.506
Mucrone mean Length	*B. huaronensis*	73.3	0.67	2.13×10^{-5}	5.118	8.71
Pheophorbide *a*		97.5	0.955	$<2 \times 10^{-16}$	11.85	39.2
Planktivory index		29.2	0.265	0.00106	1	10.71 *
Fish scales		64.4	0.547	2.6×10^{-13}	2.219	52.1 *

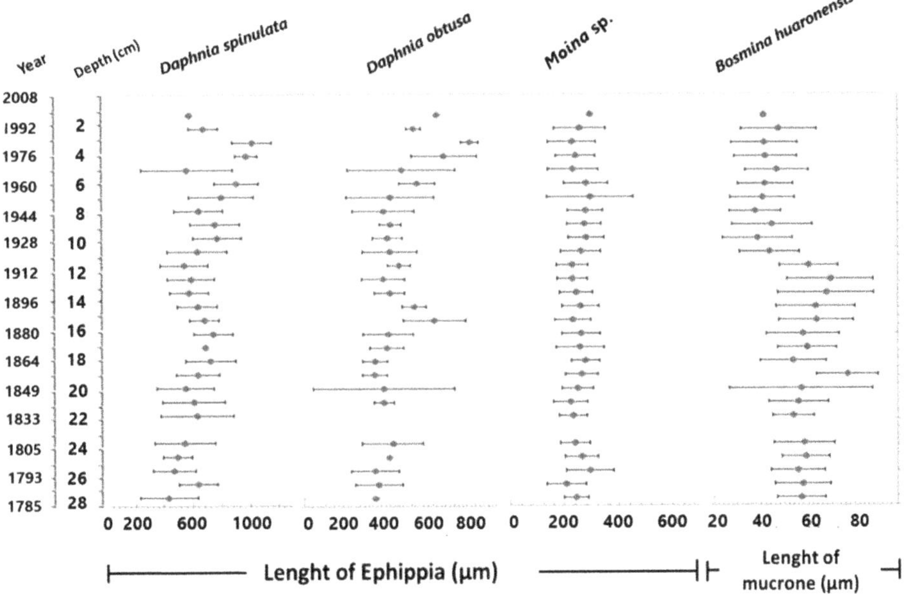

Figure 3. Size of the ephippia of *Daphnia* species and *Moina* sp. and the mucrone of *Bosmina huaronensis* in the sedimentary record of Lake Blanca Chica.

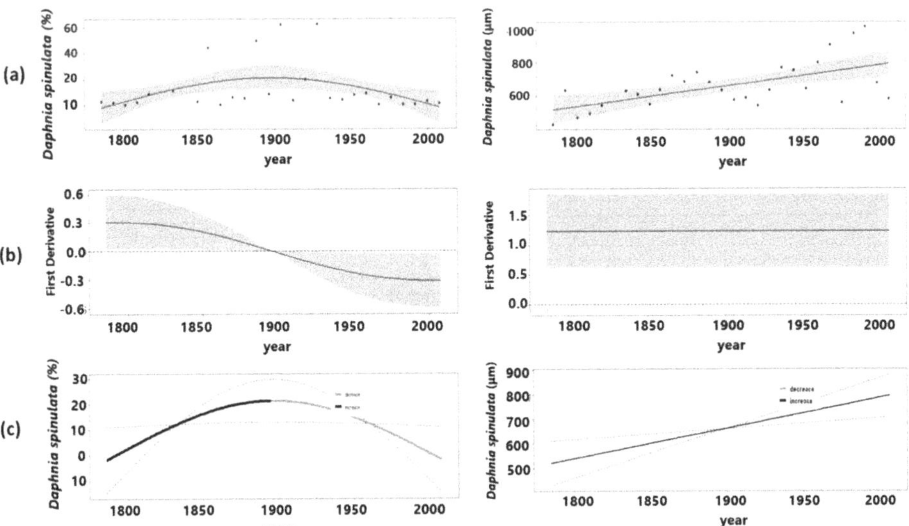

Figure 4. GAM fitted to the time series of *Daphnia spinulata* ephippia abundance (left pannel) and ephippia size (right pannel) in the sedimentary archive of Lake Blanca Chica. (**a**) Observed values, GAM-based trend fitted and its simultaneous interval; (**b**) estimated first derivative of the GAM fitted trend and the 95% simultaneous interval; (**c**) period of transition.

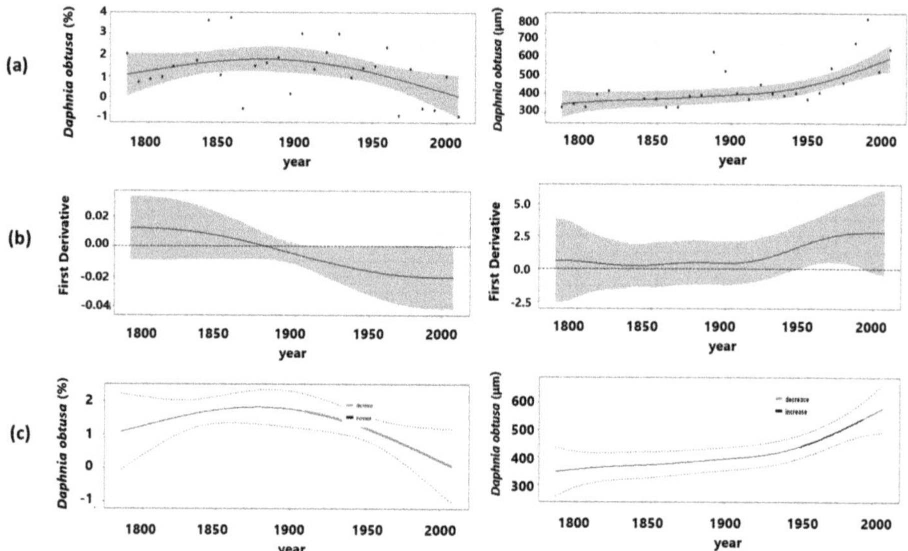

Figure 5. GAM fitted to the time series of *Daphnia obtuse* ephippia abundance (left pannel) and ephippia size (right pannel) in the sedimentary archive of Lake Blanca Chica. (**a**) Observed values, GAM-based trend fitted and its simultaneous interval; (**b**) estimated first derivative of the GAM fitted trend and the 95% simultaneous interval; (**c**) period of transition.

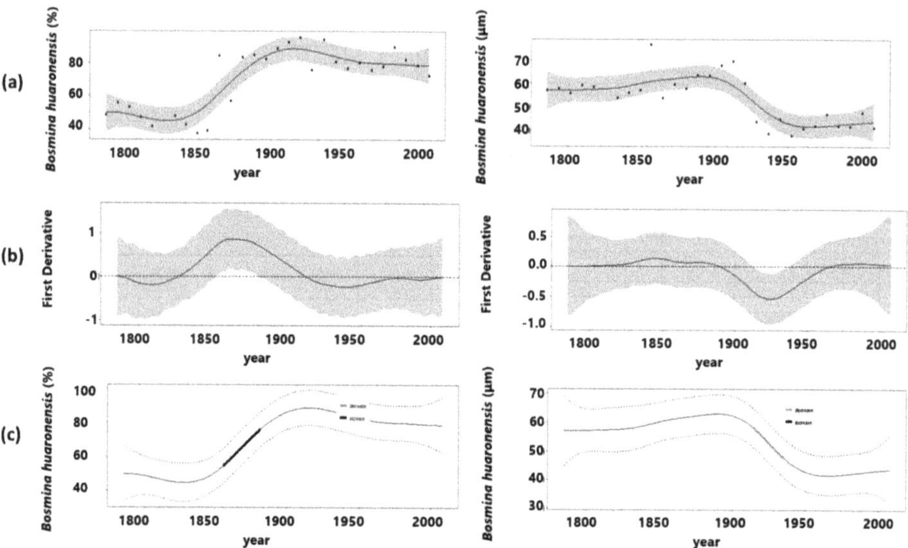

Figure 6. GAM fitted to the time series of *Bosmina huaronensis* abundance (chitinous remians) (left pannel) and mucrone size (right pannel)in the sedimentary archive of Lake Blanca Chica. (**a**) Observed values, GAM-based trend fitted and its simultaneous interval; (**b**) estimated first derivative of the GAM fitted trend and the 95% simultaneous interval; (**c**) period of transition.

Fish scale abundance, an indicator of fish abundance, showed an increasing trend since ca.1930 to 2000 according to the first derivative of the fitted GAM. The model explained a high percentage of data deviance (64.4%) (Table 1; Figure 7). In concordance, a decreasing trend in the planktivory index occurred along the time series (Table 1; Figure 8). The GAM fitted to the planktivory index explained 29.2% of the deviance, and although it showed a decreasing pattern, it can also be observed that most of the values lower than 0.5 occurred after 1900 (Figure 8).

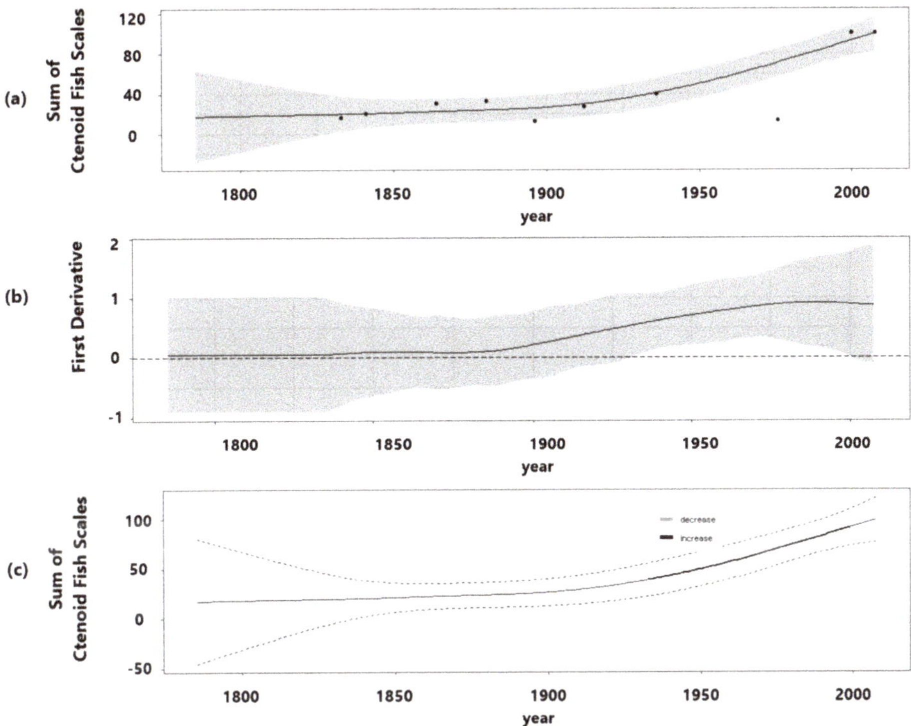

Figure 7. GAM fitted to the time series of fish scale abundance in the sedimentary archive of Lake Blanca Chica. (**a**) Observed values, GAM-based trend fitted and its simultaneous interval; (**b**) estimated first derivative of the GAM fitted trend and the 95% simultaneous interval; (**c**) period of transition.

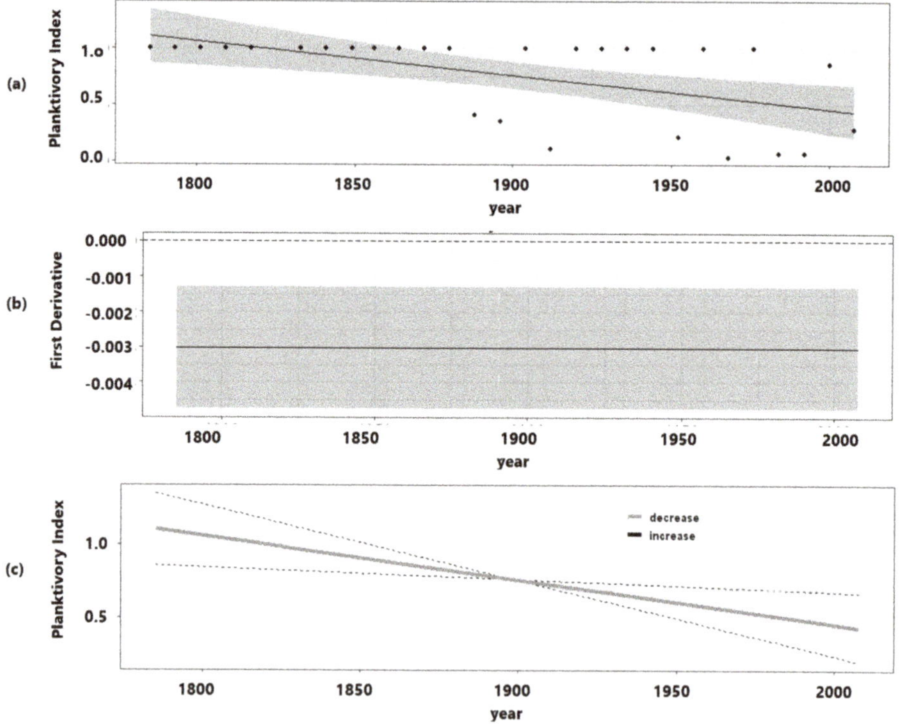

Figure 8. GAM fitted to the planktivory index (*Daphnia*/(*Daphnia* + *Bosmina*)) estimated from the ephippial assemblage in the sedimentary archive of Lake Blanca Chica. (**a**) Observed values, GAM-based trend fitted and its simultaneous interval; (**b**) estimated first derivative of the GAM fitted trend and the 95% simultaneous interval; (**c**) period of transition.

4. Discussion

Predation and grazing are the principal processes that control community composition and size structure in lake ecosystems [6]. We found major changes in planktivory and herbivory in Lake Blanca Chica over the last 250 years. The alteration of zooplankton composition and size structure, the increase in the contribution of fish scales, and the decrease of the planktivory index are indicative of a shift in the planktivory regime. On the other hand, the increase of a biomarker of invertebrate grazing supports a change in the herbivory regime. Thus, this study shows a switch in the grazing and predation pressure and, as a consequence, in the fluxes of the pelagic food web.

Bosmina and *Daphnia* species showed opposite population dynamics in relation to their abundance prior to and post ca. 1900. The cladoceran assemblage changed its composition (ca. 1880–1900), shifting from being rich in littoral, benthic, and pelagic species (including *Pleuroxus*, *Chydorus*, *Leydigia*, *Ceriodaphnia*, *Moina*, *B. huaronensis*, *D. spinulata*, and *D. obtusa*) to being dominated by *B. huaronensis*. This fact is supported by the high contribution of chitinous remains (~80%) (Figure 6) and the appearance of *Bosmina* ephippia in the lacustrine archive post ca. 1900 (Figure S1). Several mechanisms can explain the abrupt change that occurred in the zooplankton community. The switch in the chitinous remains and in the diatom assemblages in Lake Blanca Chica (ca. 1880–1900) has been associated with an increase of the lake water level that affected the physical structure of the lake [28]. This abrupt change was triggered by the increase in precipitation since ca. 1870/1880 [39] and involved a switch in the relationship between the littoral and pelagic habitat and the loss of the aquatic vegetation [28]. Pelagic cladocerans perform dial horizontal migrations to littoral areas during the daytime, finding refuge on the edge or inside submerged macrophyte stands to escape from planktivorous fish that occur in open water. However, neolimnological studies showed that the role of macrophyetesas refuge areas in subtropical temperate shallow lakes from the Southern Hemisphere differs from cold temperate shallow lakes from the Northern Hemisphere [5,12,40]. These studies were performed on eutrophic shallow lakes, and demonstrated that the lack of refuge provision in the southern lakes was due to the rich and abundant assemblage of predacious macroinvertebrates and small littoral fish that also seek refuge in macrophyte stands [5,12,40]. Despite this, the trophic status of Blanca Chica was not eutrophic before 1920–1930, according to a C/N relationship of around 25 [28]. Fish and macroinvertebrate abundance is directly correlated with primary production and nutrient level [6]. Thus, the abundance of small fish and predacious macroinvertebrates in the macrophyte stands must have been low given the C/N ratio of ~25; therefore, aquatic vegetation could have provided refuge to zooplankton, counteracting planktivory. Indeed, *D. spinulata* was favoured until 1900, as shown by an increasing trend. The increase in fish predation, evidenced by low values of the planktivory index, and the loss of the vegetation from the system (ca. 1880–1900) might have exposed zooplankton to a higher predation risk, making large-bodied species such as *D. spinulata* and *D. obtusa* more vulnerable to fish predation than small-bodied species such as *Bosmina*. Altogether, the change in the physical structure of the lake and the predation increase might have facilitated the decline of *Daphnia* species, and would have probably favoured *Bosmina* dominance, releasing it from *Daphnia* competition in addition to the benefit of its smaller size. The top-down regulation and the competition release are the most plausible mechanisms that led to *Bosmina* dominance because no enhancement of primary production (bottom-up regulation) was recorded for the period of the shift. The same mechanisms have been proposed to explain the shift from *Daphnia* to *Bosmina* in other lakes [41].

The mucrone (*Bosmina*) and ephippial (*Daphnia*) size decreases serve as independent indicators of planktivorous fish predation [3,21,42,43]. A shift in the planktivory regime occurred in the lake ca. 1930, indicated by the increase of fish scales since 1930 until present day and by the decreasing trend of the planktivory index (Figures 7 and 8). In this case, the higher top-down control on the zooplankton impacted the size composition of *Bosmina*, which showed a reduction in its body size during 1915–1930 (Figures 3 and 6). An important consideration in paleolimnological studies about *Bosmina* morphology is to obtain a statistically valid estimate of size structure [43]. A minimum sample size of 35 remains per stratigraphic level was determined as necessary to detect a statistically significant variation in size [44]. In our study, we measured between 30–60 remains per sample of sediment; thus, the mucrone size reduction results detected are valid, accurate, and act as an independent indicator in supporting the shift in the planktivory regime and the enhanced top-down control. The main driver of this shift was the eutrophication process that the lake underwent between 1920 and 1990 [28]. Our findings are in agreement with several studies that have recorded changes in the zooplankton composition and size structure as a result of an eutrophication process, leading to an increase of pelagic taxa, a replacement of large-bodied *Daphnia* by small-bodied *Bosmina* or *Ceriodaphnia*, and/or changes in the body size of pelagic taxa in different types of lakes (deep and shallow) [41,45,46].

Planktivorous fish select the larger species or individuals; thus, zooplankton display different strategies such as the reduction of its body size to counteract fish predation [4]. The increase in *Daphnia* body size or the lack of a specific trend in *Moina* sp. constitute the opposite expected patterns under an increasing fish predation scenario; then, it cannot be associated to planktivory. Even though, it can be explained by other mechanisms. *D. spinulata* increased its size over the entire time series whereas *D. obtusa* did so since ca.1950, that is, during the eutrophication phase. Jeppesen et al. [21] found that for a given CPUE (fish catch per unit of effort), *Daphnia* ephippia size apparently increased with total phosphorous (P). All *Daphnia* species are sensitive to food quality, requiring algae rich in phosphorous (P) content [47], and small *Daphnia* species, such as *D. spinulata* or *D. obtusa*, are favoured in high P lakes as they are less adversely affected by cyanobacteria filaments than large *Daphnia* species [48]. Therefore, a possible explanation is that the ephippiasize is bottom-up regulated, as no invertebrate predator occurred in this lake (such as *Chaoborus* or other cladocerans) and thus an increase in ephippia size could not be induced. In the case of *Moina*, a summer species in shallow Pampean lakes, the high inter-annual/decadal variation in abundance and size could be associated with a low selection pressure and/or a mismatch with young of the year fish during this period.

Considering the different responses found in this study and combining *Bosmina* and *Daphnia* abundance and size trends, it is clear that they cannot be explained by a specific type of control. The explanation most likely lies in the interaction of an increase in nutrient concentration in the system (rendering a higher zooplankton production and higher *Daphnia* size) and a top-down control by planktivorous fish (impacting on the composition of the pelagic assemblage and the size of the dominant pelagic cladoceran, *Bosmina*).

The pheophorbide *a*, a grazer biomarker of invertebrate herbivory, has increased over two periods in Lake Blanca Chica: ca. 1930–1950 and 1990 until the present day. The increasing trend and the detected transitions also coincide with the eutrophication process in the lake and with the increase of the total sum of chitinous remains (represented by pelagic, benthic, and littoral species) [28], implying an increase of the overall grazing pressure in the system. The production of this pigment results from a higher food availability (bottom-up regulation) and a top-down control of herbivorous invertebrates, even when in the cladoceran assemblage the large-bodied species were replaced by small-bodied species with less grazing capacity. It seems that the higher zooplankton production counteracts, at least in part, the loss of species with higher clearance rates.

In this study, we found shifts in the herbivory and the planktivory regime in Lake Blanca Chica over the last 250 years, with impacts on different trophic levels and also on different attributes of populations and communities. The shift in herbivory implied a higher control over lake primary producers and, probably, a major flux of pelagic, and also littoral/benthic, carbon to higher trophic levels. Changes in predation pressure led to compositional changes in the zooplankton, with the replacement of large-bodied *Daphnia* by small-bodied *Bosmina*, and also to a reduction of body size in the latter. Shifts in predation pressure were first (ca. 1880–1900) related to a natural driver (hydrology) and then reinforced by eutrophication (ca. 1920–1930). The first driver induced an increase in predation risk due to the loss of refuge in littoral areas and the second one by an increment in the lake fish stock. Thus, we demonstrated that human activities and climate change affect relevant lake processes such as grazing and predation, inducing profound impacts on the function, structure, as well as in fluxes across lake food webs. In addition, our findings represent the first record of the ephippial assemblage for Pampean shallow lakes, showing the baseline condition for these lakes. These results should be taken into consideration when managing restoration activities are carried out in these degraded shallow lakes. Lastly, we emphasize and agree with previous results [49], pointing out that cladocerans are a strong candidate for the single best indicator in paleoecological studies related to the alteration in food web structure and changes in trophic status in shallow lakes.

Supplementary Materials: The following are available online at http://www.mdpi.com/2073-4441/12/2/597/s1: Figure S1: Ephippial stratigraphy of Lake Blanca Chica, including the scores from the first and second axes of correspondence analysis. Figure S2: GAM fitted to the time series of scores of the second axis (CA2) of the correspondence analysis estimated for the ephippial assemblage in the sedimentary record of Lake Blanca Chica. Figure S3: GAM fitted to the time series of ephippia (total sum) recorded in the sedimentary archive of Lake Blanca Chica. Table S1: Descriptive statistics for the size of the ephippia (μm) of *Daphnia spinulata*, *Daphnia obtuse*, and *Moina* sp., as well as the mucrone of *Bosmina huaronensis* (μm).

Author Contributions: Conceptualization, M.d.l.Á.G.S.; methodology, D.C., F.E.C., and M.d.l.Á.G.S.; validation, M.d.l.Á.G.S.; formal analysis, D.C.; investigation, D.C., S.M., A.L., F.E.C., and M.d.l.Á.G.S.; resources, S.M., A.L., and M.d.l.Ã.G.S.; writing—original draft preparation, S.M., A.L., and M.d.l.Á.G.S.; writing—review and editing, D.C., S.M., A.L., F.E.C., and M.d.l.Á.G.S.; visualization, D.C.; supervision, M.d.l.Á.G.S.; project administration, A.L. and M.d.l.Á.G.S.; funding acquisition, A.L. and M.d.l.Á.G.S. All authors have read and agreed to the published version of the manuscript.

Funding: This research was supported by National Council of Scientific and Technical Research CONICET (grant: PIP 465/2013), National Research Council (CNR), Water Research Institute (IRSA), Bilateral Cooperation Project CONICET-CNR/2015 and Mar del Plata University (EXA775/2017).

Acknowledgments: We thank A. Kotov for all the material that was shared to M.d.l.Ã.G.S., J.C. Paggi for his help with Cladocera identification, to the three anonymous reviewers that contributed with their comments to improve this manuscript, and to G. Free and R. Burks for the English style edition.

Conflicts of Interest: The authors declare no conflict of interest. The funders had no role in the design of the study; in the collection, analyses, or interpretation of data; in the writing of the manuscript; or in the decision to publish the results.

References

1. Dumont, H.J.; Van de Velde, I.; Dumont, S. The dry weight estimate of biomass in a selection of Cladocera, Copepoda and Rotifera from the plankton, periphyton and benthos of continental waters. *Oecologia* **1975**, *19*, 75–97. [CrossRef]
2. Sterner, R.W. Role of Zooplankton in Aquatic Ecosystems. In *Encyclopedia of Inland Waters*; Likens, G.E., Ed.; Academic Press: Oxford, UK, 2009; pp. 678–688.
3. Nevalainen, L.; Brown, M.; Manca, M. Sedimentary record of cladoceran functionality under eutrophication and re-oligotrophication in Lake Maggiore, Northern Italy. *Water* **2018**, *10*, 86. [CrossRef]
4. Brooks, J.; Dodson, S. Predation, body size, and composition of plankton. *Science* **1965**, *150*, 28–35. [CrossRef] [PubMed]
5. González Sagrario, M.A.; Balseiro, E. The role of macroinvertebrates and fish in regulating the provision by macrophytes of refugia for zooplankton in a warm temperate shallow lake. *Freshw. Biol.* **2010**, *55*, 2153–2166. [CrossRef]
6. Jeppesen, E.; Jensen, J.P.; Søndergaard, M.; Lauridsen, T.; Pedersen, L.J.; Jensen, L. Top-down control in freshwater lakes: The role of nutrient state, submerged macrophytes and water depth. *Hydrobiologia* **1997**, *342*, 151–164. [CrossRef]
7. Korosi, J.B.; Paterson, A.M.; DeSellas, A.M.; Smol, J.P. Linking mean body size of pelagic Cladocera to environmental variables in Precambrian Shield lakes: A paleolimnological approach. *J. Limnol.* **2008**, *67*, 22–34. [CrossRef]
8. Hann, B.J. Methods in Quaternary Ecology #6. Cladocera. *Geosci. Can.* **1989**, *16*, 17–26.
9. DeSellas, A.M.; Paterson, A.M.; Sweetman, J.N.; Smol, J.P. Cladocera assemblages from the surface sediments of south-central Ontario (Canada) lakes and their relationships to measured environmental variables. *Hydrobiologia* **2008**, *600*, 105–119. [CrossRef]
10. Scheffer, M.; Jeppesen, E. Regime Shifts in Shallow Lakes. *Ecosystems* **2007**, *10*, 1–3. [CrossRef]
11. Jeppesen, E.; Leavitt, P.; De Meester, L.; Jensen, J.P. Functional ecology and palaeolimnology: Using cladoceran remains to reconstruct anthropogenic impact. *Trends Ecol. Evol.* **2001**, *16*, 191–198. [CrossRef]
12. Meerhoff, M.; Iglesias, C.; De Mello, F.T.; Clemente, J.M.; Jensen, E.; Lauridsen, T.L.; Jeppesen, E. Effects of habitat complexity on community structure and predator avoidance behaviour of littoral zooplankton in temperate versus subtropical shallow lakes. *Freshw. Biol.* **2007**, *52*, 1009–1021. [CrossRef]
13. Catalan, J.; Pla-Rabés, S.; Wolfe, A.P.; Smol, J.P.; Rühland, K.M.; Anderson, N.J.; Kopáček, J.; Stuchlík, E.; Schmidt, R.; Koinig, K.A.; et al. Global change revealed by palaeolimnological records from remote lakes: A review. *J. Paleolimnol.* **2013**, *49*, 513–535. [CrossRef]
14. Battarbee, R.W. Mountain lakes, pristine or polluted? *Limnetica* **2005**, *24*, 1–8.
15. Carpenter, S.R.; Leavitt, P.R. Temporal variation in a paleolimnological record arising from a trophic cascade. *Ecology* **1991**, *72*, 277–285. [CrossRef]
16. Daley, R.J. Experimental characterization of lacustrine chlorophyll diagenesis. II. Bacterial, viral and herbivore grazing effects. *Arch. Hydrobiol.* **1973**, *72*, 409–439.
17. Leavitt, P.R. A review of factors that regulate carotenoid and chlorophyll deposition and fossil pigment abundance. *J. Paleolimnol.* **1993**, *9*, 109–127. [CrossRef]
18. Davidson, T.A.; Sayer, C.D.; Perrow, M.; Bramm, M.; Jeppesen, E. The simultaneous inference of zooplanktivorous fish and macrophyte density from sub-fossil cladoceran assemblages: A multivariate regression tree approach. *Freshw. Biol.* **2010**, *55*, 546–564. [CrossRef]
19. Davidson, T.A.; Sayer, C.D.; Perrow, M.R.; Bramm, M.; Jeppesen, E. Are the controls of species composition similar for contemporary and sub-fossil cladoceran assemblages? A study of 39 shallow lakes of contrasting trophic status. *J. Paleolimnol.* **2007**, *38*, 117–134. [CrossRef]

20. Hann, B.J.; Leavitt, P.R.; Chang, P.S.S. Cladocera community response to experimental eutrophication in Lake 227 as recorded in laminated sediments. *Can. J. Fish. Aquat. Sci.* **1994**, *51*, 2312–2321. [CrossRef]

21. Jeppesen, E.; Jensen, J.P.; Amsinck, S.; Landkildehus, F.; Lauridsen, T.; Mitchell, S. Reconstructing the historical change in *Daphnia* mean size and planktivorous fish abundance in lakes from the size of *Daphnia* ephippia in the sediment. *J. Paleolimnol.* **2002**, *27*, 133–143. [CrossRef]

22. Verschuren, D.; Marnell, L.F. Fossil zooplankton and the historical status of westslope cutthroat trout in a headwater lake of Glacier National Park, Montana. *Trans. Am. Fish. Soc.* **1997**, *126*, 21–34. [CrossRef]

23. Paggi, J.C. Revisión de las especies argentinas del género *Bosmina* Bird agrupadas en el subenero *Neobosmina* Lieder (Crustacea: Cladocera). *Acta Zool. Lilloana* **1979**, *35*, 137–162.

24. Iriondo, M. Quaternary lakes of Argentina. *Palaeogeogr. Palaeoclimatol. Palaeoecol.* **1989**, *70*, 81–88. [CrossRef]

25. Sanzano, V.; Grosman, F.; Colasurdo, P. Estudio limnológico de Laguna Blanca Chica (Olavarría, Provincia de Buenos Aires) durante un período de sequía. *Biol. Acuát.* **2014**, *30*, 189–202.

26. Piovano, E.L.; Ariztegui, D.; Damatto-Moreira, S. Recent environmental changes in Laguna Mar Chiquita (central Argentina): A sedimentary model for a highly variable saline lake. *Sedimentology* **2002**, *49*, 1371–1384. [CrossRef]

27. Guerra, L.; Piovano, E.L.; Córdoba, F.E.; Sylvestre, F.; Damatto, S. The hydrological and environmental evolution of shallow Lake Melincué, central Argentinean Pampas, during the last millennium. *J. Hydrol.* **2015**, *529*, 570–583. [CrossRef]

28. González Sagrario, M.A.; Musazzi, S.; Córdoba, F.E.; Mendiolar, M.; Lami, A. Inferring the occurrence of regime shifts in a shallow lake during the last 250 years based on multiple indicators. *Ecol. Indic.* **2019**, submitted.

29. Lami, A.; Musazzi, S.; Marchetto, A.; Buchaca, T.; Kerna, M.; Jeppesen, E.; Guilizzoni, P. Sedimentary pigments in 308 alpine lakes and their relation to environmental gradients. *Adv. Limnol.* **2009**, *62*, 217–238.

30. Korhola, A.; Rautio, M. Cladocera and other branchiopod crustaceans. In *Tracking Environmental Change Using Lake Sediments*; Zoological Indicators; Smol, J.P., Birks, H.J.B., Last, W.M., Eds.; Kluwer Academic Publisher: Dordrecht, The Netherlands, 2001; Volume 4, pp. 5–41.

31. Kotov, A.A. Separation of *Leydigia louisi* Jenkin, 1934 from *L. leydigi* (Schoedler, 1863) (Chydoridae, Anomopoda, Cladocera). *Hydrobiologia* **2003**, *490*, 147–168. [CrossRef]

32. Smirnov, N.N.; Kotov, A.A.; Coronel, J.S. Partial revision of the *aduncus*-like species of *Pleuroxus* Baird, 1843 (Chydoridae, Cladocera) from the southern hemisphere with comments on subgeneric differentiation within the genus. *J. Nat. Hist.* **2006**, *40*, 1617–1639. [CrossRef]

33. Szeroczyńska, K.; Sarmaja-Korjonen, K. *Atlas of Subfossil Cladocera from Central and Northern Europe*; Friends of the lower Vistula Society: Świecie, Poland, 2007; p. 84.

34. Wood, S.N. *Generalized Additive Models: An Introduction with R*, 2nd ed.; Chapman and Hall/CRC: Boca Raton FL, USA, 2017.

35. Simpson, G.L. Modelling palaeoecological time series using generalised additive models. *Front. Ecol. Evol.* **2018**, *6*, 149. [CrossRef]

36. Oksanen, J.; Blanchet, F.G.; Kindt, R.; Legendre, P.; Minchin, P.R.; O'Hara, R.B.; Simpson, G.L.; Solymos, P.; Stevens, M.H.H.; Wagner, H. Vegan: Community Ecology Package. R package version 2.2-1. 2015. Available online: http://cran.r-project.org (accessed on 15 November 2018).

37. Borcard, D.; Gillet, F.; Legendre, P. *Numerical Ecology with R*, 1st ed.; Springer: New York, NY, USA, 2011; p. 319.

38. R Core Team. *R: A language and environment for statistical computing*; R Foundation for Statistical Computing: Vienna, Austria, 2017; Available online: https://www.R-project.org/ (accessed on 15 November 2018).

39. Córdoba, F.E.; Piovano, E.L.; Guerra, L.; Mulsow, S.; Sylvestre, F.; Zárate, M. Independent time markers validate ^{210}Pb chronologies for two shallow Argentine lakes in Southern Pampas. *Quatern. Int.* **2017**, *438*, 175–186. [CrossRef]

40. González Sagrario, M.A.; Balseiro, E.; Ituarte, R.; Spivak, E. Macrophytes as refuge or risky area for zooplankton: A balance set by littoral predacious macroinvertebrates. *Freshw. Biol.* **2009**, *54*, 1042–1053. [CrossRef]

41. Perga, M.-E.; Desmet, M.; Enters, D.; Reyss, J.-L. A century of bottom-up- and top-down driven changes on a lake planktonic food web: A paleoecological and paleoisotopic study of Lake Annecy, France. *Limnol. Oceanogr.* **2010**, *55*, 803–816.

42. Amsinck, S.L.; Jeppesen, E.; Landkildehus, F. Inference of past changes in zooplankton community structure and planktivorous fish abundance from sedimentary subfossils—A study of a coastal lake subjected to major fish kill incidents during the past century. *Arch. Hydrobiol.* **2005**, *162*, 363–382. [CrossRef]
43. Korosi, J.B.; Kurek, J.; Smol, J.P. A review on utilizing *Bosmina* size structure archived in lake sediments to infer historic shifts in predation regimes. *J. Plankton Res.* **2013**, *35*, 444–460. [CrossRef]
44. Brahney, J.; Routledge, R.; Bos, D.G.; Pellatt, M.G. Changes to the productivity and trophic structure of a sockeye salmon Rearing Lake in British Columbia. *N. Am. J. Fish. Manag.* **2010**, *30*, 433–444. [CrossRef]
45. Davidson, T.A.; Sayer, C.D.; Langdon, P.G.; Burgess, A.; Jackson, M. Inferring past zooplanktivorous fish and macrophyte density in a shallow lake: Application of a new regression tree model. *Freshw. Biol.* **2010**, *55*, 584–599. [CrossRef]
46. Manca, M.; Torretta, B.; Comoli, P.; Amsinck, S.L.; Jeppesen, E. Major changes in trophic dynamics in large, deep sub-alpine Lake Maggiore from 1940s to 2002: A high resolution comparative palaeo-neolimnological study. *Freshw. Biol.* **2007**, *52*, 2256–2269. [CrossRef]
47. Acharya, K.; Kyle, M.; Elser, J.J. Effects of stoichiometric dietary mixing on *Daphnia* growth and reproduction. *Oecologia* **2003**, *138*, 333–340. [CrossRef]
48. Gliwicz, Z.M. *Daphnia* growth at different concentrations of blue-green filaments. *Arch. Hydrobiol.* **1990**, *120*, 51–65.
49. Davidson, T.A.; Bennion, H.; Jeppesen, E.; Clarke, G.H.; Sayer, C.D.; Morley, D.; Odgaard, B.V.; Rasmussen, P.; Rawcliffe, R.; Salgado, J.; et al. The role of cladocerans in tracking long-term change in shallow lake trophic status. *Hydrobiologia* **2011**, *676*, 299–315. [CrossRef]

 water

MDPI

Article

Evaluation of the Egg Bank of Two Small Himalayan Lakes

Roberta Piscia [1,*], Sara Bovio [2], Marina Manca [1], Andrea Lami [1] and Piero Guilizzoni [1]

[1] CNR-IRSA, Largo Tonolli 50, 28922 Verbania (VB), Italy; marina.manca@irsa.cnr.it (M.M.); andrea.lami@irsa.cnr.it (A.L.); piero.guilizzoni@irsa.cnr.it (P.G.)

[2] Independent Researcher, Via Rimembranza 12, 28043 Bellinzago Novarese (NO), Italy; sarabovio@gmail.com

* Correspondence: roberta.piscia@irsa.cnr.it

Received: 18 December 2019; Accepted: 6 February 2020; Published: 11 February 2020

Abstract: High mountain lakes are biodiversity treasures. They host endemic taxa, adapted to live in extreme environments. Among adaptations, production of diapausing eggs allows for overcoming the cold season. These diapausing eggs can rest in the sediments, providing a biotic reservoir known as an egg bank. Here, we estimated changes in abundance of the egg bank in two lakes in the Khumbu Region of the Himalayas, during the last ca. 1100 and 500 years, respectively, by analyzing two sediment cores. We tested viability of the diapausing eggs extracted from different layers of the sediment cores under laboratory conditions. We found that only diapausing eggs of the Monogont rotifer *Hexarthra bulgarica nepalensis* were able to hatch, thus suggesting that a permanent egg bank is lacking for the other taxa of the lakes, not least for the two *Daphnia* species described from these sites. Our results confirm previous studies suggesting that in high mountain lakes, the production of diapausing is mainly devoted to seasonal recruitment, therefore leading to a nonpermanent egg bank. The different ability of different taxa to leave viable diapausing eggs in the sediments of high mountain lakes therefore poses serious constraints to capability of buffering risk of biodiversity loss in these extremely fragile environments.

Keywords: zooplankton; diapausing eggs; high mountain lakes; Himalayas

---

1. Introduction

High mountain lakes are highly vulnerable to impact of human activities and to climate change [1–3]. They host endemic populations of aquatic invertebrates, adapted to live under extreme environmental conditions using different mechanisms for optimizing life-cycles and growth during a very short ice-free period and under the ice cover [3,4].

Research on high mountain crustacean populations has mainly focused on physiological adaptation mechanisms, such as development of cuticular pigmentation in cladocerans (melanin) [5,6] and diet dependent carotenoid pigmentation, in copepods [7,8]. Less well documented are zooplankton reproductive traits and their importance as survival strategies through time. Life-cycles of high mountain zooplankton taxa are characterized by cyclical parthenogenesis, where a shift from parthenogenetic to sexual reproduction occurs at the end of the ice-free period leading to production of diapausing eggs (embryos at early developmental stages). These stages are capable of surviving the harsh seasons and under the ice cover, securing recruitment of individuals into the population during the ice-free period [9,10].

The production of diapausing eggs with delayed development (undergoing diapause or dormancy) can be regarded as a mechanism through which taxa maintain their genetic reservoirs—buffering the high risk of extinction to which their populations are exposed [11–15]. Evaluating the "egg bank" potential allows us to estimate the vulnerability of different taxa to extinction and the risk of biodiversity loss in these environments. The dormant egg bank can in fact influence the rate and the direction of population, community, and ecosystem response to climate change [12,14,15]. Under global environmental changes, the frequency of extreme conditions will likely increase and species with prolonged diapause will be able to survive extreme years of no recruitment, whereas species lacking an egg bank will not. An altered environment may change species having poor recruitment and even which species produce an egg bank, through effects on thermal diapause cues [16].

Information on remote high mountains environments is usually scant, limited in time and space, and even a list of taxa from contemporary samples is generally incomplete as plurennial surveys can be rarely accomplished. To this aim of investigating spatial and temporal variability in biodiversity and time changes, integration of paleo- and neo-limnological approaches is very useful. Analyses of sediment cores allow for reconstructing long-term changes in biodiversity of biotic communities along with changes in the environment, allowing for reconstructing changes in climate and in different types of human impacts, such as long distance transport of pollutants.

In this study, we trace long-term changes in abundance of diapausing eggs in two small lakes, LCN10 and LCN70, located in the highest mountains of the world. Diapausing eggs were extracted from two short sediment cores representing different stages of the lake life history and through different climatic conditions. Abundance and viability of diapausing eggs in the sediments are crucial for complementing contemporary biodiversity estimates with potential biodiversity. Consistency of the active egg bank in the two lakes was estimated by hatching diapausing eggs recovered from the sediment cores under controlled laboratory conditions. The aim was to understand risk of loss of endemic species and vulnerability of different taxa to impact of perturbations in these fragile environments, in which impact of climate change and of anthropogenic activities is high.

Our study was conceived in the framework of an extensive multiannual study aimed at characterizing lakes in the Sagarmatha National Park, Nepal. Both water samples and sediment cores were collected during fifteen limnological surveys carried out in the period 1992–2011. Of the 120 cataloged lakes in the region, ca. 70 were characterized for hydrochemistry, 40 of which were analyzed for plankton and benthos, providing taxonomic and biodiversity estimates in different years [17]. Subfossil Cladocera remains also allowed for understanding changes in biodiversity during different climatic phases [18].

2. Study Sites

LCN10 and LCN70 are located in the upper Khumbu Valley (Table 1; Figure 1), in the Nepalese Himalayas [19,20]. They are small and shallow lakes, with an ice-covered period of 9–10 months and a high transparency during the open water phase. LCN10 receives water primarily from LCN9 located upstream and from snowfields, while LCN70, classified as a temporary lake by Löffler [21], receives water only from snowfields. Both sites are part of the 40 lakes in which plankton has been studied [22,23] in the framework of the Ev-K2-CNR project [24] developed in collaboration with the Nepal Academy of Science and Technology (NAST). Previous information on taxonomy, reporting endemic taxa, was provided by Löffler [21,25] as a result of an extensive sampling campaign in the Khumbu Valley. In previous studies, zooplankton samples were collected with plankton nets (126 μm mesh size in 1993 and 1994; 50 μm mesh size in 1997 and 2004) by vertical and horizontal hauls at the deepest point and from the littoral. Samples were only qualitative, as the amount of filtered water could not be estimated. The samples were preserved in buffered 10% formaldehyde and counted to obtain estimates of relative abundance of the different taxa. LCN10, the deep lake close to the Pyramid laboratory, was visited in September–October of each sampling campaign while LCN70 was visited only in October 1994 and 1997. The list of taxa of LCN10 includes *Arctodiaptomus jurisovitchi* Löffler, 1968 [26], *Daphnia dentifera* Forbes, 1893 [27], *Daphnia fusca* Gurney, 1906 [28], *Alona werestschagini* Sinev, 1999 [29,30] and among rotifers, *Ascomorpha* sp. LCN70 was characterized by the prevalence of the Monogonont rotifer *Hexarthra bulgarica nepalensis* [31], forming dense agglomerates and bearing diapausing eggs. In this lake, only the melanic *Daphnia fusca* was found along with the Anostracan *Branchinecta orientalis* Sars, 1901 [22,32–34]. A few *Euchlanis* sp. were also reported, along with bottom dwelling *Ilyocriptus acutifrons* Sars, 1862 (Cladocera Mactrothricidae) and a few Harpaticoidae of the genus *Maraenobiotus* sp. [35]. However, given the restricted number of samples, the sampling method and the mesh size, the list of taxa from zooplankton samples is far from exhaustive. The Ev-K2-CNR Pyramid laboratory located in the region at 5050 m a.s.l. gave technical and logistical support for the investigations, allowing for sampling some lakes in different years between 1992 and 2011. Seasonal hydrochemistry analyses and experiments in lakes closer to the laboratory, a detailed description of the morphometry and the chemistry of lakes and the traits of their catchment basin is reported in Lami and Giussani [17]. Besides, paleolimnological and paleoclimatic reconstructions were investigated in LCN10 and LCN70 [36,37]. The region is characterized by a cold and dry climate with monthly precipitation lower than 200 mm. Monthly mean temperatures range between $-7\,°C$ (December) and $5\,°C$ (June).

Table 1. Geographic coordinates and principal morphometric characteristics of the studied lakes.

	LCN10	LCN70
Latitude	27°57′45″ N	27°53′36″ N
Longitude	86°48′56″ E	86°46′26″ E
Altitude	5067 m	4830 m
Area	1.67 ha	0.63 ha
Maximum depth	14.8 m	2 m

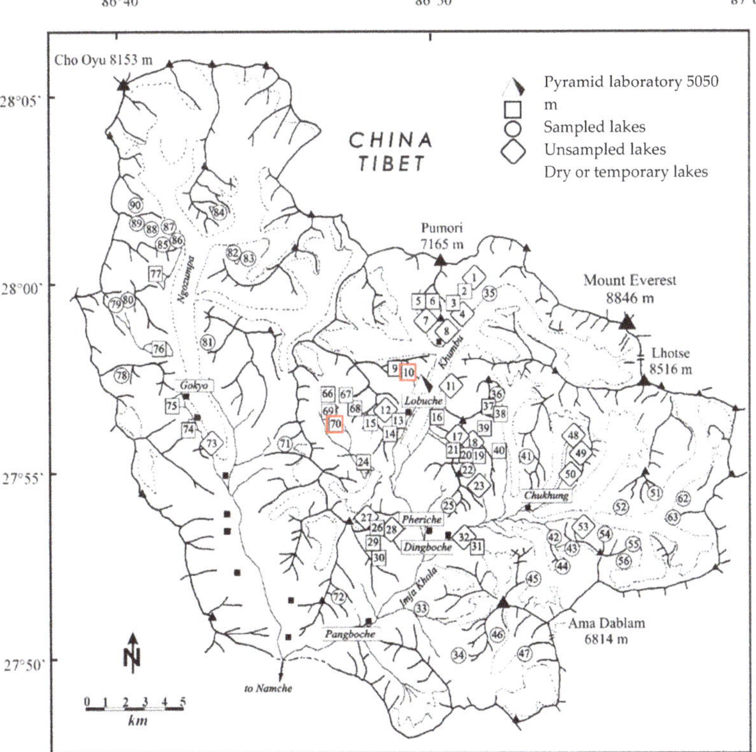

Figure 1. A map of the Khumbu and Imja Khola catchment within the Sagarmatha National Park, Nepal in which are highlighted all lakes sampled in the framework of the EV-K2-CNR project since 1987 (G.A. Tartari, pers. comm.). Black triangles refer the mountain peaks; black squares refer to localities. Red squares highlight the study sites.

3. Materials and Methods

A short sediment core (04/1, 26 cm in length) was collected near the deepest zone of LCN10 in 2004 using a gravity corer (with 6 cm diameter Plexiglas tube). The age–depth model was established by correlation with a previous core 02/3, dated by ^{14}C method. A detailed description is provided in Lami et al. [37]. About 200 years were represented in the topmost six-centimeter sections of the core (resulting in a sedimentation rate of 0.03 cm y^{-1}), and the deepest section was dated as corresponding to ca. 900 CE. The topmost 5 cm were sliced into sections of 1 cm, while for the remaining core a section of 1 cm was taken every 5 cm. In total, nine subsamples were extracted from the core.

Sediment samples were stored in the dark at 4 °C until microscopic analysis, in order to preserve the material and stimulate subsequent hatching [38]. All the diapausing eggs' cases found in the sediment core sections were counted, including those which, being open, were empty. The latter may indicate changes in recruitment from diapausing eggs over time. Taxonomic identification was performed, comparing images of rotifer diapausing eggs cases, which are often species-specific [39], to photographic material from previous studies [23,40] and on the basis of a guide on rotifers' diapausing eggs [41], while Anostraca diapausing eggs was based on a study [35] and literature [42–44]. Diapausing eggs were extracted following the sugar flotation method [45] modified for rotifer diapausing eggs by Garcia-Roger [46]. Taxonomic identification, counting, measurements and sorting of ephippia and rotifer diapausing egg cases were performed using a Zeiss Stemi 2000-C microscope (Carl Zeiss™, Oberkochen, Germany), equipped with software for image analysis (Image-Pro Express 5.1 (Media Cybernetics Inc., Silver Spring, USA)). The presence of eggs in dark ephippia was verified only after viability experiments, as opening ephippia cases prior to hatching might have damaged the eggs. Extracted ephippia and rotifer diapausing egg cases were individually transferred into multiwell plates (with well oxygenated culture medium of 4 mL oligomineral water) and stored in a thermostatic cell at 17 °C, with light intensity of about 50×10^3 Lux and L:D photoperiod of 16:8 h, reproducing summer environmental conditions. Every 24 hours we recorded hatchlings and renewed the culture medium.

Ephippia and diapausing egg cases from LCN70 were extracted from a sediment core collected in October 1994 (core 94/2) with a gravity corer (of 6 cm diameter Plexiglas tube) in the deepest zone of the lake. The core, 11.5 cm in length, was cut longitudinally and sectioned into 2–1.5 cm slices. Chronology had been established by ^{210}Pb; sedimentation rate was 0.1 cm y^{-1} in the topmost 6 cm and 0.02 cm y^{-1} between 6 and 12 cm [36].

No attempt was made to hatch cysts of the Anostracan *Branchinecta orientalis*.

Duration of incubation was 21 days. Eggs that did not hatch during this time period were considered as unviable.

4. Results

The analysis of Cladocera ephippia in core LCN10 04/1 sections revealed periods during which only one subgenus was present and periods during which both *Daphnia* (*Ctenodaphnia*) and *Daphnia* (*Daphnia*) coexisted. In particular, in sections 3–4 cm and 4–5 cm only *D.* (*Ctenodaphnia*) ephippia were present, while in the topmost section only *D.* (*Daphnia*) ephippia were found (Figure 2A). As expected from previous studies on this environment, the few ephippia found in the superficial section belonged exclusively to a *D.* (*Daphnia*). Contemporary zooplankton samples had identified the species as *Daphnia dentifera* [47]. *D.* (*Ctenodaphnia*) ephippia were of the melanic species of the region.

Apart from the deepest section, ephippia were much more abundant from 2 to 5 cm sections, while in sections 10–11, 1–2 and 0–1 cm, their abundance was low. The highest abundance of *D.* (*Ctenodaphnia*) ephippia (12,385 m^{-2}) was found in the section 4–5 cm corresponding to ca. 1800 CE; a second peak was detected in the section 2–3 cm where both *D.* (*Ctenodaphnia*) and *D.* (*Daphnia*) ephippia were found (1415 ephippia m^{-2} and 7431 ephippia m^{-2}, respectively). The two *Daphnia* species co-occurred also in sections 10–11 and 1–2 cm. Ephippia were lacking from deeper sections (15–26 cm) of the core.

After 21 days of incubation, no hatchlings were recovered from the ephippia extracted from core LCN10 04/1. Opening and inspection of their content revealed that ephippia cases were either empty or with only one egg which we classified as nonviable.

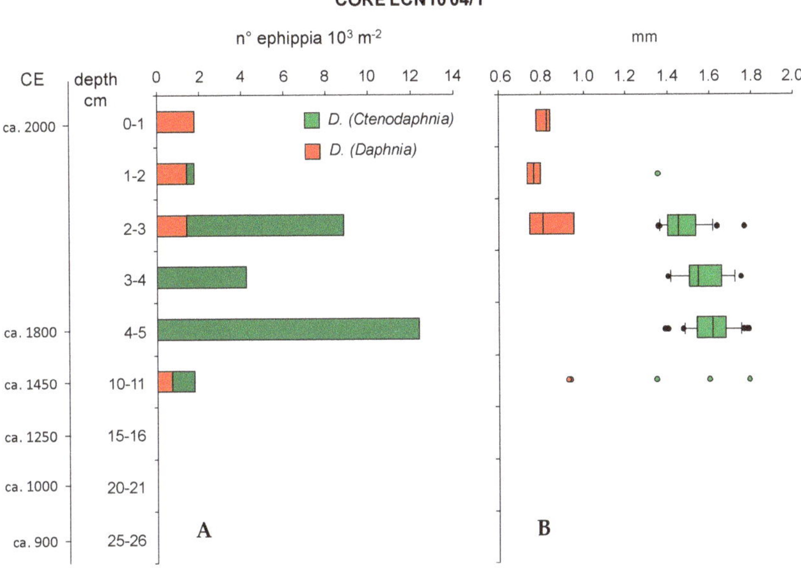

Figure 2. Abundance of ephippia in the analyzed sections of the sediment core 04/1 of LCN10 (**A**) and range of variation of their length (**B**). Vertical lines represent median values while horizontal whiskers represent the 95th percentile.

The two types of ephippia had different sizes (lengths), with those of *D.* (*Daphnia*) always smaller than those of *D.* (*Ctenodaphnia*). The latter decreased in size between the 3–5 and 2–3 cm sections and, in the latter section, co-occurred with those of *D.* (*Daphnia*). *D.* (*Daphnia*) ephippia at 2–3 cm section were in a relatively large range of sizes, along with smaller *D.* (*Ctenodaphnia*) ephippia size. Five ephippia in total were found in the 10–11 cm section, corresponding to ca. 1450 CE; two were from *D.* (*Daphnia*) and their length was 0.9 mm, corresponding to the largest value of the 2 cm section. The other three pertained to *D.* (*Ctenodaphnia*), and their range of size did not differ from those detected in upper sediment sections. The relatively small size of the only *D.* (*Ctenodaphnia*) ephippium in section 1–2 cm (1.35 mm) seems to confirm the trend towards decreasing size observed between deeper and more surficial sections of the sediment core. The decrease in size of *D.* (*Ctenodaphnia*) ephippia in recent times was statistically significant (F = 11.803, $p < 0.001$; Figure 2B). In particular, mean ephippia length was statistically different between sections 4–5 and 2–3 cm ($t = 4.855$, $p < 0.001$; Table 2) and between sections 3–4 and 2–3 cm ($t = 2.467$, $p < 0.016$), but the difference between sections 4–5 and 3–4 cm was nonsignificant (Table 2). Changes in size of *D.* (*Daphnia*) ephippia in different sections of the core were not statistically significant.

Table 2. Results of all pairwise multiple comparison procedures (Holm–Sidak method, overall significance level = 0.05) on size of *D.* (*Ctenodaphnia*) ephippia extracted from sediment core LCN10 04/1. Significant comparisons are highlighted in bold.

Comparison	Difference of Means	*t*	Unadjusted *p*
4–5 cm vs. 2–3 cm	135.724	4.855	**<0.001**
3–4 cm vs. 2–3 cm	90.405	2.467	**0.016**
4–5 cm vs. 3–4 cm	45.319	1.338	0.186

Microscopical analysis of qualitative planktonic samples taken in LCN70 in 1994 revealed that both crustaceans and rotifers were present. Rotifers dominated in terms of numbers, the most common taxon being *Hexarthra bulgarica nepalensis* in lakes of the region [25,31]. The sample was very rich in rotifer eggs (Figure 3), with young and ovigerous females forming dense agglomerates on detritus and filamentous green algae. Few *Euchlanis* sp. specimens were also found [35].

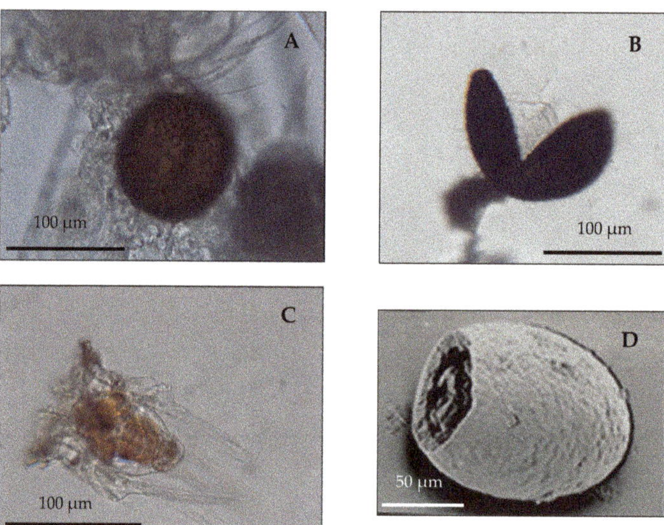

Figure 3. (**A**) *Hexarthra bulgarica nepalensis* from a plankton sample of LCN70 carrying a diapausing egg; (**B**) Hatching of a diapausing egg of *Hexarthra bulgarica nepalensis* recovered from sediment core LCN70 94/2; (**C**) Newborn of *Hexarthra bulgarica nepalensis* hatched from the diapausing egg; (**D**) SEM image of a *Brachionus* sp. open egg case recovered from sediment core LCN70 94/2.

The profile of core LCN70 94/2 (Figure 4) shows the presence of diapausing eggs of three Rotifera taxa in all the sections: the dominant *Hexarthra bulgarica nepalensis* and, in low abundance, *Brachionus* of two morphotypes as reported in Sudzuki [48] and Piscia et al. [49]. The latter two were not found in the zooplankton samples. *Hexarthra* diapausing eggs were abundant in all sections but tended to increase in deeper sections of the core, with a peak of 30.4×10^6 m^{-2} at the 8–10 cm section. They were between 1% and 2.5% of the total resting eggs found between the 10–11.5 and 6 cm sections and represented 23% and 15% of the total abundance at the 2 and at 4–6 cm sections, respectively. Cladocera ephippia of *D.* (*Ctenodaphnia*) and of *Alona werestschagini*, gradually increased from bottom to top layers of the sediment core and were 4.5 times more abundant in the upper 2 cm section than in the deepest one (10–11.5 cm). Anostracan cysts were also found to vary in abundance (min = 1.8×10^5 cysts m^{-2}; max = 5.1×10^5 cysts m^{-2}). *Brachionus* resting egg cases were all open, therefore revealing that diapausing eggs produced might serve as seasonal recruitment of the species at the end of the ice cover phase. No attempt was made of hatching Anostracan cysts, as we had no indication of conditions for promoting their hatching.

Only *Hexarthra* diapausing eggs from the topmost 2 cm were still viable, with a high rate of hatching (63.6%). Given the estimated mean sedimentation rate of 0.06 cm y^{-1} [36], diapausing eggs in this section were deposited in a timeframe of ca. 33 years before date of collection of the core. Individual hatching timing varied between 4 and 7 days.

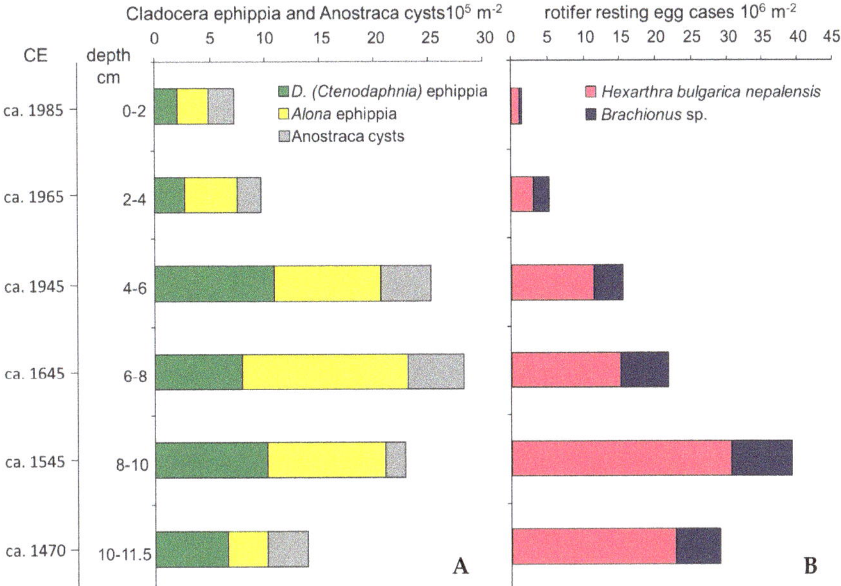

Figure 4. Abundance of Cladocera ephippia and Anostraca cysts (**A**) and rotifer resting eggs (**B**) recovered from sediment core LCN70 94/2.

5. Discussion and Conclusions

The presence of diapausing eggs in different sections of the two sediment cores confirmed that their production is a fundamental survival strategy in high-altitude lakes. The accumulation in sediments of diapausing eggs of different species, generations and genotypes, represents a potential reserve of biodiversity [12,16], which buffers the consequent extinction risk from anthropogenic impacts and makes the population resilient to the settlement of invasive species [50,51]. Organisms able to survive in extreme environments, such as LCN10 and LCN70, belong to a small number of highly specialized species; nevertheless, the reservoir of diapausing eggs in the sediments was of the same order of size (10^3–10^5 eggs per m^{-2}) of those reported from less remote environments [16].

Zooplankton can produce different types of diapausing eggs: from embryos at early developmental stages of monogonont rotifers, Branchiopods and Calanoid Copepods, to advanced larval stages (Cyclopoid Copepods) and finally to adults of Harparcticoids and Bdelloid rotifers [16]. In previous studies, microscopic analysis of contemporary (plankton) samples revealed that Rotifera, Cladocera and Anostraca produced diapausing eggs at the end of the ice-free period [23,35]. Anostracan cysts recovered from sediments of LCN70 were of the species *Branchinecta orientalis* previous found in LCN70 and LCN30 zooplankton samples, in which the presence of diapausing eggs was also reported [35]. Among cladocerans, we found ephippia of the endemic *Alona werestschagini* in LCN70, *D.* (*Daphnia*) ephippia in LCN10, and *D.* (*Ctenodaphnia*) ephippia in both LCN70 and LCN10. In LCN10, in particular, an alternation in the occurrence of *D.* (*Daphnia*) and *D.* (*Ctenodaphnia*) ephippia was conspicuous and a co-occurrence of the two in the transition periods was observed. In a previous investigation, the presence of melanic *D.* (*Ctenodaphnia*) and pale *D.* (*Daphnia*) was reported [52]. Ephippia of the former were dark because of cuticular pigmentation (melanic) while those of the latter were transparent. The two also differed in size and morphology; ephippia morphology is in fact a diagnostic trait for distinguishing between *Daphnia* of the *D.* (*Ctenodaphnia*) type and those of the *D.* (*Daphnia*) type. The former belong to an ancestral type clearly separated from the other during the phylogeny ca. 200 million years ago [53].

Ephippia isolated from the surface layers of the sediment core indicate the substitution in recent times of the subgenus *D.* (*Ctenodaphnia*) with the subgenus *D.* (*Daphnia*) in agreement with plankton samples previously analyzed [23]. As suggested by Nevalainen et al. [18], the presence of *D.* (*Ctenodaphnia*) ephippia in the sediment core of LCN10 could be related to phases of enhanced connectivity with LCN9, the source of a cascade system in which *D.* (*Ctenodaphnia*) is permanent. Such changes in connectivity may in turn, be related to glacial advance or retreat driven by climate change [36]. Ephippia size is positively correlated with *Daphnia* body size, larger females producing larger ephippia resulting in larger neonates [54,55]. Such correspondence is the basis for reconstructing changes in *Daphnia* body size with changes in the environment [56]. In fishless, high mountain lakes, a wide range of *Daphnia* size (reconstructed from fossils, such as postabdomens, head shields, mandibles and, of course, from ephippia [52,57]) is regarded as indicative of a longer growing season, i.e., a longer ice-free season, securing a larger number of molts during the parthenogenetic phase which precedes the appearance of males and ephippial females at the end of the growing season [58,59]. In a study on LCN40, the substitution of the melanic *D.* (*Ctenodaphnia*) by the pale *Daphnia dentifera* in the most recent period was interpreted as related to changes in climate, with an increase in precipitation resulting in increase of dissolved organic carbon. In the case of a sediment core with low accumulation rates, however, multiple seasons in variable numbers are integrated in a 1 cm section during different time periods. Results of hatching experiments indicate that ephippial eggs recovered from the sediments of both lakes were nonviable, thus suggesting that an active egg bank is lacking for this important component of the zooplankton population. The apparent lack of a permanent *Daphnia* egg bank has been reported also from other alpine lakes (e.g., in *D. umbra* from a lake in southern Norway [60] and from Lake Paione Inferiore in the central Alps [58,61]. Ephippia bearing only one egg (i.e., lacking the bilocular structure typical of *Daphnia* ephippia [58]) or having only one diapausing egg left, are often reported as nonviable, also in the case of lakes in temperate regions [62]. As suggested in the case of lakes in the Alps forming a cascade system, we might hypothesize that the Himalayan upper Lake Piramide Superiore might represent the biotic reservoir for *D.* (*Ctenodaphnia*) *fusca*, as suggested in previous studies [33].

Daphnia is an important member of pond and lake systems in the Himalayas and Tibet [63–66]. Most lakes of this region were sampled during the 1960s by Löffler [21], who reported the presence of melanic *Daphnia tibetana* Sars, 1903 [63], endemic to the region, and pale *Daphnia longispina* Müller, 1776 [67] (classified as var. *aspina* by Werestschagin [30] because of the lack of a carapace spine at the adult stage). The former swims during the day in bottom waters and the latter apparently is restricted to lakes with limited transparency. Such distinction between transparent, "*D.* (*Ctenodaphnia*) lakes" and turbid "*D.* (*Daphnia*) lakes" was not confirmed by multiannual studies and by sediment core analyses. Our results on ephippia abundance in core 04/1 of LCN10 therefore represent a further evidence of co-occurrence of the two *Daphnia* species in lakes of the region. Co-occurrence of the large, melanic *D.* (*Ctenodaphnia*) and the smaller, transparent *Daphnia dentifera* in LCN10, is likely enhanced by different habitats: while the former is usually found swimming near the littoral, the latter tends to swim in deeper water, minimizing effects of UV damage. Our observation confirms what was reported by Löffler [21], who found the melanic *Daphnia* swimming in the littoral and the pale one swimming in deeper lakes and/or in turbid ones [68]. Largely differing in size, the two species might be regarded as able to feed on different food particles. In a very interesting and pioneer study, however, Geller and Müller [69] demonstrated that, while the distances between the setulae of filtering combs increase with individual body length in different Cladocera taxa, they do not increase with body length in *Daphnia magna*, a *D.* (*Ctenodaphnia*) species. Sharing the same food particle size, therefore, might result in a strong competition between the two *Daphnia* species; literature suggests that the larger one would be favored, being able to grow and reproduce at lower food concentrations [70]. Therefore, the fact that the two tend to live in different habitats appears as the most reasonable explanation.

Since the first studies carried out during the Yale North India expedition [66] of 1932–1933, endemic populations were reported from lakes in the Khumbu Region of the Himalayas, particularly among the zooplankton: *Arctodiaptomus jurisovitchi* [19] and *Daphnia tibetana*. Among monogonont rotifers, *Hexarthra bulgarica nepalensis* [31] were described. *Hexarthra bulgarica* has been reported from several high altitude lakes: the Rila Mountains, Bulgaria [71], the Italian Alps [72,73], the Himalayas [74], the Khumbu area (East Nepal) in lakes within the subnival belt [21,25], the Rocky Mountains (Grouse Lake, Alberta, Canada) [31], the Andes [75–77] and the Sierra Nevada [78].

The taxonomy of the melanic *D. (Ctenodaphnia)* from this region is problematical, largely because incomplete descriptions have been published to date. Originally, three species were identified: *Daphnia pamirensis* Rylov, 1930 [64], *Daphnia fusca* and *Daphnia tibetana* (first described as *Daphniopsis tibetana*). These species have been synonymized in every pairwise combination, or all three merged in one species, over the last century [65,79]. In his *Daphnia* taxonomic revision, Benzie [80] synonymized the genus *Daphniopsis* with the *genus Daphnia*, and *D. pamirensis* with *D. fusca*. He also synonymized *Daphnia* specimens from the Khumbu Region in Nepalese Himalayas, referred to as *D. tibetana* by Löffler [21] and as *D. himalaya* by Manca et al. [33]. More recently, a detailed analysis and taxonomic revision of diagnostic traits of the four *D. (Ctenodaphnia)* species (*D. tibetana*, *D. fusca*, *D. pamirensis* and *D. himalaya*) from their original sites (Kyrgyzstan, Gokyo and Khumbu, Nepal), revealed that morphology of populations of *D. fusca*, *D. pamirensis* and *D. himalaya* did not substantially differ from each other [47]. Results on morphology were also confirmed by genetic analyses (based on three mitochondrial genes, 12S, 16S COI). According to the latter, Himalayan and Tian Shan populations are conspecific, while *D. tibetana* clearly differs from the other two. The latter belongs to a wider group of *D. (Ctenodaphnia)* species that include *D. atkinsoni* Baird, 1859 [81] and *D. mediterranea* Alonso, 1985 [82] complexes as well as *D. fusca*. While *D. tibetana* and *D. fusca* are two valid, well differentiated but distantly related species, *D. pamirensis* and *D. himalaya* are younger synonyms of *D. fusca*. Populations of *D. fusca* so far occur only in high mountain glacial lakes, while *D. tibetana* inhabits saline lakes of Central Asian mountain ranges and high altitude plateau. Molecular systematics of the non-pigmented *D.* gr. *longispina*, presenting phylogenetic analyses for individuals collected from the Himalaya (Nepal) and the Pamir (Tajikistan) mountain ranges [47] recently suggested that populations from Nepalese Himalayas actually constitute a lineage within the *D. dentifera* clade. *Daphnia* from Lake Rangkul (Tajikistan) pertained to *D. longispina* clade. Results suggest that the two morphotypes represented two different species, among the most widespread ones in the *D. longispina* complex. Each species was confirmed genetically in at least two biogeographic regions (Palearctic and Holarctic in case of *D. dentifera*; [83] and Palearctic and Ethiopian in case of *D. longispina* [84]). According to these results, distribution of both species is therefore extended: the Himalayan LCN10 represents not only the westernmost, but also the highest locality reported for *D. dentifera*. Similarly, Lake Rangkul is the southeastern most and also the highest known locality for *D. longispina*, as evidenced by molecular data. A recent study on *Daphnia* of the Tibetan Plateau based on molecular analysis (COI gene) revealed a higher species richness with respect to those described in the past, however six of the taxonomic units identified by DNA analysis agreed very well to classical taxonomy descriptions [85].

In LCN70, Rotifera dormancy does not seem to persist over multiple seasons, except for the species *Hexarthra bulgarica nepalensis*. The relatively high numbers of *Hexarthra* diapausing eggs' cases may be indicative of a high production of diapausing eggs is this lake. In rotifers, the ability to produce diapausing eggs varies from species to species, and not necessarily with environmental conditions [86]. Differently from most Monogononta, *Hexarthra* stem females can in turn be mictic in temporary ponds from Texas [87]. Mictic females tend to appear very early in the season in lakes of the Sierra Nevada [78] and in temporary ponds of the Chihuahuan desert [87]. Fertilized eggs in dried sediment probably regularly hatch in large numbers soon after flooding in LCN70. They hatch during a short period that marks the beginning of population growth. The timing of this hatching period could be controlled by temperature at the sediment surface, by physical factors that promote resuspension of diapausing eggs into an oxygenated and illuminated water and by biotic factors [88–91].

Our results indicate that some of these diapausing eggs remain unhatched in their egg cases and preserve well in the sediments deposited at the bottom of LCN70, year after year, forming an active egg bank [92]. If the eggs are buried in the sediments, their ability to hatch probably depends on physical and biological processes that mix the sediments and bring them to the surface where they can experience hatching cues. Sediment mixing is more pronounced in shallow waters, and therefore, buried diapausing eggs should have a greater chance of hatching there [68]. In the Himalayan temporary lake studied, during the ice-free period, the water level is subject to wide fluctuations and the life cycle may start after the summer monsoon inundation provides a hatching environment with well-oxygenated water and light. Changes in the proportion of diapausing eggs in different sections of the core might therefore also reflect changes in environmental conditions related to the monsoon.

Viability tests revealed that 64% of *Hexarthra* diapausing eggs of the uppermost two centimeters of the sediment core were able to hatch. This result is in agreement with previous studies which reported a range of hatching success between 10% and 88% [51,93–95]. Given the low sedimentation rate of this lake, the deposition time of diapausing eggs spans over three decades, thus the large variation of hatching time we observed (between 4 and 7 days) may reflect differences in age of the eggs. Consistent with our results, viable zooplankton diapausing eggs are recorded in the upper 4–6 cm of sediments [49,92,96–101]. Differences between environments can be related not only to differences in sedimentation rates but also to differences in the mortality rates of dormant stages, that can range between 1% y^{-1} and 64% y^{-1} [51,100,102,103]. In shallow ponds, as well as in our lakes, diapausing eggs are more exposed to stressing conditions (e.g., desiccation, UV radiation) than in deep lakes, and this exposure can accelerate the rate of degradation processes [100]. Nevertheless, there is evidence in literature of a high longevity, up to a maximum of 100 years [12,49,100,104,105].

Brachionus hatched resting eggs were found in the sediment core of LCN70. While reported from nearby lakes [22], *Brachionus* was not found in zooplankton samples of LCN70 at the time of sampling. It might have grown earlier in the season, thus leaving in the sediments the egg resting cases we found, although in lower numbers that those of *Hexarthra*. We expect that in these high mountain lakes, seasonal succession is similar to that already reported from similar systems in the Alps, in which small zooplankton (monogonont rotifers and immature stages of copepods) grows earlier in the season than large *Daphnia* [58]. The chance of finding taxa depends on time and space, and on the method used for sampling. Species rare at the time of sampling might be overlooked, and such might be the case for *Brachionus* in plankton samples of LCN70, which is shallow and with a rocky, heterogeneous bottom. Fossils from sediment samples generally complement taxa detection in zooplankton contemporary samples (see [18] for a list of Cladocera taxa found from the same sediment core from LPI that we used in the present study). Unfortunately, rotifers do not leave fossil remains in the sediments, and copepods are also very rarely reported, as they tend to quickly degrade. Therefore, the only chance for detecting their presence is by means of their diapausing stages. Production of diapausing eggs in copepods is rare and finding of viable diapausing eggs in lake sediments, namely of *Diaptomus sanguineus* Forbes, 1876 [106], as old as 350 years, is deserving of topmost journals [94]. On the contrary, production of diapausing eggs in monogonont rotifers, particularly *Brachionus*, forming an egg bank in the sediments is commonly reported. So, our inability to find unhatched eggs in the sediments appears contrasting with the standard very long diapause of most monogonont rotifers. Of course, it might be that, given rarity, these viable eggs were not found in our sediment samples.

In conclusion, despite producing diapausing eggs in non-negligible amounts, the Cladocera zooplankton taxa investigated in two Himalayan lakes do not seem leave a biotic reservoir in the sediments by means of viable diapausing eggs. *Daphnia* ephippia were in fact empty and/or with only one, nonviable egg. Similar results were obtained in previous studies, reporting that when a bilocular ephippium is partly empty, the remaining egg is usually nonviable. The presence of unilocular ephippium is also reported from high mountain lakes [58], and infertility of the ephippial egg inside is attributed to UV damage [107]. Therefore, in these high mountain lakes, the egg bank is restricted to the seasonal recruitment of active individuals into the population when the ice cover melts. The biotic reservoir of an active eggs bank seems restricted to one monogonont rotifer species, *Hexarthra bulgarica nepalensis*, whose diapausing eggs were successfully resurrected (hatched) in large numbers from relatively recent sediments.

We cannot, of course, exclude that we were truly unable to study them at best; we might simply think that, as suggested by Ruggiu et al. 1998: "Any information on freshwater organisms from a truly remote area could be intrinsically interesting" [108].

If our interpretation is true, we might be very concerned about the future of biodiversity of these lakes, so beautiful, so fragile. Given the low connections of these lakes with other aquatic environments, and their remoteness, our results highlight their high vulnerability to loss of biodiversity and of taxa which are unique and precious, super-exposed to impact of climate change and long distance transport of pollutants [109–111].

Author Contributions: Conceptualization, M.M.; investigation, M.M., R.P., S.B., A.L. and P.G.; formal analysis: R.P. and S.B.; writing—original draft preparation, M.M. and R.P.; writing—review and editing, M.M., R.P., A.L. and P.G.; visualization, funding acquisition, M.M. and A.L. All authors have read and agreed to the published version of the manuscript.

Funding: The present research was funded to M.M. by CNR RSTL (Basic research funds on special projects) n°552 on "Evolution of biodiversity and evaluation of the egg bank of aquatic organisms of remote lakes". Sampling and basic sediment core analyses were supported by funds to A.L. from CNR, Research Project on "Impact of Global Change in high elevation remote sites".

Acknowledgments: We wish to thank Simona Musazzi for helping in sampling of the sediment core of LCN10 in 2004 and G.A. Tartari for sampling the core of LCN70 in 1994. We wish to express our gratitude to the two anonymous referees: their contribution was crucial for substantially improving a previous version of the manuscript.

Conflicts of Interest: The authors declare no conflict of interest.

References

1. Battarbee, R.W.; Grytnes, J.A.; Thompson, R.; Appleby, P.G.; Catalan, J.; Korhola, A.; Birks, H.J.B.; Heegaard, E.; Lami, A. Comparing palaeolimnological and instrumental evidence of climate change for remote mountain lakes over the last 200 years. *J. Paleolimnol.* **2002**, *28*, 161–179. [CrossRef]

2. Kong, L.; Yang, X.; Kattel, G.; Anderson, N.J.; Hu, Z. The response of Cladocerans to recent environmental forcing in an Alpine Lake on the SE Tibetan Plateau. *Hydrobiologia* **2017**, *784*, 171–185. [CrossRef]

3. Catalan, J.; Ninot, J.M.; Aniz, M.M. (Eds.) High mountain conservation in a changing world. In *Advances in Global Change Research*; Springer International Publishing AG: Basel, Switzerland, 2017; p. 413. [CrossRef]

4. Kořínek, V.; Villalobos, L. Two South American endemic species of *Daphnia* from high Andean lakes. *Hydrobiologia* **2003**, *490*, 107–123. [CrossRef]

5. Herbert, P.D.N.; Emery, C.J. The adaptive significance of cuticular pigmentation in *Daphnia. Funct. Ecol.* **1990**, *4*, 703–710. [CrossRef]

6. Hessen, D.O. Competitive trade-off strategies in Arctic *Daphnia* linked to melanism and UV-B stress. *Polar Biol.* **1996**, *16*, 573–579. [CrossRef]

7. Ringelberg, J. Aspects of red pigmentation in zooplankton, especially copepods. In *Ecology and Evolution of Zooplankton Communities*; Kerfoot, W.C., Ed.; University Press of New England: Lebanon, NH, USA, 1980; pp. 91–97.

8. Moeller, R.E.; Gilroy, S.; Williamson, C.E.; Grad, G.; Sommaruga, R. Dietary acquisition of photoprotective compounds (mycosporine-like amino acids, carotenoids) and acclimation to ultraviolet radiation in a freshwater copepod. *Limnol. Oceanogr.* **2005**, *50*, 427–439. [CrossRef]

9. Cáceres, C.E. Interspecific variation in the abundance, production, and emergence of *Daphnia* diapausing eggs. *Ecology* **1998**, *79*, 1699–1710. [CrossRef]

10. Pérez-Martínez, C.; Barea-Arco, J.; Conde-Porcuna, J.M.; Morales-Baquero, R. Reproduction strategies of *Daphnia pulicaria* population in a high mountain lake of Southern Spain. *Hydrobiologia* **2007**, *594*, 75–82. [CrossRef]

11. De Stasio, B.T. The seed bank of a freshwater crustacean: Copepodology for the plant ecologist. *Ecology* **1989**, *70*, 1377–1389. [CrossRef]

12. Hairston, N.G., Jr. Zooplankton egg banks as biotic reservoirs in changing environments. *Limnol. Oceanogr.* **1996**, *41*, 1087–1092. [CrossRef]

13. De Meester, L.; Gómez, A.; Okamura, B.; Schwenk, K. The Monopolization Hypothesis and the dispersal–gene flow paradox in aquatic organisms. *Acta Oecol.* **2002**, *23*, 121–135. [CrossRef]

14. Dupuis, A.P.; Hann, B.J. Climate change, diapause termination and zooplankton population dynamics: An experimental and modelling approach. *Freshw. Biol.* **2009**, *54*, 221–235. [CrossRef]

15. Jones, N.T.; Gilbert, B. Changing climate cues differentially alter zooplankton dormancy dynamics across latitudes. *J. Anim. Ecol.* **2016**, *85*, 559–569. [CrossRef]

16. Brendonck, L.; De Meester, L. Egg banks in freshwater zooplankton: Evolutionary and ecological archives in the sediment. In *Recent Developments in Fundamental and Applied Plankton Research*; Van Donk, E., Spaak, P., Boersma, M., Eds.; Kluwer Academic Publishers: Dordrecht, The Netherlands, 2003; Volume 491, pp. 65–84. [CrossRef]

17. Lami, A.; Giussani, G. (Eds.) Limnology of high altitude lakes in the Mt. Everest Region (Nepal). *Mem. Ist. Ital. Idrobiol.* **1998**, *57*, 1–235.

18. Nevalainen, L.; Lami, A.; Luoto, T.P.; Manca, M. Fossil cladoceran record from Lake Piramide Inferiore (5067 m asl) in the Nepalese Himalayas: Biogeographical and paleoecological implications. *J. Limnol.* **2014**, *73*. [CrossRef]

19. Tartari, G.A.; Panzani, P.; Adreani, L.; Ferrero, A.; De Vito, C. Lake cadastre of Khumbu Himal Region: Geographical-geological-limnological data base. Limnology of high altitude lakes in the Mt Everest Region (Nepal). *Mem. Ist. Ital. Idrobiol.* **1998**, *57*, 151–235.

20. Tartari, G.A.; Tartari, G.; Mosello, R. Water chemistry of high altitude lakes in the Khumbu and Imja Kola valleys (Nepalese Himalayas). *Mem. Ist. Ital. Idrobiol.* **1998**, *57*, 51–76.

21. Löffler, H. High altitude lakes in Mt. Everest region. *Verh. Int. Ver. Limnol.* **1969**, *17*, 373–385. [CrossRef]

22. Manca, M.; Nocentini, A.M.; Ruggiu, D.; Panzani, P.; Bonardi, M.; Cammarano, P.; Spagnuolo, T. Observations on plankton and macrobenthic fauna of some high altitude Himalayan lakes. *Mem. Ist. Ital. Idrobiol.* **1995**, *53*, 17–25.

23. Manca, M.; Ruggiu, D.; Panzani, P.; Asioli, A.; Mura, G.; Nocentini, A.M. Report on a collection of aquatic organisms from high mountain lakes in the Khumbu Valley (Nepalese Himalayas). *J. Limnol.* **1998**, *57*, 77–98.

24. Baudo, R.; Tartari, G.; Munawar, M. (Eds.) Top of the world environmental research: Mount Everest-Himalayan ecosystem. In *Ecovision World Monograph Series*; Schweizerbart Science Publishers: Stuttgart, Germany, 1998; pp. 1–294.

25. Daems, G.; Dumont, H.J. Rotifers from Nepal, with the description of a new species of *Scaridium* and a discussion of the Nepalese representatives of the genus *Hexarthra*. *Biol. Jb. Dodonaea* **1974**, *42*, 61–81.

26. Löffler, H. *Diaptomus* (*Arctodiaptomus*) *jurisowitchi* nov. spec. aus dem Khumbu—gebiet (Nepal). *Khumbu Himal.* **1968**, *3*, 9–16.

27. Forbes, S.A. A preliminary report on the aquatic invertebrate fauna of the Yellowstone National Park, Wyoming, and the Flathead Region of Montana. *US Fish. Comm. Bull.* **1893**, *11*, 207–256.

28. Gurney, R. On some freshwater Entomostraca in the collection of the Indian Museum, Calcutta. *J. Asia. Soc. Bengal* **1906**, *2*, 273–281.

29. Sinev, A.Y. *Alona werestschagini* sp., a new species of the genus *Alona* Baird, 1843 related to *A. guttata* Sars, 1862 (Anomopoda: Chydoridae). *Arthropoda Sel.* **1999**, *8*, 23–30.

30. Werestschagin, G. *K faune Cladocera Kavkaza (The Caucasian Fauna of Cladocera). Raboty laboratorii zoologicheskogo kabineta Imeratorskogo Varshavskogo Universtiteta*; University of Warsaw: Warsaw, Poland, 1911; pp. 1–17.

31. Dumont, H.J.; Coussement, M.; Anderson, R.S. An examination of some *Hexarthra* species (Rotatoria) from western Canada and Nepal. *Can. J. Zool.* **1978**, *56*, 440–445. [CrossRef]

32. Sars, G.O. Contributions to the Knowledge of the Fresh-water "Entomostraca" of South America, as Shown by Artificial Hatching from Dried Material, by GO Sars. Part II. "Copepoda-Ostracoda". *Arch. Mat. Nat.* **1901**, *23*, 1–101.

33. Manca, M.; Martin, P.; Peñalva-Arana, D.C.; Benzie, J.A. Re-description of *Daphnia* (*Ctenodaphnia*) from lakes in the Khumbu Region, Nepalese Himalayas, with the erection of a new species, *Daphnia himalaya*, and a note on an intersex individual. *J. Limnol.* **2006**, *65*, 132–140. [CrossRef]

34. Manca, M.; Cammarano, P.; Spagnuolo, T. Notes on Cladocera and Copepoda from high altitude lakes in the Mount Everest Region (Nepal). *Hydrobiologia* **1994**, *287*, 225–231. [CrossRef]

35. Manca, M.; Mura, G. On *Branchinecta orientalis* Sars (Anostraca) in the Himalayas. *Hydrobiologia* **1997**, *356*, 111–116. [CrossRef]

36. Lami, A.; Guilizzoni, P.; Marchetto, A.; Bettinetti, R.; Smith, D.J. Palaeolimnological evidence of environmental changes in some high altitude Himalayan lakes (Nepal). *Mem. Istit. Ital. Idrobiol.* **1998**, *57*, 130–131.

37. Lami, A.; Marchetto, A.; Musazzi, S.; Salerno, F.; Tartari, G.; Guilizzoni, P.; Rogora, M.; Tartari, G.A. Chemical and biological response of two small lakes in the Khumbu Valley, Himalayas (Nepal) to short-term variability and climatic change as detected by long-term monitoring and paleolimnological methods. *Hydrobiologia* **2010**, *648*, 189–205. [CrossRef]

38. Cousyn, C.; De Meester, L. The vertical profile of resting egg banks in natural populations of the pond-dwelling cladoceran *Daphnia magna* Straus. *Arch. Hydrobiol.* **1998**, *52*, 127–139.

39. Guerrero-Jiménez, G.; Ramos-Rodríguez, E.; Silva-Briano, M.; Adabache-Ortiz, A.; Conde-Porcuna, J.M. Analysis of the morphological structure of diapausing propagules as a potential tool for the identification of rotifer and cladoceran species. *Hydrobiologia* **2019**, *847*, 1–24. [CrossRef]

40. Suzduki, M. New systematical approach to the Japanese planktonic Rotatoria. *Hydrobiologia* **1964**, *23*, 1–124.

41. Koste, W. Rotatoria. Die Rädertiere Mitteleuropas (Überordnung Monogononta). In *Bestimmugswerk Begründet von Max Voigt*; Gebrüder Borntraeger: Stuttgart, Germany, 1978; Volume 2, p. 234.

42. Mura, G. SEM morphological survey on the egg shell in the Italian anostracans (Crustacea, Branchiopoda). *Hydrobiologia* **1986**, *134*, 273–286. [CrossRef]

43. Mura, G. SEM morphology of resting eggs in the species of the genus *Branchinecta* from North America. *J. Crustac. Biol.* **1991**, *11*, 432–436. [CrossRef]

44. Fugate, M. *Branchinecta sandiegonensis*, a new species of fairy shrimp (Crustacea: Anostraca) from western North America. *Proc. Biol. Soc. Wash.* **1993**, *106*, 296.

45. Onbe, T. Sugar flotation method for the sorting the resting eggs of marine cladocerans and copepods from sea-bottom sediment. *Bull. Jpn. Soc. Sci. Fish.* **1978**, *44*, 1411. [CrossRef]

46. García-Roger, E.M.; Carmona, M.J.; Serra, M. Deterioration patterns in diapausing egg banks of *Brachionus* (Müller, 1786) rotifer species. *J. Exp. Mar. Biol. Ecol.* **2005**, *314*, 149–161. [CrossRef]

47. Moest, M.; Petrusek, A.; Sommaruga, R.; Juračka, P.J.; Slusarczyk, M.; Manca, M.; Spaak, P. At the edge and on the top: Molecular identification and ecology of *Daphnia dentifera* and *D. longispina* in high-altitude Asian lakes. *Hydrobiologia* **2013**, *715*, 51–62. [CrossRef]

48. Sudzuki, M. Intraspecific variability of *Brachionus plicatilis*. *Hydrobiologia* **1987**, *147*, 45–47. [CrossRef]

49. Piscia, R.; Guilizzoni, P.; Fontaneto, D.; Vignati, D.A.; Appleby, P.G.; Manca, M. Dynamics of rotifer and cladoceran resting stages during copper pollution and recovery in a subalpine lake. *Ann. Limnol. Int. J. Lim.* **2012**, *48*, 151–160. [CrossRef]

50. Latta, L.C.; Fisk, D.L.; Knapp, R.A.; Pfrender, M.E. Genetic resilience of *Daphnia* populations following experimental removal of introduced fish. *Conserv. Genet.* **2010**, *11*, 1737–1745. [CrossRef]

51. García-Roger, E.M.; Lubzens, E.; Fontaneto, D.; Serra, M. Facing Adversity: Dormant Embryos in Rotifers. *Biol. Bull.* **2019**, *237*, 119–144. [CrossRef] [PubMed]

52. Manca, M.; Comoli, P. Reconstructing long-term changes in *Daphnia*'s body size from subfossil remains in sediments of a small lake in the Himalayas. *J. Paleol.* **2004**, *32*, 95–107. [CrossRef]

53. Colbourne, J.K.; Hebert, P.D. The systematics of North American Daphnia (Crustacea: Anomopoda): A molecular phylogenetic approach. *Philos. Trans. R. Soc. Lond. B Biol. Sci.* **1996**, *351*, 349–360.

54. Green, J. Growth, size and reproduction in *Daphnia* (Crustacea: Cladocera). *Proc. Zool. Soc. Lond.* **1956**, *126*, 173–204. [CrossRef]

55. Arbačiauskas, K. Seasonal phenotypes of *Daphnia*: Post-diapause and directly developing offspring. *J. Limnol.* **2004**, *63*, 7–15. [CrossRef]

56. Jeppesen, E.; Jensen, J.P.; Amsinck, S.; Landkildehus, F.; Lauridsen, T.; Mitchell, S.F. Reconstructing the historical changes in *Daphnia* mean size and planktivorous fish abundance in lakes from the size of *Daphnia* ephippia in the sediment. *J. Paleolimnol.* **2002**, *27*, 133–143. [CrossRef]

57. Manca, M.; Comoli, P. Reconstructing population size structure in Cladocera by measuring their body remains. *Mem. Ist. Ital. Idrobiol.* **1995**, *54*, 61–68.

58. Cammarano, P.; Manca, M. Studies on zooplankton in two acidified high mountain lakes in the Alps. *Hydrobiologia* **1997**, *356*, 97–109. [CrossRef]

59. Manca, M.; Comoli, P. Studies on zooplankton of Lago Paione Superiore. *J. Limnol.* **1999**, *58*, 131–135. [CrossRef]

60. Larsson, P.; Wathne, I. Swim or rest during the winter–what is best for an alpine daphnid? *Arch. Hydrobiol.* **2006**, *167*, 265–280. [CrossRef]

61. Guilizzoni, P.; Lami, A.; Manca, M.; Musazzi, S.; Marchetto, A. Palaeoenvironmental changes inferred from biological remains in short lake sediment cores from the Central Alps and Dolomites. *Hydrobiologia* **2006**, *562*, 167–191. [CrossRef]

62. Marková, S.; Černý, M.; Rees, D.; Stuchlík, E. Are they still viable? Physical conditions and abundance of *Daphnia pulicaria* resting eggs in sediment cores from lakes in the Tatra Mountains. *Biologia* **2006**, *61*, S135–S146. [CrossRef]

63. Sars, G.O. *An Account of the Crustacea of Norway, with Short Descriptions and Figures of all the Species: IV. Copepoda Calanoida*; Compilation, Ed.; Bergens Museum: Bergen, Germany, 1903; Volume 171.

64. Rylov, W.M. *The Fresh-Water Calanoids of the U.S.S.R. Keys to Determination of Fresh-Water Organism of the U.S.S.R. A Fresh-Water Fauna*; Institute Fisheries and Science Explorations: Leningrad, Russia, 1930; pp. 1–288.

65. Wagler, E. Die Systematik und geographische Verbreitung des Genus *Daphnia* O.F. Müller mit besonderer Berucksichtigung des sudafrikanischen Arten. *Arch. Hydrobiol.* **1936**, *30*, 505–556.

66. Hutchinson, G. Limnological studies in Indian Tibet. *Int. Rev. Gesamten Hydrobiol.* **1937**, *35*, 134–177. [CrossRef]

67. Müller, O.F. *Zoologiæ Danicæ Prodromus, seu Animalium Daniæ et Norvegiæ indigenarum characteres, nomina, et synonyma imprimis popularium*; Hallageri: Copenhagen, Denmark, 1776; p. 274.

68. Manca, M.; Comoli, P.; Margaritora, F.G. An unusual type of *Daphnia* head shields from plankton and sediments of Himalayan lakes. *J. Limnol.* **1999**, *58*, 29–32. [CrossRef]

69. Geller, W.; Müller, H. The filtration apparatus of Cladocera: Filter mesh-sizes and their implications on food selectivity. *Oecologia* **1981**, *49*, 316–321. [CrossRef]

70. Gliwicz, Z.M. Food thresholds and body size in cladocerans. *Nature* **1990**, *343*, 638–640. [CrossRef]

71. Wiszniewski, J. Un nouveau rotifere du genre *Pedalia* habitant les lacs des hautes montagnes. *Int. Rev. Ges. Hydrobiol. Hydrogr.* **1933**, *29*, 229–236. [CrossRef]

72. Tonolli, V.; Tonolli, L. Osservazioni sulla biologia ed ecologia di 170 popolamenti zooplanctonici di laghi italiani de alta quota. *Mem. 1st Ital. Idrobiol.* **1951**, *6*, 53–136.

73. Ferrari, I. Notes on the dynamics of the reproductive activity of *Arctodiaptomus bacillifer* in high altitude alpine lakes. *Boll. Zool.* **1971**, *38*, 221–235. [CrossRef]

74. Edmondson, W.T.; Hutchinson, G.E. Report on Rotatoria. *Mem. Conn. Acad. Arts Sci.* **1934**, *10*, 153–186.

75. Ruttner-Kolisko, A. Plankton Rotifers, biology and taxonomy. *Binnengewässer* **1974**, *26*, 146.

76. Modenutti, B.E. Summer population of *Hexarthra bulgarica* in a high elevation lake of south Andes. *Hydrobiologia* **1993**, *259*, 33–37. [CrossRef]

77. Zagarese, H.E.; Diaz, M.; Pedrozo, F.; Ubeda, C. Mountain lakes in northwestern Patagonia. *Int. Ver. Angew. Limnol. Verh.* **2000**, *27*, 533–538. [CrossRef]

78. Cruz-Pizarro, L.; Morales, R. Taxonomic and ecological notes on *Hexarthra bulgarica* from high mountain lakes and ponds in the Sierra Nevada (Spain). *Hydrobiologia* **1987**, *147*, 91–95. [CrossRef]

79. Brehm, V.; Woltereck, R. Die Daphniden der Yale-Northindia-Expedition. *Int. Rev. Gesamten Hydrobiol. Hydrograp.* **1939**, *39*, 1–19. [CrossRef]

80. Benzie, J.A. The genus *Daphnia* (including *Daphniopsis*) (anomopoda: Daphniidae). In *Guides to the Identification of the Microinvertebrates of the Continental Waters of the World*; Kenobi Productions: Grant Pass, OR, USA, 2005; Volume 21, pp. 1–376.

81. Baird, W. Description of some new recent Entomostraca from Nagpur, collected by the Rev. S. Hislop. *Proc. Zool. Soc. Lond.* **1859**, *27*, 231–234.

82. Alonso, M. *Daphnia (Ctenodaphnia)* Mediterranea: A new species of hyperhaline waters, long confused with *D. (C.) Dolichocephala* Sars, 1895. *Hydrobiologia* **1985**, *128*, 217–228. [CrossRef]

83. Ishida, S.; Taylor, D.J. Mature habitats associated with genetic divergence despite strong dispersal ability in an arthropod. *BMC Evol. Biol.* **2007**, *7*, 52. [CrossRef] [PubMed]

84. Petrusek, A.; Hobæk, A.; Nilssen, J.P.; Skage, M.; Černý, M.; Brede, N.; Schwenk, K. A taxonomic reappraisal of the European *Daphnia longispina* complex (Crustacea, Cladocera, Anomopoda). *Zool. Scr.* **2008**, *37*, 507–519. [CrossRef]

85. Xu, L.; Lin, Q.; Xu, S.; Gu, Y.; Hou, J.; Liu, Y.; Dumont, H.J.; Han, B.P. *Daphnia* diversity on the Tibetan Plateau measured by DNA taxonomy. *Ecol. Evol.* **2018**, *8*, 5069–5078. [CrossRef]

86. Gilbert, J.J. Timing of diapause in monogonont rotifers: Mechanisms and strategies. In *Diapause in Aquatic Invertebrates Theory and Human Use*; Alekseev, V.R., de Stasio, B.T., Gilbert, J.J., Eds.; Springer: Dordrecht, Germany, 2007; pp. 11–27.

87. Schröder, T.; Howard, S.; Arroyo, M.L.; Walsh, E.J. Sexual reproduction and diapause of *Hexarthra* sp. (Rotifera) in short-lived ponds in the Chihuahuan Desert. *Freshw. Biol.* **2007**, *52*, 1033–1042. [CrossRef]

88. Vandekerkhove, J.; Declerck, S.; Brendonck, L.; Conde-Porcuna, J.M.; Jeppesen, E.; Meester, L.D. Hatching of cladoceran resting eggs: Temperature and photoperiod. *Freshw. Biol.* **2005**, *50*, 96–104. [CrossRef]

89. Gilbert, J.J.; Schröder, T. Rotifers from diapausing, fertilized eggs: Unique features and emergence. *Limnol. Oceanogr.* **2004**, *49*, 1341–1354. [CrossRef]

90. Gilbert, J.J. Resting-egg hatching and early population development in rotifers: A review and a hypothesis for differences between shallow and deep waters. *Hydrobiologia* **2017**, *796*, 235–243. [CrossRef]

91. Bozelli, R.L.; Tonsi, M.; Sandrini, F.; Manca, M. A Big Bang or small bangs? Effects of biotic environment on hatching. *J. Limnol.* **2008**, *67*, 100–106. [CrossRef]

92. Cáceres, C.E.; Hairston, N.G. Benthic-pelagic coupling in planktonic crustaceans: The role of the benthos. *Ergeb. Limnol.* **1998**, *52*, 163–174.

93. Marcus, N.H.; Lutz, R.; Burnett, W.; Cable, P. Age, viability, and vertical distribution of zooplankton resting eggs from an anoxic basin: Evidence of an egg bank. *Limnol. Oceanogr.* **1994**, *39*, 154–158. [CrossRef]

94. Hairston, N.G., Jr.; Van Brunt, R.A.; Kearns, C.M.; Engstrom, D.R. Age and survivorship of diapausing eggs in a sediment egg bank. *Ecology* **1995**, *76*, 1706–1711. [CrossRef]

95. Franch-Gras, L.; García-Roger, E.M.; Serra, M.; José Carmona, M. Adaptation in response to environmental unpredictability. *Proc. R. Soc. Lond. B Biol. Sci.* **2017**, *284*, 2017.0427. [CrossRef]

96. Herzig, A. Resting eggs—A significant stage in the life cycle of crustaceans *Leptodora kindtii* and *Bythotrephes longimanus*. *Int. Ver. Angew. Limnol. Verh.* **1985**, *22*, 3088–3098.

97. Carvalho, G.R.; Wolf, H.G. Resting eggs of lake-*Daphnia*, I. Distribution, abundance and hatching of eggs collected from various depths in lake sediments. *Freshwat. Biol.* **1989**, *22*, 459–470. [CrossRef]

98. Hairston, N.G.; Van Brunt, R.A. Diapause dynamics of two diaptomid copepod species in a large lake. *Hydrobiologia* **1994**, *292*, 209–218. [CrossRef]

99. García-Roger, E.M.; Carmona, M.J.; Serra, M. Hatching and viability of rotifer diapausing eggs collected from pond sediments. *Freshw. Biol.* **2006**, *51*, 1351–1358. [CrossRef]

100. García-Roger, E.M.; Carmona, M.J.; Serra, M. Patterns in rotifer diapausing egg banks: Density and viability. *J. Exp. Mar. Biol. Ecol.* **2006**, *336*, 198–210. [CrossRef]

101. Piscia, R.; Tabozzi, S.; Bettinetti, R.; Nevalainen, L.; Manca, M.M. Unexpected increases in rotifer resting egg abundances during the period of contamination of Lake Orta. *J. Limnol.* **2016**, *75*, 76–85. [CrossRef]

102. García-Roger, E.M.; Ortells, R. Trade-offs in rotifer diapausing egg traits: Survival, hatching, and lipid content. *Hydrobiologia* **2018**, *805*, 339–350. [CrossRef]

103. Gabaldón, C.; Serra, M.; Carmona, M.J.; Montero-Pau, J. Life-history traits, abiotic environment and coexistence: The case of two cryptic rotifer species. *J. Exp. Mar. Biol. Ecol.* **2015**, *465*, 142–152. [CrossRef]

104. King, C.E.; Serra, M. Seasonal variation as a determinant of population structure in rotifers reproducing by cyclical parthenogenesis. *Hydrobiologia* **1998**, *387*, 361–372. [CrossRef]

105. Kotani, T.; Ozaki, M.; Matsuoka, K.; Snell, T.W.; Hagiwara, A. Reproductive isolation among geographically and temporally isolated marine *Brachionus* strains. *Hydrohiologia* **2001**, *446*, 283–290. [CrossRef]

106. Forbes, S.A. List of Illinois Crustacea. *Bull. Iii Mus. Nat. Hist.* **1876**, *1*, 3–76.

107. Raikow, D.F.; Reid, D.F.; Blatchley, E.R., III; Jacobs, G.; Landrum, P.F. Effects of proposed physical ballast tank treatments on aquatic invertebrate resting eggs. *Environ. Toxicol. Chem. Int. J.* **2007**, *26*, 717–725. [CrossRef]

108. Ruggiu, D.; Bertoni, R.; Callieri, C.; Manca, M.; Nocentini, A. Assessment of biota in lakes form the Khumbu Valley, high Hiamalayas. In *Top of the World Environmental Research: Mount Everest-Himalayan Ecosystem*; Baudo, R., Tartari, G., Munawar, M., Eds.; Blackhuys Publishers: Leiden, The Netherlands, 1998; pp. 219–233.

109. Fernandez, P.; Carrera, G.; Grimalt, J. Persistent organic pollutants in remote freshwater ecosystems. *Aquat. Sci.* **2005**, *67*, 263–273. [CrossRef]

110. Schmid, P.; Bogdal, C.; Blüthgen, N.; Anselmetti, F.S.; Zwyssig, A.; Hungerbühler, K. The missing piece: Sediment records in remote mountain lakes confirm glaciers being secondary sources of persistent organic pollutants. *Environ. Sci. Technol.* **2011**, *45*, 203–208. [CrossRef]

111. Guzzella, L.; Salerno, F.; Freppaz, M.; Roscioli, C.; Pisanello, F.; Poma, G. POP and PAH contamination in the southern slopes of Mt. Everest (Himalaya, Nepal): Long-range atmospheric transport, glacier shrinkage, or local impact of tourism? *Sci. Total Environ.* **2016**, *544*, 382–390. [CrossRef]

Article

Zooplankton Size as a Factor Determining the Food Selectivity of Roach (*Rutilus Rutilus*) in Water Basin Outlets

Robert Czerniawski [1,2,*] and Tomasz Krepski [1,2]

[1]   Department of Hydrobiology and General Zoology, University of Szczecin, ul. Felczaka 3c, 71-412 Szczecin, Poland; tomasz.krepski@usz.edu.pl

[2]   Centre of Molecular Biology and Biotechnology, University of Szczecin, ul. Wąska 13, 70-415 Szczecin, Poland

*   Correspondence: robert.czerniawski@usz.edu.pl

Received: 21 May 2019; Accepted: 19 June 2019; Published: 19 June 2019

Abstract: Fish occurring in the outlets of water basins reduce the abundance of zooplankton. The study was performed at the outlet sections of the lake and waste stabilization pond of a sewage treatment plant. The aim of the study was to determine which zooplankton is chosen more often by the roach (*Rutilus rutilus*), those drifting from the waste stabilization pond or from the lake. The zooplankton from the pond was dominated by *Daphnia pulex* while zooplankton from the lake was dominated by small planktonic rotifers. We observed that the larger the plankter-victim's size, the faster the reduction of its number. The fish were more likely to feed on zooplankton drifting from the waste stabilization pond than from the lake. It was influenced by *D. pulex* individuals, attractive for fish due to their largest body size among the analyzed zooplankton. The significance of riverine zooplankton in the downstream food web may render this data even more important.

Keywords: *Daphnia pulex*; stream ecology; river dispersion; live organic matter; fish feeding

---

1. Introduction

Fish make the most significant contribution to reduction of zooplankton communities [1–5]. This is especially visible in lake outlets where drifting zooplankton constitutes a rich food base for fish [6,7]. Even though the phenomenon of cyprinids feeding on zooplankton in outlets of stagnant basins is very common, it remains poorly investigated and documented [6,7]. Flow-through lakes and reservoirs provide a rich source of zooplankton for river ecosystems because organisms are washed out of the upper sections of lakes into river outlets. There is scarce literature regarding the efficacy and food selectivity of fish feeding in outlets of stagnant water basins. Certain authors speculate that fish are the main factor reducing zooplankton biomass in outlets [6–8], while others prove the point by showing zooplankton in fish stomach contents [9]. However, there is a lack of studies showing the stomach content of fish in regard to the zooplankton food-base in a river and the size of zooplankton victims.

This can be verified by checking how fish in the outlets of water basins reduce the abundance of zooplankton ranging in size. However, in the natural environment, it is difficult to find lake outlets with a high amount of large plankters such as *Daphnia magna* or *D. pulex* because they are immediately eaten by fish in the lake or at the beginning of a lake outlet, and therefore cannot reach the lower sections of a river outflow from the lake [6,7]. Therefore, it was crucial to find reservoirs without fish that allow large plankters to drift downstream in the outlet where their abundance is reduced by fish. One such reservoir is the retention pond of sewage treatment plants, in which biological purification of treated sewage takes place. This pond may abound with large planktonic crustaceans such as *D. pulex*, which reproduce freely if they do not encounter their greatest stressor, that is, fish [10]. Such conditions were observed in a sewage treatment plant in a small town. An abundant amount of zooplankton was found in the pond and its outlet, where *D. pulex* was identified as the quantitative dominant. Compared to other plankters, this species reaches a large size and therefore is probably more susceptible to fish predation. The greatest reductions were observed in the largest plankters, adult cladocerans, and adult copepods [5,9]. Zooplankton that drift from the waste stabilization pond find their way to the Drawa river, and mix with zooplankton drifting from Adamowo lake. A high quantity of zooplankton attracts a great cyprinid biomass to the outlet of Drawa, including the roach (*Rutilus rutilus*), which feeds on this zooplankton type. This is a good field of research to determine which zooplankton is chosen more often by the fish, either the one drifting from the waste stabilization pond or the one from the lake, the factors that characterize the fish (e.g., their size and age), and the factors that characterize the zooplankton (e.g., body size and abundance) and influence the fish's choice. Answering the above questions was the aim of the study. The significance of riverine zooplankton in the downstream food web may render this data even more important.

2. Methods

2.1. Study Area

The study was performed at the outlet sections of the waste stabilization pond in the town of Drawno and at the outlet of Adamowo lake (GPS: 53.2089568; 15.7502837; NW Poland) (Figure 1). The samples were collected at 5 sites: the Adamowo lake outlet into the Drawa river (RLO), the waste stabilization pond (POND), the outlet from the waste stabilization pond (PO), a site in the Drawa river 20 m below the pond outlet (R1), and a site in the Drawa river 40 m below the pond outlet (R2). There were 2 control samples from 2 sites, i.e., RLO and POND, as a source of dispersed zooplankton, with river current in other sites, i.e., PO, R1, and R2. The biological treatment of sewage, previously purified by physical and chemical methods and with the use of activated sludge, is carried out in the retention pond of the sewage treatment plant (POND). The pond is densely covered with submerged vegetation such as *Ceratophyllum demersum*. Great feeding conditions and no fish presence in the pond create the optimal conditions for the development of *D. pulex* [1–5,11,12]. Phytoplankton communities and chlorophyll *a* concentration were not determined during the study, although these findings could have been useful for determining feeding conditions for *D. pulex*. Chlorophyll *a* measurements, taken the following year during 4 water discharges, indicated that in the summer chlorophyll *a* concentration was 38–84 µg L^{-1}. Consistent with other study results, the analysis of our findings indicated that the pond provided great feeding conditions for *D. pulex*. Several authors have indicated that there is a high positive correlation between chlorophyll *a* concentration and the density of *Daphnia* species [13,14].

The other sites (in river) were dominated by pelagic rotifers; such a pattern is common in lotic waters due to several reasons. Rotifers are smaller in size and thus may be avoided by fish [9]. Also, their sedimentation is reduced due to their smaller specific weight, and consequently the rotifers may drift long distances, carried by the river current [9,15]. Additionally, the short development cycle allows rotifers to reproduce successfully under favorable conditions even in running waters [16]. However, in the case of this study, the definite quantitative prevalence of rotifers is determined by the presence of a highly eutrophicated lake that provides favorable conditions for the development of zooplankton.

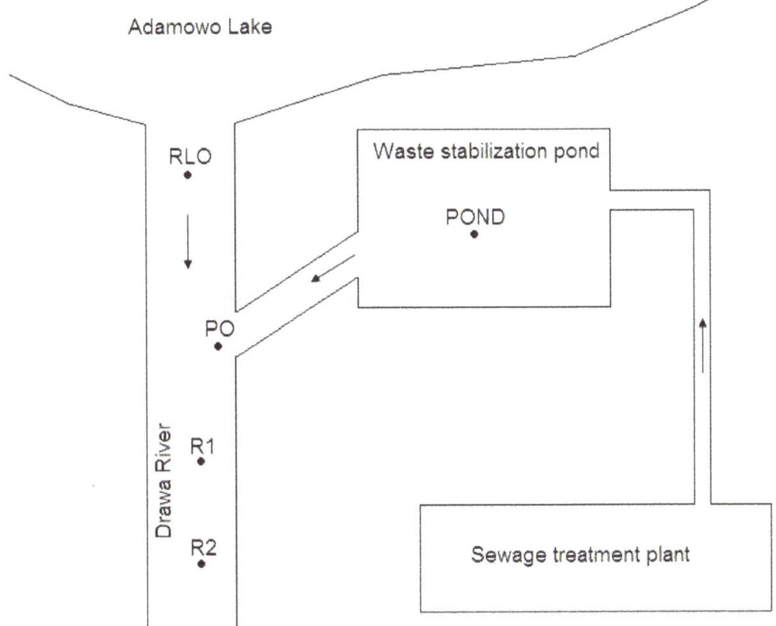

Figure 1. Scheme of the study area. RLO—river outflow from the lake; POND—waste stabilization pond of sewage treatment plant; PO—pond outflow; R1—site in the river, 20 m below the pond outflow; R2—site in the river, 40 m below the pond outflow.

2.2. Sampling Methods

The zooplankton samples were collected for 4 days when the biologically purified water was being discharged from the retention pond by the sewage treatment plant. The sewage treatment plant discharged water for 30 min daily. Zooplankton samples were collected at the above-mentioned sites 10 min before the water discharge was around midday.

At each site, 50 L of water was collected from the river current. The samples were collected using a Van Dorn 5-L water sampler (KC Denmark) at five depths: 20%, 40%, 60%, 80%, and at the surface. At each depth level, 10 L of water was collected to obtain 50 L of water. The water was filtered through a plankton net with a mesh of 30 mm. The samples were then concentrated to 150 mL and fixed in a 4–5% formalin solution. Five sub-samples (2 mL) from the stirred total sample were pipetted into a glass Sedgewick Rafter Counting Chamber. The samples were identified using a Nikon Eclipse 50i microscope. Similarly to methods applied in our research on the relationship between the reduction rate in drifting zooplankton abundance and the type of fish present in the pond, in the course of this study we divided zooplankton into 8 groups, according to the body size and environmental preferences [7]: (i) benthic rotifers (Bdelloidea, Colurellidae, *Mytilina* sp., *Euchlanis* sp., *Lecane* sp., *Cephalodella* sp.), (ii) planktonic rotifers (*Pompholyx* sp., Filinidae, Trichocercidae, Brachionidae, *Polyarthra* sp., *Synchaeta* sp.), (iii) *Asplanchna* sp., (iv) small cladocerans (Bosminidae, Chydoridae), (v) large cladocerans (Dapnidae and *Diaphanosoma* sp. except *D. pulex*), (vi) adult *D. pulex*, (vii) cyclopid nauplii, and (viii) copepods (adult copepods and copepodites). Because of their large size and their dominance in the pond and pond outlet, adult individuals of *D. pulex* were not included in the large cladoceran group, but instead were categorized as a separate group. Planktonic rotifers, benthic rotifers, and cyclopoid nauplii were considered the smallest plankters. *Asplanchna* sp., bosminids, and chydorids were considered medium-sized plankters, whereas Dapnidae (except for *D. pulex*), *Diaphanosoma* sp., and copepods were considered large plankters. The largest plankters observed were *D. pulex* individuals [17–19].

In the attempt to verify food selectivity of fish, 30 roach were caught from the pond outlet site (PO) on the same day when the zooplankton samples were collected. The fish were caught 5 min before the end of the water discharge from the waste stabilization pond by 3 anglers of the Polish Angling Association. The anglers cast fishing rods into the pond outlet, straight into the area where the fish gathered. The fish took the bait immediately. The number of fish needed for the study was caught in a range of 1 min 15 s to 2 min 45 s. None of the fish were discarded. The fish sampling was random, and the fish caught differed in size, mass, and age. We also intended to determine zooplankton selectivity by fish varying in size and age. Scales were used to estimate fish age. The fish were measured and weighed. Total length and mass values were used to calculate the fish condition factor. The means of biomasses and the total body length of fish are shown in Table 1.

Table 1. Mean values ± SD of total length, mass and condition factor of fish.

Date	Fish Age	Length (cm)	Mass (g)	Condition Factor
9th August	6+ ($n = 2$)	13.70 ± 0.57	35.05 ± 5.16	1.36 ± 0.03
	5+ ($n = 10$)	10.88 ±1.05	17.25 ± 4.92	1.31 ± 0.08
	4+ ($n = 7$)	10.31 ± 0.39	13.80 ± 1.91	1.25 ± 0.11
	3+ ($n = 11$)	8.69 ± 0.34	8.09 ± 1.58	1.22 ± 0.13
16th August	6+ ($n = 4$)	12.58 ± 0.62	23.90 ± 3.63	1.20 ± 0.10
	5+ ($n = 6$)	10.57 ± 1.05	14.78 ± 3.55	1.24 ± 0.10
	4+ ($n = 13$)	9.14 ± 0.82	9.30 ± 2.99	1.17 ± 0.11
	3+ ($n = 5$)	7.42 ± 0.22	4.72 ± 0.69	1.15 ± 0.08
	2+ ($n = 2$)	6.20 ± 0.14	2.55 ± 0.07	1.07 ± 0.10
20th August	7+ ($n = 1$)	14.30	39.40	1.35
	6+ ($n = 7$)	12.37 ± 0.76	24.63 ± 3.93	1.29 ± 0.09
	5+ ($n = 7$)	11.03 ± 0.73	17.30 ± 3.96	1.27 ± 0.07
	4+ ($n = 5$)	9.40 ± 0.72	10.74 ± 2.44	1.28 ± 0.05
	3+ ($n = 9$)	7.42 ± 0.36	5.19 ± 0.77	1.26 ± 0.07
	2+ ($n = 1$)	5.60	2.20	1.25
4th September	6+ ($n = 9$)	12.16 ± 0.30	23.14 ± 1.90	1.29 ± 0.04
	5+ ($n = 9$)	11.38 ± 0.31	18.88 ± 1.71	1.28 ± 0.05
	4+ ($n = 3$)	8.47 ± 0.46	7.50 ± 1.68	1.22 ± 0.09
	3+ ($n = 9$)	7.49 ± 0.33	5.31 ± 0.94	1.25 ± 0.10

The anglers caught 3 fish species—the roach (*Rutilus rutilus*), the white bream (*Abramis bjoerkna*), and the bleak (*Alburnus alburnus*). However, since roach constituted most of the catches (96%) the study focused only on this species. Given the above, we could objectively determine fish foraging efficiency. Once caught, the fish were stunned and killed. Afterwards, fish digestive tracts were preserved in 4% concentration of formaldehyde and taken to the laboratory where their contents were identified.

Given the site locations, no special field sampling permits were needed to conduct our study. We confirm that the land owner (the Polish Angling Association) gave us the permission to conduct the study at the examined sites. All fish were caught in the presence of the fish lake owner. Therefore, no specific permissions were needed for these locations and activities, and no approval of the Institutional Animal Care and Use Committee (IACUC) or an equivalent animal ethics committee was required. The field studies did not involve endangered or protected species. All sampling procedures and experimental manipulations required for our study have been reviewed and specifically approved under the field permit.

2.3. Statistical Analyses

To check for significant differences in abundance between the zooplankton groups found at the sites, the Kruskal–Wallis test was used ($p < 0.05$). The same test was used to check for significant differences in zooplankton amount between all zooplankton groups found in fish stomachs ($p < 0.05$). To determine significant differences between sites in terms of abundance and zooplankton number in fish stomachs, we performed post-hoc multiple comparisons of mean ranks for all groups ($p < 0.05$). The correlations between fish age and amount of zooplankton in fish stomachs were checked using

Spearman's rank correlation ($p < 0.05$). The aforementioned test was used to calculate the correlations between zooplankton abundance at sites and the zooplankton amount in fish stomachs ($p < 0.05$).

We checked the model with the Poisson distribution for overdispersion and found that it did not fit the data ($p < 0.05$). Then we used a generalized linear mixed model with negative binomial error distribution and log link function. The measurement values were added as a random effect. Zooplankton group abundance in the pond (POND) and the lake outlet (RLO) were included as covariates. We hypothesized that species abundance in the pond should increase the abundance at the sampling sites in the river (PO, R1 and R2). For modelling we used the "Mass" package [20] in the R package [21] for the analyses.

3. Results

We observed each zooplankton group at all sites except for the PO (Figure 2). The greatest zooplankton density was observed on 20 August. On every day of the data collection, *D. pulex* were predominant in the pond and in the pond outlet (Figure 2). Other sampling sites were dominated by planktonic Rotifera. A significant reduction of this species below the lake outlet indicates that the fish can consume almost the total amount of *D. pulex* drifting from the pond (Figure 2, Table 2).

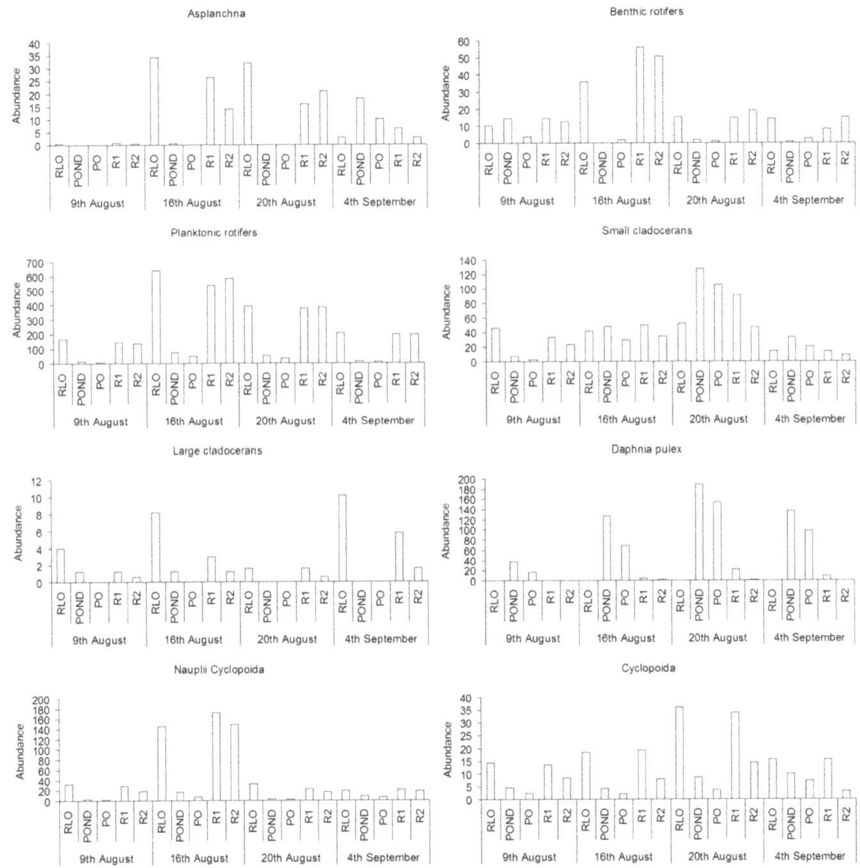

Figure 2. Abundance of zooplankton at examined sites.

The greatest and most significant reduction in the abundance of drifting zooplankton groups between the PO and the R2 was reported for *D. pulex* ($p < 0.05$) (Table 2). Such great reductions were not reported for other zooplankton groups drifting from the lake ($p > 0.05$) or found between the RLO and the R2 sites, which is a section longer than between the PO and the R2. Even though no significant differences were reported, we noted that the larger the plankter size, the greater the differences in zooplankton abundance between subsequent sites (Figure 2, Table 2).

The model with the Poisson distribution showed that the abundance of zooplankton in the RLO did not affect its abundance at other sites ($p > 0.05$) (Table 3). We found that abundance in the pond affected the abundance in other sites ($p < 0.05$) (Table 3). We found that sites R1 and R2 differed significantly from the PO site in terms of *D. pulex* abundance reduction ($p < 0.05$) (Table 3).

Table 2. Mean ± SD of the zooplankton groups abundance at examined sites ($n = 4$). Different letters in rows show significant differences between sites (post-hoc multiple comparisons of mean ranks for all groups $p < 0.05$).

Zooplankton Group	RLO	POND	PO	R1	R2
Asplanchna	17.4 ± 18.2 a	4.7 ± 8.9 ab	2.5 ± 5.0 b	12.4 ± 11.2 ab	9.6 ± 9.6 ab
Benthic rotifers	18.9 ± 11.6 a	4.3 ± 6.8 a	2.4 ± 1.1 a	23.3 ± 22.0 a	24.5 ± 17.9 a
Planktonic rotifers	351.5 ± 216.7 a	37.7 ± 30.3 ab	24.9 ± 21.2 b	313.6 ± 178.8 ab	325.4 ± 202.5 ab
Small cladocerans	38.6 ± 16.8 ab	54.2 ± 51.8 ab	39.7 ± 45.4 ab	46.8 ± 32.8 ab	28.3 ± 15.8 ab
Large cladocerans	6.0 ± 3.9 ab	0.6 ± 0.7 ab	-	2.9 ± 2.1 ab	1.0 ± 0.5 ab
Daphnia pulex	-	122.9 ± 62.5 a	84.4 ± 57.5 a	9.1 ± 9.6 b	0.8 ± 1.0 b
Nauplii Cyclopoida	57.5 ± 59.3 a	7.7 ± 6.5 ab	4.5 ± 3.1 b	61.4 ± 73.8 ab	50.5 ± 65.3 ab
Cyclopoida	21.2 ± 10.2 a	6.9 ± 2.9 b	3.9 ± 2.4 b	20.6 ± 9.3 a	8.5 ± 4.6 ab

Table 3. Estimates of fitted model with species abundance as dependent variable. Significant results are marked with bold.

Fixed Effects	Estimate	Std. Error	z Value	*p*
(Intercept)	0.803	0.426	1.883	0.060
R1 vs PO	**1.492**	**0.492**	**3.035**	**0.002**
R2 vs PO	**1.248**	**0.500**	**2.496**	**0.013**
Benthic rotifers	−0.311	0.602	−0.516	0.606
Cyclopoida	0.336	0.538	0.625	0.532
Daphnia pulex	**1.825**	**0.581**	**3.142**	**0.002**
Large cladocerans	−21.717	42.375	−0.512	0.608
Nauplii Cyclopoida	0.251	0.538	0.466	0.641
Planktonic rotifers	**1.167**	**0.594**	**1.965**	**0.049**
Small cladocerans	**1.678**	**0.502**	**3.344**	**0.001**
RLO	0.001	0.001	1.209	0.227
POND	**0.013**	**0.003**	**4.384**	**<0.001**
R1 vs PO: Benthic rotifers	0.951	0.714	1.333	0.182
R2 vs PO: Benthic rotifers	**1.273**	**0.719**	**1.769**	**0.077**
R1 vs PO: Cyclopoida	0.240	0.663	0.362	0.717
R2 vs PO: Cyclopoida	−0.400	0.682	−0.586	0.558
R1 vs PO: *Daphnia pulex*	**−3.903**	**0.635**	**−6.149**	**<0.001**
R2 vs PO: *Daphnia pulex*	**−6.017**	**0.850**	**−7.075**	**<0.001**
R1 vs PO: Large cladocerans	20.295	42.376	0.479	0.632

Table 3. *Cont.*

Fixed Effects	Estimate	Std. Error	z Value	p
R2 vs PO: Large cladocerans	18.908	42.377	0.446	0.655
R1 vs PO: Nauplii Cyclopoida	1.137	0.656	1.733	0.083
R2 vs PO: Nauplii Cyclopoida	1.148	0.663	1.730	0.084
R1 vs PO: Planktonic rotifers	**1.282**	**0.614**	**2.087**	**0.037**
R2 vs PO: Planktonic rotifers	**1.527**	**0.621**	**2.458**	**0.014**
R1 vs PO: Small cladocerans	−0.993	0.614	−1.619	0.106
R2 vs PO: Small cladocerans	−1.173	0.624	−1.879	0.060

Examination of roach stomach contents showed that their diet consists mainly of *D. pulex* (Figure 3) because it had the highest percentage contribution of the total plankters across all individuals (Table 4). Even though *D. pulex* dominated in the stomachs of fish of all ages, we observed that older fish consumed considerably more *D. pulex* ($p < 0.05$) than younger fish (Table 4).

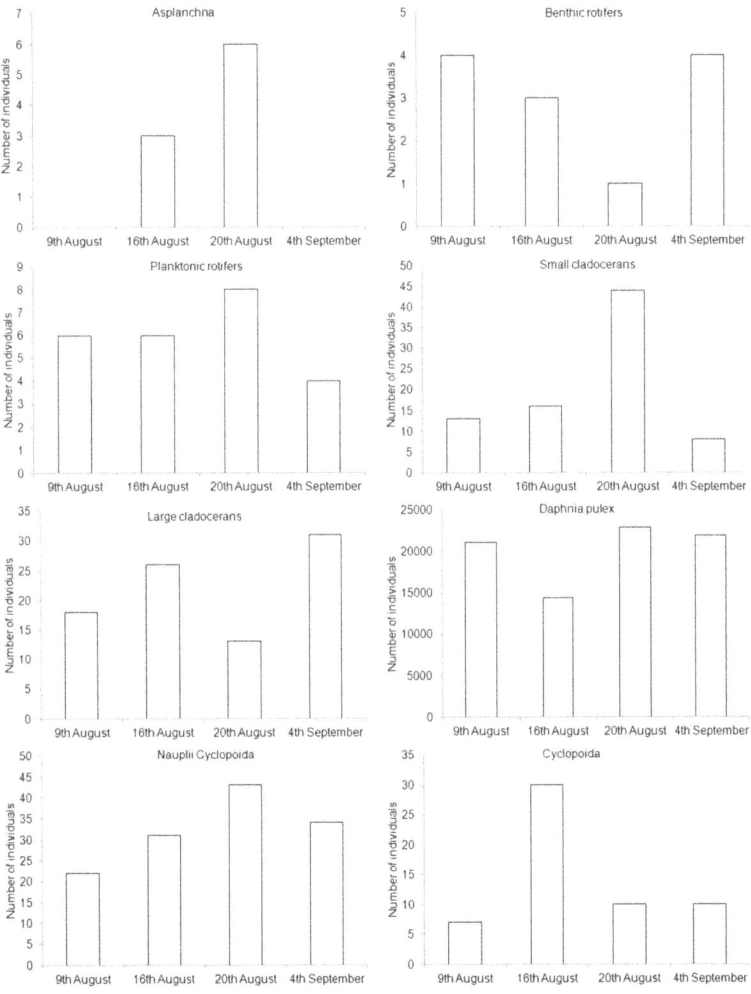

Figure 3. Total number of zooplankton individuals in stomachs of total fish.

A significant correlation between the fish age and the number of zooplankton group individuals in fish stomachs was observed only for *D. pulex* ($p < 0.05$) (Figure 4). Correlations between fish age and number of zooplankton individuals from other groups found in fish stomachs were insignificant ($p > 0.05$).

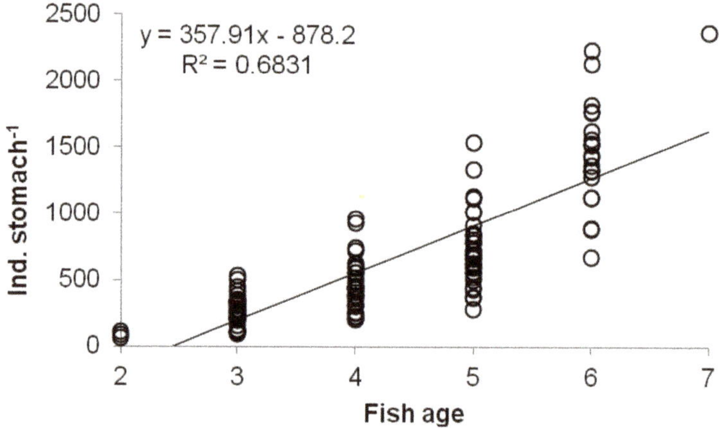

Figure 4. Significant Spearman correlations between fish age and abundance of *D. pulex* in fish stomach ($R = 0.86$, $p < 0.0001$).

We did not observe any relationship between the increase in abundance of zooplankton groups in the RLO or in the PO, and a significant increase in the number of individuals of these groups in fish stomachs ($p > 0.05$) (Table 5). The aforementioned correlation was significant only in the case of large cladocerans in the RLO.

Table 4. Mean values ± SD of percentage contribution of zooplankton abundance in fish stomachs. Different letters in rows show significant differences between fish-age samples (post-hoc multiple comparisons of mean ranks for all groups $p < 0.05$).

	Fish age	2+	3+	4+	5+	6+	7+
	n	3	34	28	32	22	1
	Asplanchna	-	0.03 ± 0.10 a	0.02 ± 0.09 a	0.01 ± 0.04 a	0.01 ± 0.03 a	-
	Benthic rotifers	0.88 ± 1.52 a	0.10 ± 0.31 a	0.02 ± 0.06 a	-	-	-
	Pelagic rotifers	1.02 ± 0.91 a	0.23 ± 0.47 a	0.03 ± 0.10 a	0.02 ± 0.08 a	-	-
Zooplankton group	Small cladocerans	2.50 ± 0.22 a	0.40 ± 0.83 ab	0.16 ± 0.35 ab	0.05 ± 0.12 b	0.04 ± 0.07 b	0.25 ab
	Large cladocerans	0.88 ± 1.52 a	0.26 ± 0.47 a	0.15 ± 0.23 a	0.10 ± 0.16 a	0.09 ± 0.12 a	0.08 a
	Daphnia pulex	90.96 ± 7.01 a	98.11 ± 2.13 a	99.16 ± 1.13 ab	99.65 ± 0.32 ab	99.74 ± 0.27 b	99.58 ab
	Nauplii	3.32 ± 3.29 a	0.39 ± 0.62 ab	0.18 ± 0.40 ab	0.02 ± 0.07 b	0.01 ± 0.03 b	-
	Cyclopoida	0.44 ± 0.76 a	0.47 ± 0.80 a	0.27 ± 0.52 a	0.15 ± 0.20 a	0.12 ± 0.20 a	-

Table 5. Spearman correlations between zooplankton group abundance in RLO and PO sites versus number of zooplankton in fish stomachs ($p < 0.05$). Ab.—abundance in water, Stom.—number in stomach. Significant results are marked in bold.

Site RLO	n	r	p
Ab. *Asplanchna* vs. Stom. *Asplanchna*	4	0.74	0.262
Ab. Benthic rotifers vs. Stom. Benthic rotifers	4	−0.74	0.262
Ab. Planktonic rotifers vs. Stom. Pelagic rotifers	4	0.32	0.683
Ab. Small cladocerans vs. Stom. Small cladocerans	4	0.80	0.200
Ab. Large cladocerans vs. Stom. Large cladocerans	4	0.99	0.001
Ab. *Daphnia pulex* vs. Stom. *Daphnia pulex*	1	-	-
Ab. Nauplii Cyclopoida vs. Stom. Nauplii	4	0.63	0.367
Ab. Cyclopoida vs. Stom. Cyclopoida	4	0.80	0.200
Site PO			
Ab. *Asplanchna* vs. Stom. *Asplanchna*	1	-	-
Ab. Benthic rotifers vs. Stom. Benthic rotifers	4	0.94	0.051
Ab. Planktonic rotifers vs. Stom. Pelagic rotifers	4	0.32	0.684
Ab. Small cladocerans vs. Stom. Small cladocerans	4	0.80	0.200
Ab. Large cladocerans vs. Stom. Large cladocerans	0	-	-
Ab. *Daphnia pulex* vs. Stom. *Daphnia pulex*	4	0.80	0.200
Ab. Nauplii Cyclopoida vs. Stom. Nauplii	4	0.83	0.167
Ab. Cyclopoida vs. Stom. Cyclopoida	4	0.60	0.400

4. Discussion

The present study shows that the reduction in abundance of several large plankters in a short river section may vary from a few percent to almost 100%. Naturally, the larger the zooplankton individuals, the greater the reduction of their abundance [6,7,22]. It would seem that the body size of plankters was a key factor that affected the amount of reduction in species abundance [5,7,9,19,23]. Fish substantially reduced large invertebrate prey, i.e., *D. pulex*. This species dominates in fish stomachs, and therefore proves that *D. pulex* abundance has been reduced by fish. Several papers have indicated that fish have a significant impact on the reduction of drifting zooplankton abundance in rivers [6,7,9]. Our study showed that fish are a crucial factor that reduces the abundance of drifting zooplankton in rivers, and their reduction efficiency increases with zooplankton size primarily because preying fish prefer large-bodied zooplankton.

Fish of all ages presented with high condition values due to the consumption of large amounts of *D. pulex*, which constituted an easily accessible food source. Low standard deviation scores for the fish condition factor indicate small differences in this parameter. This means that all the fish had equal and unlimited access to *D. pulex* as a food base.

Examination of fish stomach contents showed that older fish consumed less small planktonic organisms and more large-bodied ones, so these findings were not surprising [24]. However, given the results of the present study, it is difficult to prove the same for organisms other than *D. pulex*, because this species dominated in the stomach content of all fish of all ages. For this reason, when analyzing the plankter abundance reduction it is worth focusing on that specific species.

It is worth considering the correlations between fish age and number of zooplankton group individuals in fish stomachs, which was significant only in the case of *D. pulex*. Perhaps the correlation would have also been significant in the case of relatively large plankters such as large cladocerans or copepods. However, the fish had a large density of large *D. pulex* individuals to choose from, and clearly selected for this species. Hence, a conclusion can be drawn that fish made clear food selectivity decisions amongst the biggest plankters, which confirms the assumptions and results of other authors [5,7,9,25]. Even though other large planktonic crustaceans were available, the fish preferred *D. pulex*. We observed a significant reduction in Copepoda only in the further river sections, which means that fish consume other larger plankters once *D. pulex* is no longer available.

The results of the present work also show that the higher the abundance of zooplankton groups in a lake outlet or in a pond outlet, the higher the number of these groups observed in fish stomachs. However, there are no significant correlations between these variables to unambiguously support this statement. Perhaps, had the *D. pulex* been unavailable, the fish would have consumed other zooplankton groups and consequently such a correlation would have been significant. Nonetheless, the fish avoided feeding on small plankters because they provide little energetic gain [9]. The results of our present study also confirmed this hypothesis.

The abundance increase of *D. pulex* in a pond outlet does not affect the number of individuals in fish stomachs. This means that each day the abundance of *D. pulex* in the PO was sufficient for the fish biomass who almost entirely reduced the abundance of *D. pulex* in the river each day. Gliwicz [1] documented a case of continuous zooplankton consumption by fish in Cahora Bassa, Africa, and found that a great amount of easily accessible large zooplankton biomass was consumed by fish even though their stomachs were full. According to Gliwicz [1], it was a natural reaction for fish who had easy and short access to energetic gain. Therefore, despite having a full stomach, fish consume as much food as possible. Presumably, fish in the pond outlet consumed all the available large zooplankton biomass. It also confirms the foraging efficiency of fish examined in our experiment. The results of *D. pulex* abundance for the examined sites show that fish can consume almost the total amount of this species. This means that the pond outlet has a small effect on the movement of significant *D. pulex* biomass into the river, and thus on the entry of organic nutrients into it. Another factor is the incorporation of nutrients, eaten by zooplankton, into the fish tissues, and which eventually take part in the circulation of biogenic compounds [24].

The presented work shows the importance in limiting the abundance of drifting zooplankton in rivers by fish and the role of the plankter body size. The larger the plankter-victim's size, the faster the reduction of its number. In future research, it would be worth focusing on the issue of the efficiency and feeding potential of one fish to determine what size of fish biomass can eat the entire biomass of zooplankton carried away from the pond. This is an important problem, because the analysis of the results of this work does not indicate whether the zooplankton was continuously eaten by the same fish, or if the already fed fish were replaced by hungry fish. Finally, it can be concluded that the fish were more likely to feed on zooplankton drifting from the waste stabilization pond than from the lake. This was influenced by *D. pulex* individuals, attractive for fish due to them having the largest body size among the analyzed zooplankton, irrespective of the *D. pulex* percentage in the total abundance of zooplankton. Moreover, zooplankton are a valuable source of nutrients [26].

Author Contributions: Conceptualization, R.C.; methodology, R.C. and T.K.; software, R.C. and T.K.; validation, R.C. and T.K.; formal analysis, R.C. and T.K.; investigation, R.C. and T.K.; resources, R.C. and T.K.; data curation, R.C. and T.K.; writing—original draft preparation, R.C.; writing—review and editing, R.C. and T.K.; visualization, R.C. and T.K.

Funding: This research received no external funding.

References

1. Gliwicz, Z.M. A lunar cycle in zooplankton. *Ecology* **1986**, *67*, 883–897. [CrossRef]
2. Christoffersen, K.; Riemann, B.; Klysner, A.; Søndergaard, M. Potential role of fish predation and natural populations of zooplankton in structuring a plankton community in eutrophic lake water. *Limnol. Oceanogr.* **1993**, *38*, 561–573. [CrossRef]

3. Wissel, B.; Boeing, W.J.; Ramcharan, C.W. Effects of water color on predation regimes and zooplankton assemblages in freshwater lakes. *Limnol. Oceanogr.* **1998**, *48*, 1965–1976. [CrossRef]

4. Hoffman, J.C.; Smith, M.E.; Lehman, J.T. Perch or plankton: Top-down control of Daphnia by yellow perch (*Perca flavescens*) or *Bythotrephes cederstroemi* in an inland lake? *Freshw. Biol.* **2001**, *46*, 759–775. [CrossRef]

5. Jack, J.D.; Thorp, J.H. Impacts of fish predation on an Ohio River zooplankton community. *J. Plankton Res.* **2002**, *24*, 119–127. [CrossRef]

6. Czerniawski, R.; Domagała, J. Reduction of zooplankton communities in small lake outlets in relation to abiotic and biotic factors. *Oceanol. Hydrobiol. Stud.* **2013**, *42*, 123–131. [CrossRef]

7. Czerniawski, R.; Sługocki, Ł.; Kowalska-Góralska, M. Diurnal changes of zooplankton community reduction rate at lake outlets and related environmental factors. *PLoS ONE* **2016**, *11*, e0158837. [CrossRef]

8. Walks, D.J.; Cyr, H. Movement of plankton through lake-stream systems. *Freshw. Biol.* **2004**, *49*, 745–759. [CrossRef]

9. Chang, K.H.; Doi, H.; Imai, H.; Gunji, F.; Nakano, S.I. Longitudinal changes in zooplankton distribution below a reservoir outfall with reference to river planktivory. *Limnology* **2008**, *9*, 125–133. [CrossRef]

10. Mehner, T. Influence of spring warming on the predation rate of underyearling fish on daphnia—A deterministic simulation approach. *Freshw. Biol.* **2000**, *45*, 253–263. [CrossRef]

11. Kuczyńska-Kippen, N.M.; Nagengast, B. The influence of the spatial structure of hydromacrophytes and differentiating habitat on the structure of rotifer and cladoceran communities. *Hydrobiologia* **2006**, *559*, 203–212. [CrossRef]

12. Estlander, S.; Nurminen, L.; Olin, M.; Vinni, M.; Horppila, J. Seasonal fluctuations in macrophyte cover and water transparency of four brown-water lakes: Implications for crustacean zooplankton in littoral and pelagic habitats. *Hydrobiologia* **2009**, *620*, 109–120. [CrossRef]

13. Kamarainen, A.M.; Rowland, F.E.; Biggs, R.; Carpenter, S.R. Zooplankton and the total phosphorus-chlorophyll a relationship: Hierarchical bayesian analysis of measurement error. *Can. J. Fish. Aquat. Sci.* **2008**, *65*, 2644–2655. [CrossRef]

14. Gołdyn, R.; Kowalczewska-Madura, K. Interactions between phytoplankton and zooplankton in the hypertrophic swarzędzkie Lake in western Poland. *J. Plankton Res.* **2008**, *30*, 33–42.

15. Pourriot, R.; Rougier, C.; Miquelis, A. Origin and development of river zooplankton: Example of the Marne. *Hydrobiologia* **1997**, *345*, 143–148. [CrossRef]

16. Ejsmont-Karabin, J.; Kruk, M. Effects of contrasting land use on free-swimming rotifer communities of streams in Masurian Lake District, Poland. *Hydrobiologia* **1998**, *387/388*, 241–249. [CrossRef]

17. Radwan, S. *Wrotki (Rotifera)*; Wydawnictwo Uniwersytetu Łódzkiego: Łódź, Polska, 2004; pp. 1–447.

18. Rybak, J.I.; Błędzki, L.A. *Planktonic Crustaceans of Freshwaters*; Wydawnictwo Uniwersytetu Warszawskiego: Warszawa, Poland, 2010; pp. 1–366.

19. Czerniawski, R.; Domagała, J. Small dams profoundly alter the spatial and temporal composition of zooplankton communities in running waters. *Int. Rev. Hydrobiol.* **2014**, *99*, 300–311. [CrossRef]

20. Venables, W.N.; Ripley, B.D. *Modem Applied Statistics with S. Fourth Edition*; Springer: New York, USA, 2002; pp. 1–487.

21. R Core Team. *A Language and Environment for Statistical Computing—R Foundation for Statistical Computing*; Vienna, Austria. 2018. Available online: https://www.R-project.org (accessed on 20 December 2018).

22. Czerniawski, R.; Domagała, J. Zooplankton communities of two lake outlets in relation to abiotic factors. *Open Life Sci.* **2010**, *5*, 240–255. [CrossRef]

23. Czerniawski, R.; Domagała, J. Similarities in zooplankton community between river drawa and its two tributaries (Polish part of River Odra). *Hydrobiologia* **2010**, *638*, 137–149. [CrossRef]

24. Bone, Q.; Moore, R. *Biology of Fishes*; Taylor and Francis: New York, NY, USA, 2008; pp. 1–497.

25. Akopian, M.; Garnier, J.; Pourriot, R. A large reservoir as a source of zooplankton for the river: Structure of the populations and influence of fish predation. *J. Plankton Res.* **1999**, *21*, 285–297. [CrossRef]

26. Meeren, T.; Olsen, E.R.; Hamre, K.; Fyhn, J.H. Biochemical composition of copepods for evaluation of feed production of juvenile marine fish. *Aquaculture* **2008**, *274*, 375–379. [CrossRef]

Article

Species Richness and Taxonomic Distinctness of Zooplankton in Ponds and Small Lakes from Albania and North Macedonia: The Role of Bioclimatic Factors

Giorgio Mancinelli [1,2,3], **Sotir Mali** [4] **and Genuario Belmonte** [1,5,*]

1   CoNISMa, Consorzio Nazionale Interuniversitario per le Scienze del Mare, 00196 Roma, Italy; giorgio.mancinelli@unisalento.it
2   Laboratory of Ecology, Department of Biological and Environmental Sciences and Technologies (DiSTeBA), University of Salento, 73100 Lecce, Italy
3   National Research Council (CNR), Institute of Biological Resources and Marine Biotechnologies (IRBIM), 08040 Lesina, Italy
4   Department of Biology, Faculty of Natural Sciences, "Aleksandër Xhuvani" University, 3001 Elbasan, Albania; sotirmali@hotmail.com
5   Laboratory of Zoogegraphy and Fauna, Department of Biological and Environmental Sciences and Technologies (DiSTeBA), University of Salento, 73100 Lecce, Italy
*   Correspondence: genuario.belmonte@unisalento.it

Received: 13 October 2019; Accepted: 11 November 2019; Published: 14 November 2019

Abstract: Resolving the contribution to biodiversity patterns of regional-scale environmental drivers is, to date, essential in the implementation of effective conservation strategies. Here, we assessed the species richness S and taxonomic distinctness Δ+ (used a proxy of phylogenetic diversity) of crustacean zooplankton assemblages from 40 ponds and small lakes located in Albania and North Macedonia and tested whether they could be predicted by waterbodies' landscape characteristics (area, perimeter, and altitude), together with local bioclimatic conditions that were derived from Wordclim and MODIS databases. The results showed that a minimum adequate model, including the positive effects of non-arboreal vegetation cover and temperature seasonality, together with the negative influence of the mean temperature of the wettest quarter, effectively predicted assemblages' variation in species richness. In contrast, taxonomic distinctness did not predictably respond to landscape or bioclimatic factors. Noticeably, waterbodies' area showed a generally low prediction power for both S and Δ+. Additionally, an in-depth analysis of assemblages' species composition indicated the occurrence of two distinct groups of waterbodies characterized by different species and different precipitation and temperature regimes. Our findings indicated that the classical species-area relationship hypothesis is inadequate in explaining the diversity of crustacean zooplankton assemblages characterizing the waterbodies under analysis. In contrast, local bioclimatic factors might affect the species richness and composition, but not their phylogenetic diversity, the latter likely to be influenced by long-term adaptation mechanisms.

Keywords: crustacean zooplankton; species richness; phylogenetic diversity; bioclimate; freshwater ponds

---

1. Introduction

Lentic freshwaters are acknowledged to play a crucial role in regulating the global ecosystem functions e.g., carbon cycle [1] and they are among the Earth's most threatened habitats in terms of intensity of anthropogenic pressures, biodiversity loss, and non-indigenous species introduction [2–4].

They include an extreme variety of habitats differing in ecological characteristics and fragility [5,6]. Surface area represents one of the most apparent differentiating properties: indeed, lentic environments

(304 million water bodies; 4.2 million km^2 in total area [7]) include the lake Superior (82,000 km^2), together with small ponds, i.e., waterbodies less than 0.05 km^2 in area [8].

Ponds and small lakes (hereafter PSL) have significant ecological functions [9,10]: among others, they provide a considerable contribution to inland water CO_2 and CH_4 emissions [11]. In addition, they are, to date, recognized as important biodiversity hotspots, especially in mountainous regions, supporting a high species richness and contributing a high degree of rare species to regional pools ([12–15]; see also [16] for an example on planktonic Calanoida). Noticeably, PSL are threatened by a number of anthropogenic pressures, including nutrient loading, contamination, acid rain, and invasion of exotic species [17]. In addition, infilling (both natural and caused by direct habitat destruction), land drainage, decline in many of their traditional uses, and changes of function determine at a regional scale the drastic reduction in PSL number and connectivity [12].

In the last decade, several investigations have focused on the diversity of benthic invertebrates, as they are excellent bio-indicators of PSL ecological integrity [18,19]. Local factors that are related with e.g., hydroperiod, environmental harshness, water chemistry, spatial connectivity, habitat heterogeneity, and presence of predators, have been recognized to influence the biodiversity of macroinvertebrate assemblages ([20] and literature cited). At a regional scale, attention has been primarily given to the influence of waterbodies area [21–23], while assuming, within the general theoretical background provided by the species-area relationship (SAR) hypothesis [24], that basins' size correlates with the number of microhabitats within the basin itself and with populations' abundance, and thence inversely correlated with the likelihood of random extinctions. However, resolving the contribution to biodiversity patterns of environmental factors acting at a regional scale is, to date, essential to the implementation of effective conservation strategies in the face of e.g. deforestation and climate change ([25,26] and literature cited). Accordingly, several attempts have been made to model biodiversity of freshwater environments by means of regional bioclimatic factors [27,28].

In the present study, we focused on the diversity of crustacean zooplankton assemblages in 40 ponds and small lakes differing remarkably in terms of origin, extension, and altitude from a relatively wide region comprising part of Albania and North Macedonia. A recent faunal inventory focusing on ponds and lakes in the area [29] provided the starting reference information on the taxonomic characteristics of the assemblages.

A number of studies have generally indicated a positive relationship between the surface area of lacustrine environments and zooplankton diversity (e.g., [30–32]; but see [13]); accordingly, crustacean zooplankton has been shown to have higher species richness in small ponds as compared with lakes [16,33,34]. This notwithstanding, we hypothesized that area alone may not be an adequate predictor, and that local bioclimatic conditions may ultimately contribute in explaining diversity variations across waterbodies by influencing their physical-chemical characteristics, as observed in recent investigations on freshwater macroinvertebrates and macrophytes [27,28,35]. This could be particularly true for waterbodies in mountainous habitats, where temperature and precipitation regimes intensely reflect the chemical-physical characteristics and hydroperiod of the waterbodies themselves [36], regulating the harshness and stability of the aquatic environments and, in turn, the diversity of the biota living in them ([22] and literature cited).

To verify the hypothesis and test whether bioclimatic factors can predict assemblages' diversity, we identified a minimum adequate model (MAM) predicting assemblages' diversity across the different waterbodies by means of a heuristic multiple regression approach and Bayesian Information Criterion model selection method while using satellite-derived bioclimatic variables as predictors. Multiple indices are, to date, available to quantify different aspects of biodiversity [37]. Here, we identified predictive MAMs estimating the diversity of crustacean zooplankton assemblages in terms of species richness and average taxonomic distinctness.

Species richness is the most classical measure of biodiversity across ecosystems that has been extensively used in studies on lentic habitats (see references cited above). This index provides an incomplete understanding of biological variability, because it neglects information on the identity and taxonomic relationship among species, and it is hampered by a number of critical limitations [38,39]. Accordingly, we used the average taxonomic distinctness Δ+ [40] to compare the taxonomic relatedness of species in the crustacean assemblages of every water body. In addition, we tested the influence of bioclimatic factors on crustacean assemblages in terms of species composition. To this end, multivariate approaches that are based on a canonical analysis of principal coordinates were used to model the changes in the structure of the assemblages as affected by bioclimatic variables, and identify relationships between the latter and specific groups of zooplankton taxa.

2. Material and Methods

2.1. Sampling Sites and Collection Procedures

A total of 40 sites were selected among those (53) that were surveyed between 2005 and 2017 by Belmonte and colleagues [29] in an area comprised between 39°55′22″–42°04′30″ N, and 19°24′30″–20°47′36″ E. (Figure 1, Table 1).

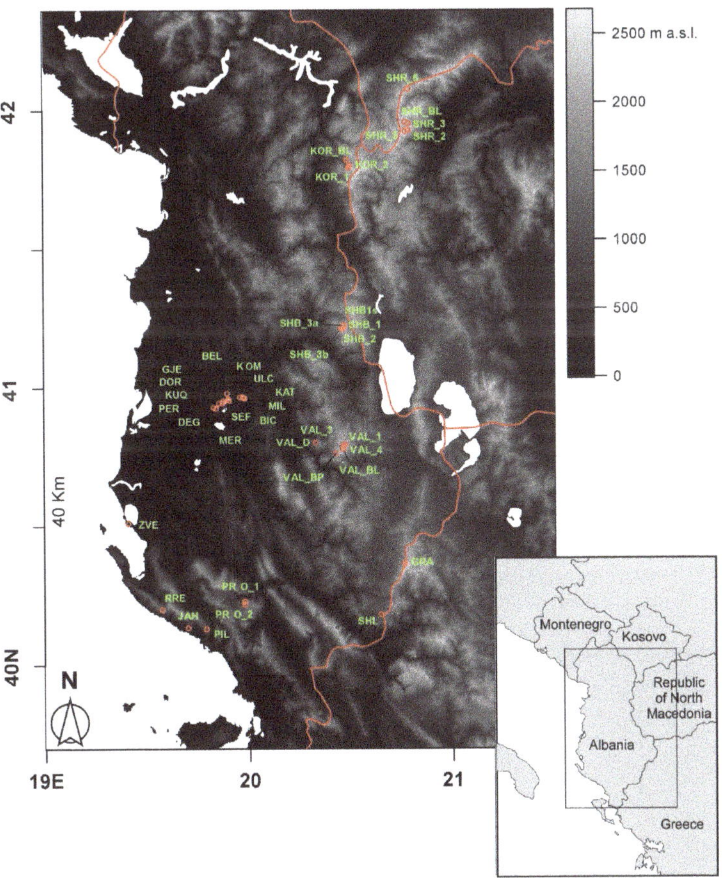

Figure 1. Location of the 40 waterbodies in Albania and North Macedonia.

Table 1. List of the waterbodies included in the study. The code used in Figure 1 is included, together with information of the basins origin (N = natural, A = artificial), typology (L= lake, P = pond (area < 0.05 km^2)), elevation (in m), coordinates, number of samples collected, year of collection, species richness S, and taxonomic distinctness Δ+.

Waterbody	Code	Origin	Typology	Altitude	Latitude	Longitude	Samples	Year	S	Δ+
Bellsh	BEL	N	L	150	40°58′47″	19°53′37″	2 *	2008	6	76.67
Bici	BIC	N	P	77	40°57′41″	19°58′46″	2 *	2009	8	74.11
Dega	DEG	N	L	105	40°55′32″	19°50′25″	2 *	2009	8	60.27
Dorbi	DOR	N	L	133	40°57′07″	19°52′30″	3 *	2008	9	69.1
Gjeluar	GJE	N	P	127	40°57′32″	19°54′02″	3 *	2010	8	74.11
Gramoz	GRA	N	P	2364	40°21′52″	20°47′26"	2	2008	4	81.25
South Coast Jahl	JAH	A	P	701	40°07′53″	19°47′16″	2	2012	3	75
Katund	KAT	N	P	100	40°57′54″	19°58′27″	2 *	2009	8	75.45
Komnjec	KOM	N	P	135	40°57′58″	19°57′26″	3 *	2009	5	80
Korab Hapave 1	KOR1	N	P	1786	41°47′59″	20°30′04″	1	2017	14	67.86
Korab Hapave 2	KOR2	N	P	1779	41°47′55″	20°30′03″	1	2017	11	70.68
Korab black lake	KOR BL	N	P	1470	41°49′13″	20°29′14″	1	2017	7	69.05
I Kuq	KUQ	N	L	135	40°56′38″	19°51′14″	3 *	2008	8	75
Merohjes	MER	N	L	112	40°56′39″	19°52′19″	2 *	2008	11	72.27
Milosh	MIL	N	L	70	40°57′49″	19°58′50″	2 *	2009	8	65.63
Pernaska	PER	N	L	107	40°55′45″	19°49′40″	3 *	2008	6	80.83
South Coast Pilur	PIL	A	P	280	40°08′08″	19°41′59″	1	2012	3	75
Progonat 1	PRO1	A	P	1325	40°13′06″	19°58′35″	1	2011	4	66.67
Progonat 2	PRO2	A	P	1262	40°13′50″	19°58′35″	1	2011	4	72.92
Karaburun Rreza	RRE	A	P	1333	40°12′03″	19°34′17″	2	2012	3	79.17
Seferan	SEF	N	L	124	40°57′12″	19°54′12″	1 *	2011	9	71.53
Shebenik 1	SHB1	N	P	1903	41°13′30″	20°28′21″	1	2015	6	75.83
Shebenik 2	SHB2	N	P	2006	41°12′50″	20°28′04″	1	2015	5	60
Shebenik 3a	SHB3a	N	P	2054	41°12′44″	20°27′43″	1	2015	4	64.58
Shebenik 3b	SHB3b	N	P	2005	41°12′45″	20°27′31″	1	2015	6	75.83
Shebenik 1s	SHB1s	N	P	1905	41°13′29″	20°28′21″	1	2015	5	78.75
Sheleguri	SHL	A	L	1002	40°10′55″	20°38′49″	1	2012	4	72.92
Sharr 2	SHR2	N	P #	2280	41°57′21″	20°46′34″	2	2008	5	76.25
Sharr 3	SHR3	N	L	1945	41°57′02″	20°47′36″	2	2008	11	68.41
Sharr 5	SHR5	N	P	2435	41°55′23″	20°46′32″	1	2016	4	68.75
Sharr 6	SHR6	N	P	2190	42°04′26″	20°47′32″	2	2016	8	71.43

Table 1. *Cont.*

Waterbody	Code	Origin	Typology	Altitude	Latitude	Longitude	Samples	Year	S	Δ+
Sharr black lake	SHR BL	N	P #	2170	41°55′34″	20°47′34″	2	2016	7	66.67
Ulca	ULC	N	L	107	40°57′57″	19°58′17″	2 *	2009	7	71.43
Valamare 1	VAL1	N	P	2051	40°47′40″	20°28′36″	1	2016	7	73.81
Valamare 3	VAL3	N	P	2062	40°47′38″	20°28′29″	1	2016	8	70.98
Valamare 4	VAL4	N	P	2121	40°47′19″	20°28′05″	1	2016	9	71.53
Valamare black lake	VAL BL	N	P	1698	40°45′43″	20°25′50″	1	2016	7	73.21
Valamare green bun pine	VAL BP	N	P	2005	40°46′40″	20°28′00″	1	2016	8	69.2
Valamare Dushq teke	VAL D	N	L	1115	40°48′06″	20°19′47″	2	2010	6	70
Narte Zvernec	ZVE	N	P	2	40°30′41″	19°24′24″	1	2012	6	71.67

temporary pond. * sample collection performed while using a canoe.

Sampling procedures are described in detail elsewhere [29]. As the water bodies varied in terms of area, altitude, origin (natural, artificial), as well as in hydrology (permanent, temporary), banks morphology, and degree of aquatic vegetation cover, it was not possible to apply a standardized sampling protocol across all of the sites.

In the selected sites, sampling operations were carried out in spring–early summer (i.e., between April and July) always by the same operators, while using a hand-held plankton net (200 µm mesh-size, mouth diameter, 30 cm). The collection (by plankton net towing from two opposite edges of the pond) covered the whole water body when its diameter was smaller than 100 m. For larger water bodies, sample collection was carried while using a canoe. In the case of small ponds, the collection procedure was repeated three times (each sample derived from the execution of three collections). In the case of larger water bodies, a sample collection was carried out in three different stations; the three different samples were ultimately cumulated.

After collection, the samples were fixed in situ in 90–96% ethanol. In the laboratory, taxa were identified to the species level while using a compound microscope (×30–×300 magnifications) that was equipped with a *camera lucida*.

The quantification of the abundance of each taxon was not performed, and only presence/absence data were considered due to the huge variability of water volumes in each pond, which made impossible the comparison among the concentrations of plankton of different sites.

2.2. Landscape-Climate Variables

Together with altitude, we used area and perimeter of the water bodies together with bioclimatic factors (i.e., temperature and precipitation) to represent the landscape-climate variables. Water bodies were geo-referenced in Google Earth Pro version 7.3.2., where their surface (in km^2) and perimeter (in km) were measured while using the software tools. Measurements were performed by preferentially choosing images that were taken in spring or summer between 2008 and 2017, assuming that negligible variations in the water bodies size and morphology occurred during this period.

Nineteen climatic layers with a 30-second spatial resolution ($0.93 \times 0.93 = 0.86$ km^2 at the equator; approximately $0.92 \times 0.70 = 0.64$ km^2 within the study area), including temperature and precipitation variables, were extracted from the WorldClim v2 data set [41] (Table A1 in online material). Besides climate, there is a growing recognition of the importance of vegetation cover in characterizing the spatial environmental heterogeneity in a given area at the meso- and topo-scales, in turn affecting climate, soil composition, hydrology and geomorphology, and, ultimately, biological processes that were related with species richness and community complexity [42,43]. Accordingly, two vegetation variables (i.e., percent tree cover and percent non-tree cover) were extracted from the Terra MODIS Vegetation Continuous Field (VCF) product (available as MOD44B v006 https://lpdaac.usgs.gov/products/mod44bv006/). Percent tree cover included all forest types and age classes, while percent non-tree cover included meadows, regeneration areas, and clear-cut areas. MODIS tiles of the study area were re-projected and re-sampled to meet the coordinate system and resolution of WorldClim layers; percent cover data were subsequently obtained for the years from 2006 to 2017, and averaged. In addition, the third VCF component of ground cover, i.e., percent bare soil (including bare soils and rocks) was extracted according to the aforementioned procedures, and was used together with tree and non-tree vegetation percent cover data to estimate the Shannon's diversity index (H) as a proxy of habitat heterogeneity.

2.3. Data Analysis

The values in the text are expressed as averages ± 1SE; for parametric statistical analysis, data were tested for conformity to assumptions of variance homogeneity (Cochran's C test) and normality (Shapiro–Wilks test) and transformed when required.

The taxonomic diversity of crustacean assemblages in each water body was estimated in terms of species richness S and average taxonomic distinctness Δ+. The index can be used as a proxy for phylogenetic diversity and it measures the mean path length through a taxonomic tree connecting every species [40]. Here, mean taxonomic distinctness values were calculated assigning equal weighting to branch lengths from a linear Linnaean classification while using eight taxonomic levels (i.e., class, subclass, order, suborder, infraorder, family, genus, and species). The taxonomic classification tree was built according with the World Register of Marine Species (WoRMS, available at https://www.marinespecies.org) and the Integrated Taxonomic Information System (ITIS, available at https://www.itis.gov).

For the sake of completeness, other taxonomic diversity indices were calculated, including the total taxonomic distinctness $s\Delta+$ and the variance in taxonomic distinctness $\Lambda+$ [44,45], the average phylogenetic diversity $\Phi+$, and the total phylogenetic diversity $s\Phi+$ [46]. S resulted in being significantly related with $s\Delta+$, $\Phi+$, and $s\Phi+$ (r = 0.98, −0.89, and 0.95, respectively; P always < 0.01, 38 degrees of freedom), while $\Delta+$ scaled negatively with $\Lambda+$ (r = −0.44, P = 0.004); conversely, S and $\Delta+$ were characterized by a non significant negative correlation (r = −0.26, P = 0.12, 38 d.f.); thus, the two indices were chosen for further analyses.

We verified the influence on both indices of potential artefacts, due to (i) possible differences in sampling procedures and (ii) the different number of total collected samples per water body. To this end, we performed a one way permutational analysis of variance (PERMANOVA; [47]) based on Euclidean distances and 999 permutations with "sampling procedure" as a fixed factor (two levels, "hand", or "canoe") and the number of collected samples as the covariate (P and N hereafter). Both factors exerted negligible influences on the S and $\Delta+$ estimations (S: Pseudo–F_P = 4.42, P(perm)$_P$ = 0.08, Pseudo–F_N = 0.48, P(perm)$_N$ = 0.47, Pseudo–$F_{P \times N}$ = 0.12, P(perm) $_{P \times N}$ = 0.72; $\Delta+$: Pseudo–F_P = 1.67, P(perm)$_P$ = 0.21, Pseudo–F_N = 0.01, P(perm)$_N$ = 0.94, Pseudo–$F_{P \times N}$ = 1.07, P(perm) $_{P \times N}$ = 0.31). Consequently, the S and $\Delta+$ values were assumed to provide robust estimation of planktonic Crustacea diversity across the studied water bodies. PERMANOVA was further used to test the effects of the factors "origin" (two levels, "natural" and "artificial") and "typology" (two levels, "pond" and "lake") on the diversity indices. As water bodies varied greatly in elevation (Table 1), the latter was included in the analyses after log-transformation as a continuous covariate.

The georeferenced locations of the sampling sites were used to extract climatic and vegetation data from environmental layers. The final data set included 19 climatic, two vegetation (% tree cover, % non-tree cover), four geomorphological (elevation, perimeter, surface, and perimeter/surface ratio), and one habitat heterogeneity variable (Table A1). They were log-transformed and z-scaled; subsequently, their original number (25) was reduced while using an iterative variance inflation factor (VIF) analysis [48,49]. In brief, if a strong linear relationship links a variable x with at least another variable y, the correlation coefficient would be close to 1, and the VIF for x would be large. Here, diversity measures with VIF factors that were larger than 10 were excluded. Variables with VIF factors larger than 10 were discarded. The identification of a minimum adequate model (MAM hereafter; [50]) linking diversity measures with environmental variables was based on the heuristic generation of alternative regression models (see [51] for complete details on the procedure). Model selection was performed while adopting an Information Theoretic criterion [52]; the second-order Akaike Information Criterion *AICc* [53,54] was calculated for each combination of n explanatory variables and used to identify the best MAM among the alternative regression models that were generated by the procedure. For model comparison, *AICc* values were used to estimate a set of positive Akaike weights w_i summing 1:

$$w_i(AIC) = \frac{\exp[-1/2(AIC_i - minAIC)]}{\sum_1^K \exp[-1/2(AIC_i - minAIC)]} \tag{1}$$

With K = number of models. The model showing the highest w_i was accepted as the best candidate; other candidate models were accepted if characterized by w_i values within 12.5% of the highest [55–57]. The model building and MAM identification procedures were performed while adopting Fox and Weisberg [58] as a general reference.

Given the non-conclusive outcomes of the analyses that were performed on taxonomic distinctness (see Results section), we verified whether bioclimatic factors influenced planktonic Crustacea assemblages in terms of species composition. Species incidence data were used to calculate a Jaccard similarity matrix across the different waterbodies. In addition, a similarity matrix that was based on Euclidean distances was constructed for bioclimatic variables, and the consistency of the two matrices was tested while using the Kendall coefficient of concordance (W). Subsequently, a canonical analysis of principal coordinates (CAP) [59] was performed to model changes in assemblages composition, as affected by bioclimatic factors. The appropriate number of principal coordinates m was chosen as to minimise the P value from the permutation test based upon the trace statistic and maximizing the leave-one-out allocation success [60]. Post-hoc PERMANOVA tests were performed to confirm the results of the ordination for both bioclimatic factors and species; SIMPER analyses were performed on the Euclidean distance matrix of the former to assess the percentage contribution of each factor to the dissimilarity between the groups of waterbodies that were identified by the CAP procedure. Furthermore, the Spearman's rank correlations were estimated to identify the bioclimatic variables and the crustacean species that most effectively described the groups of waterbodies that were identified by the CAP procedure. Only the variables with a Spearman rank correlation coefficient $r > 0.55$ were considered.

All of the analyses were implemented in the R statistical environment v3.6.1 [61] while using a suite of packages including *taxize* (for taxonomic information retrieval from online databases) *vegan* (for diversity measures and multivariate analyses), *raster*, *rgdal*, and *maptools* (for environmental layers manipulation), *usdm* (for VIF analysis), *car*, *leaps*, and *HH* (for MAM identification).

3. Results

3.1. General Features

The 40 water bodies analysed, varied remarkably in terms of altitude, area, and perimeter (Figure 2). They showed an average altitude of 1168.3 m a.s.l. (± 141.9 m SE), ranging from 2 m a.s.l. (Narte Zvernec pond, ZVE in Figure 1) to 2,435 m a.s.l. (SHR 5). The average surface extension was 0.09 km^2 ± 0.03 SE, ranging from 9.1×10^{-4} to 0.86 km^2. The average perimeter was $1.01 ± 0.21$ km, ranging between 0.035 and 6.1 km. For both of the variables, the minimum and maximum values corresponded with a small artificial pond in the karst highlands of Progonat (PRO1) and the Lake Seferan in the Dumre region (SEF).

In the 40 water bodies, 79 Crustacea species were identified in total, being almost equally distributed between the classes Branchiopoda and Hexanauplia (41 and 38 species respectively).

Among Branchiopoda, Anomopoda outnumbered the other two orders Ctenopoda and Haplopoda (38 vs. 2 and 1 species). *Daphnia*, *Ceriodaphnia*, and *Moina* were the genera that were characterized by the highest number of species (nine *Daphnia* species, four *Moina* species, and three *Ceriodaphnia* species) together representing the majority of all the sampled Anomopoda species. Ctenopoda were represented by the congeneric *Diaphanosoma brachyurum* and *D. lacustris* while Haplopoda by the single species *Leptodora kindtii*. The class Hexanauplia (*alias* Copepoda) was dominated by species belonging to the order Cyclopoida (31) and to a minor extent Calanoida (7). The genus *Cyclops* (seven species) together with *Acanthocyclops*, *Paracyclops* (four species each), and *Mesocyclops* (three species) constituted to the majority of the species in the order; Calanoida were represented by the genera *Eudiaptomus* (three species), *Mixodiaptomus* (two species each), and by *Arctodiaptomus salinus* and *Neodiaptomus schmackeri*.

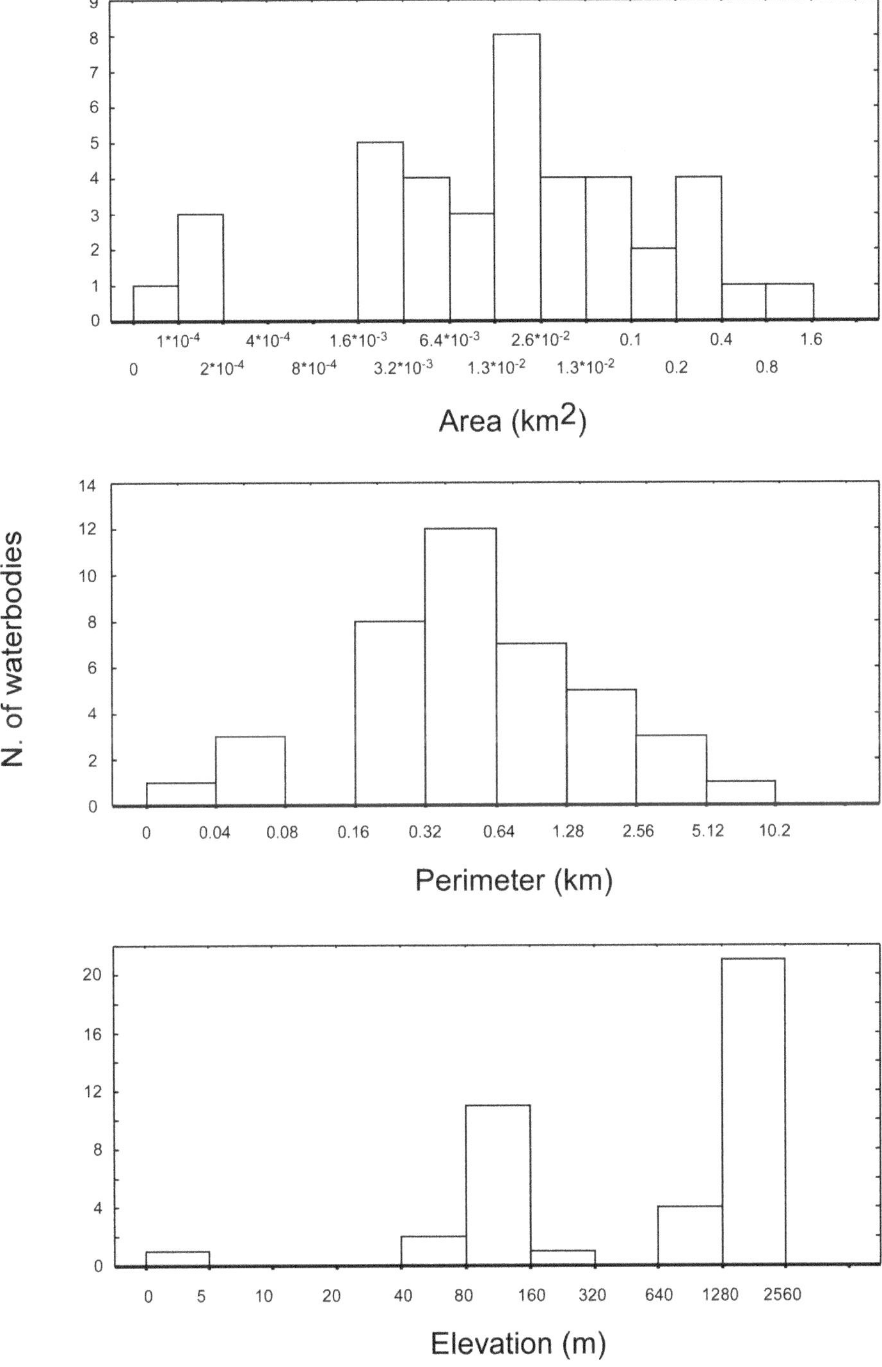

Figure 2. Frequency distribution of waterbodies area, perimeter, and altitude. The histograms for first variables adopt geometric scale increments (×2), while for altitude (elevation) a linear scale is used.

The species showing the widest distributions were the Anomopoda *Bosmina longirostris* (16 sites), *Chydorus sphaericus* (15 sites), and *Daphnia longispina* (14 sites) (Table 2); in addition, the Cyclopoida *Mesocyclops leuckarti*, the Calanoida *Mixodiaptomus tatricus*, and the Ctenopoda *Diaphanosoma brachyurum* occurred in 12 sampled sites.

Table 2. Summary of PERMANOVAs on species richness S and taxonomic distinctness + of crustacean zooplankton assemblages testing for the effects of waterbodies' origin and typology including elevation as a continuous covariate **: $p < 0.01$.

Response Variable	S			$\Delta+$	
Source of variation	df	MS	Pseudo-F	MS	Pseudo-F
Altitude (1)	1	1.93	0.41	9.43	0.35
Origin (2)	1	71.68	15.16 **	18.83	0.69
Hydrology (3)	1	11.19	2.37	29.73	1.09
1×2	1	0.16	3.4×10^{-2}	1.01	3.7×10^{-2}
1×3	1	0.11	2.2×10^{-2}	1.39	5.2×10^{-2}
2×3	1	0.82	0.17	1.11	4.1×10^{-2}

As regarding high level taxa, only Anomopoda (Cladocera) were present in every site. Cyclopoida (Hexanauplia) were present in 38 sites (95% of the total), Calanoida (Hexanauplia) in 30 sites (75%), Ctenopoda (Cladocera) in four sites (10%), and Haplopoda (Cladocera) in three sites (7.5%).

3.2. Diversity Patterns and Bioclimatic Correlates

On average, 6.7 ± 0.4 species per water body were found, ranging between three (JAH, PIL, RRE) and 14 species (KOR 1). The taxonomic distinctness $\Delta+$ of the different planktonic assemblages was on average 72.1 ± 0.78, ranging between 60 (SHB 2) and 88.9 (GRA).

The factor "origin" was the only exerting significant effects of the species richness of waterbodies (Table 2), with the six artificial water bodies being included in the study characterized by lower S values as compared with natural basins (3.5 ± 0.22 vs. 7.29 ± 0.38, respectively). Conversely, negligible effects were generally observed for the taxonomic distinctness $\Delta+$ (Table 2).

The 25 predictor variables (Table A1) were reduced to a set of nine characterized by negligible collinearity (Table A2). They included five climatic variables (i.e., Isothermality, Temperature Seasonality, Mean Temperature of Wettest Quarter, Annual Precipitation, and Precipitation of Coldest Quarter), % tree and % non-tree vegetation cover, habitat heterogeneity, and water body surface. Besides water body surface, the variable Mean Temperature of Wettest Quarter showed the greatest among-water bodies variability (Table A2), ranging from a minimum of -5.32 °C (SHB 2) to a maximum of 11 °C (ZVE). It was followed by % tree cover (varying between 1 and 70%, BEL and DEG, respectively) and % non-tree cover (ranging between 20 and approx. 81%, BEL-DEG and SHR 2, respectively).

The heuristic search procedure identified a Minimum Adequate Model (MAM) predicting the variation of species richness S across the different water bodies relying on the three explanatory variables % non tree cover, Temperature Seasonality, and Mean Temperature of Wettest Quarter (Figure 3; multiple $r = 0.58$, $P = 0.002$, d.f. $= 3, 36$). The MAM was characterized by the lowest *AICc* value, and by an Akaike weight w_i approximately eight times larger than the second-best candidate, based on the variables Temperature Seasonality, Mean Temperature of Wettest Quarter, and Habitat heterogeneity (Table 3). All of the predictors provided significant contributions to S variation across water bodies (minimum absolute *t* value $= 2.34$, $P = 0.02$ for the variable Mean Temperature of Wettest Quarter); the contributions of both % non-tree cover and Temperature Seasonality were positive ($b = 0.27 \pm 0.12$ and 0.64 ± 0.15, respectively), while the Mean Temperature of the Wettest Quarter provided a negative contribution ($b = -0.26 \pm 0.14$). Noticeably, none of the first ten best-performing models included water body area (Table 3).

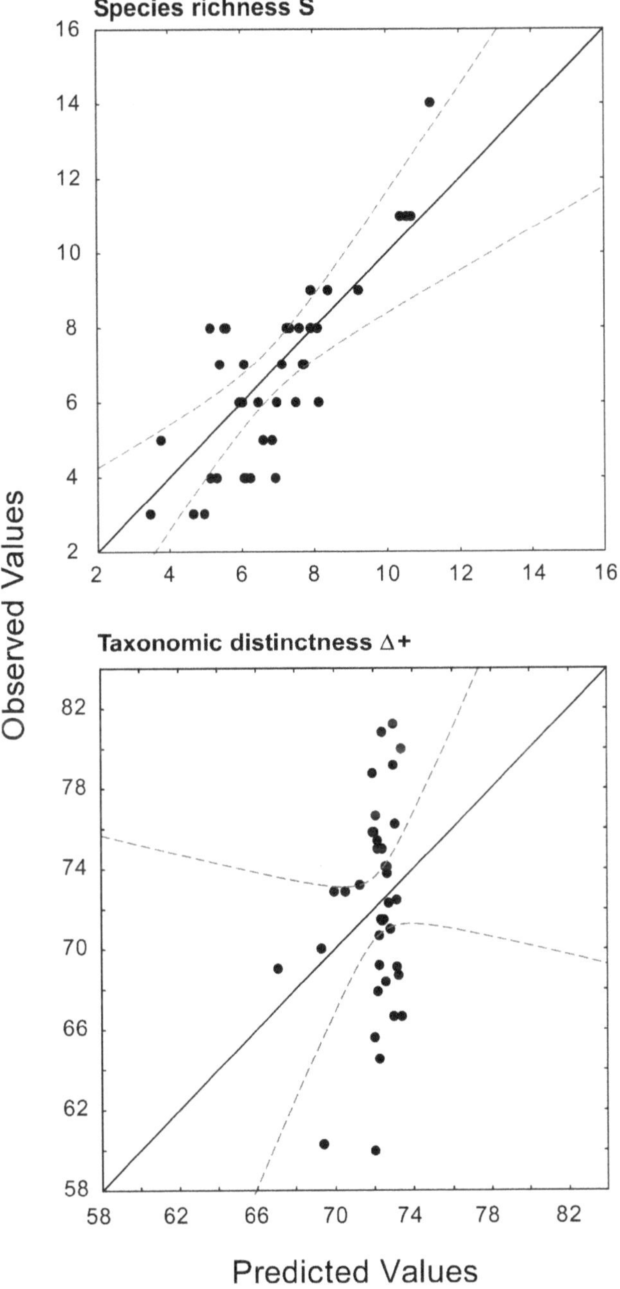

Figure 3. Species richness S of crustacean zooplankton assemblages in the 40 waterbodies under analysis plotted against the values predicted by the best Minimum Adequate Model (MAM; see Table 3 for predictor variables) identified adopting a multiple regression procedure (top). Dashed curves are 95% CI. For the sake of completeness the predicted taxonomic distinctness values Δ+ by the best MAM are also reported (bottom).

Table 3. Summary of the heuristic multiple regression analysis followed by a parsimonious selection procedure of the Minimum Adequate Model (MAM) predicting species richness (S) and of crustacean zooplankton assemblages by means of bioclimatic variables; only the first 10 best MAMs are reported. For the sake of completeness, results of MAM analysis are reported also for taxonomic distinctness (Δ+), even though the statistical power of the models was negligible (see Results). K: number of predictors included in the model; AICc: second-order Akaike Information Criterion; w_i: Akaike weight. For predictor abbreviations see Table A1.

Species Richness S			
K	**Predictors**	*AICc*	w_i
3	%nontr–TempSea–MeanTwet	62.84	0.485
3	H–TempSea–MeanTwet	67.05	0.058
3	%nontr–TempSea–PrecColdQ	67.07	0.057
4	H–Iso–TempSea–MeanTwet	67.12	0.046
2	TempSea–PrecColdQ	67.57	0.044
4	%nontr–TempSea–MeanTwet–Prec	67.62	0.044
4	%nontr–TempSea–MeanTwet	67.71	0.043
3	TempSea–MeanTwet–Prec	67.73	0.042
1	Iso–PrecColdQ	67.74	0.041
4	%nontr–Iso–TempSea–MeanTwet	67.96	0.04

Taxnomic DistinctnessΔ+			
K	**Predictors**	*AICc*	w_i
1	%tr	130.92	0.19
2	%tr–MeanTwet	131.88	0.12
1	PrecColdQ	131.98	0.11
2	%tree–PrecColdQ	132.08	0.11
1	MeanTwet	132.42	0.09
3	%tr–MeanTwet–SUR	132.58	0.08
2	MeanTwet–SUR	132.67	0.08
1	%nontr	132.71	0.08
1	SUR	132.74	0.08
2	%tr–SUR	132.79	0.07

In contrast with species richness, the heuristic search procedure was unable to identify a MAM with a significant predictive power for taxonomic distinctness. The single variable % tree cover, resulted the best predictor of Δ+ (Table 3); however, the correlation resulted in being non-significant ($r = 0.25$, $P = 0.11$, d.f. 1,38; see also Figure 3). Other models showed an even worst performance, independently from the number of variables involved (Table 3), indicating, in turn, that the taxonomic distinctness of the crustacean assemblages cannot be predicted by bioclimatic, landscape-scale factors.

Canonical analysis of principal coordinates (CAP), followed by a confirmatory PERMANOVA test identified two main groups of waterbodies significantly different in terms of species composition (Figure 4; Pseudo–F = 7.2, P(perm) = 0.001). The variables Isothermality, Mean Temperature of Wettest Quarter, Annual Precipitation, and Precipitation of Coldest Quarter showed a correlation ($r > 0.65$) with the canonical axis 1 (Figure 4) and significantly differed between the two groups of waterbodies (PERMANOVA, Pseudo–F = 27.4, P(perm) = 0.001). A Simper procedure indicated that the variable Mean Temperature of Wettest Quarter contributed by 32.2% to inter-group differences, followed by Isothermality and Annual Precipitation (28.6 and 24.2%, respectively).

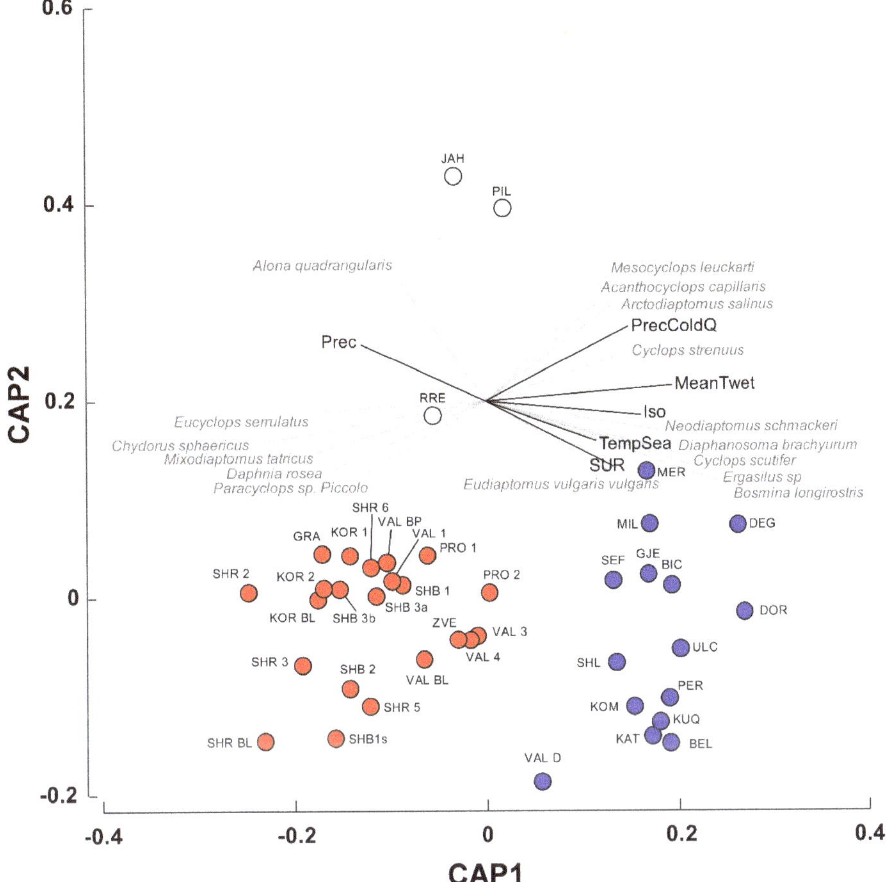

Figure 4. Canonical analysis of principal components (CAP) analysis ($m = 6$, misclassification = 13%, $P = 0.0001$) testing the differences in crustacean zooplankton assemblages across the 40 waterbodies included in the study as affected by bioclimatic factors. *Markers* represent waterbodies labelled with the respective code (Table 1); their color categorizes the two groups (i.e., red for group1 and blue for group2) showing significant differences in species composition (post hoc PERMANOVA, P(perm) < 0.001). Vector overlay are Spearman correlations of bioclimatic factors and species with canonical axes with $r > 0.55$.

In group1, the Mean Temperature of Wettest Quarter was remarkably lower than that determined for group2 (0.06 ± 0.69 vs. 8.44 ± 0.09 °C; *t*-test for separate variances: $t = 9.41$, $P < 0.0001$, 25.53 d.f.). Similarly, Isothermality showed lower values in group1 (33.91 ± 0.23 vs. 38.42 ± 0.24, $t = 9.19$, $P < 0.0001$, 29.52 d.f.), while the Annual Precipitation showed an inverse pattern (1098 ± 10.42 vs. 1012 ± 1.71 mm, $t = -6.43$, $P < 0.001$, 26.09 d.f.).

The analysis of the relationships between species occurrences and the canonical axis 1 (see Table A2 for Spearman correlations for the complete list of species) indicated that *Eucyclops serrulatus*, *Chydorus sphaericus*, *Mixodiaptomus tatricus*, and *Daphnia rosea* in the first group were correlated mainly with Precipitation. The occurrences of *Neodiaptomus schmackeri*, *Diapahnosoma brachyurum*, *Cyclops scutifer*, *Ergasilus* sp, *Bosmina longirostris*, and, to a lesser extent, *Mesocyclops leuckarti* in the group2 were related with the variables Isothermality, Mean Temperature of Wettest Quarter, and Precipitation of Coldest Quarter.

4. Discussion

The earth is undergoing an accelerated rate of native ecosystem conversion and degradation and there is increased interest in measuring and modelling biodiversity while using landscape-scale, remotely-sensed predictors. Here, we made an attempt towards this direction while using crustacean zooplankton. This group of organisms, ubiquitous in lentic habitats, has been recently subjected to renewed interest as an effective bio-indicator of the environmental status of ponds and small lakes [62–65]. The heuristic procedure was used here to identify minimum adequate models predicting the diversity the crustacean zooplankton assemblages across the 40 waterbodies under analysis provided non-univocal results. On one hand, they showed that a subset of bioclimatic variables could effectively predict the variation in species richness and composition across the different waterbodies. On the other hand, they also indicated that the assemblages' taxonomic distinctness Δ+ is unrelated with landscape-scale environmental drivers.

Before discussing these results, it must be considered that the emphasis we put on landscape and bioclimatic drivers of zooplankton diversity by no means imply that other chemical, physical, and biotic characteristics of the waterbodies, such as nutrient concentration, pH, depth, predators, aquatic vegetation, etc. are to be considered of secondary importance. A number of studies have unequivocally indicated that they can directly affect zooplankton species richness ([66,67] and references cited in the introduction; but see also further in this section). In addition, lake age [68,69], connectivity, and, in general, the spatial arrangement of the habitat have been acknowledged to play an important structuring role (e.g., [70]; see also [71] for a marine example). However, in the present study, a hierarchy of effects at different spatial and environmental scales is implicitly assumed, with landscape and bioclimatic drivers indirectly affecting the characteristics of the biota (including crustacean zooplankton) by affecting the chemical and physical conditions of the waterbodies. Indeed, the limited extension of the basins that were included in our study (Figure 1) actually implies for them a low thermal inertia, and thus the ability to rapidly respond to the external climatic conditions [72]. An indirect support to this view is also provided by the negligible predictive power of waterbodies' area for assemblages' S, Δ+, and species composition, confirming the results of investigations performed on the benthic fauna of high-altitude ponds [22].

The best MAM included as predictors the degree of non-arboreal vegetation cover of the land areas neighboring the waterbodies, temperature seasonality, and mean temperature of the wettest quarter. The positive influence of the non-arboreal vegetation cover on species richness is consistent with the results of studies that were focused on macrobenthos in lotic habitats (e.g., [73] and literature cited). This could be ascribed to a positive, indirect effect of lateral trophic enrichment on aquatic primary producers, increasing zooplankton diversity by a phytoplankton-mediated bottom up effect or, alternatively, promoting habitat heterogeneity through an increase in aquatic vegetation [66,74,75].

Noticeably, the two temperature variables had contrasting effects on species richness: the lowest number of species was predicted to occur in waterbodies that were subjected to minimum temperature variability during the year and to maximum temperatures during the wettest months, i.e., in winter. The positive influence of temperature variability on patterns of species diversity has been acknowledged for zooplankton and other freshwater invertebrates [76,77], and can generally be ascribed to an effect of habitats environmental heterogeneity on empty niches availability, and, in turn, on species richness [78]. The negative influence of the mean temperature of the wettest quarter (MeanTwet) on species richness can be explained while considering that the former scales negatively with the altitude of the water bodies ($r = -0.95$, $P < 0.001$, 38 d.f.). Thus, MeanTwet maximum values were observed for basins that were located at low altitudes, generally in highly anthropized areas. Accordingly, the lower species richness characterizing these environments might be actually determined by the interplay of a spectrum of anthropogenic perturbations, such as pollution, water caption, or introduction of predatory fish. Indeed, until 1990, the vast majority of low-altitudes waterbodies in Albania have been stocked with both native and non-indigenous fish species [79], and fish predation has been repeatedly recognized to influence the species richness of zooplankton in lentic habitats [80,81].

The "assemble first, predict later" modelling strategy that was used in the present study was successful in predicting species richness and is generally acknowledged to have several advantages, among others an enhanced capacity to synthesize complex data into a form more readily interpretable by scientists and decision-makers [82]. However, it is apparent that it presents important limitations, as testified by the failure in modeling taxonomic distinctness. Δ+ resulted in being not predictable, confirming the results of several investigations that have found weak or negligible relationships of the taxonomic distinctness of macrobenthic communities with environmental factors [27,83,84]. Species richness S and taxonomic distinctness Δ+ are not conceptually (or mechanistically) related, and they behave differently [44]. The lack of congruence between S and Δ+ and the negligible predictability of the latter is because S is likely to respond to short-term environmental changes in the waterbodies under analysis, while Δ+ is a proxy for phylogenetic diversity, reflecting a complex set of intrinsic and extrinsic traits and expressing evolutionary long-term adaptations to local environmental conditions [85]. The canonical analysis of principal coordinates allowed for us to partially overcome these limitations, applying an "assemble and predict together" strategy [82] in order to model changes in the species composition of planktonic assemblages and provide an advanced resolution of species-specific relationships with bioclimatic factors. The CAP analysis (Figure 4) distinguished two distinct group of waterbodies, showing different climatic characteristics in terms of isothermality, mean temperature of the wettest quarter, and annual precipitation. The first (red circles in Figure 4) was mainly constituted by high-altitude ponds and lakes (elevation 1725.3 ± 123.7 m, mean ± 1SE) distributed throughout the study area characterized by Crustacea species (e.g., *Eucyclops serrulatus*, *Mixodiaptomus tatricus*) that are peculiar of pristine alpine environments [86]. The second group (blue circles in Figure 4) comprised low-altitude karst waterbodies that were located in the Dumre area (Figure 1; elevation 239.9 ± 86.2 m, mean ± 1SE), where they are subjected to several anthropogenic pressures, including agricultural and urban pollution, eutrophication, and the introduction of non-indigenous fish species [87]. Accordingly, the group is characterized by the occurrence of *Neodiaptomus schmackeri*, an Australasian species of Chinese origin that was recently introduced in Albanian lentic habitats through fish stocking [88] and the copepod *Ergasilus* sp., whose adult females are ectoparasites of fish [89].

Regarding the three isolated waterbodies in Figure 4 (i.e., JAH, PIL, and RRE), they are artificial reservoirs located at 1330, 701, and 280 m a.s.l., respectively, being heavily affected by cattle frequentation (Belmonte, personal observation). Their isolation in the CAP diagram is due to the low species richness (three species in all the waterbodies), that might be ascribed to cattle-induced eutrophication conditions [90]. However, it is worth noting that copepods vary their body size (at the community level and even for single species) inversely with the eutrophication level [91,92]. Thus, by using a plankton net with a mesh size of 200 μm, we may have underestimated the smaller component of the planktonic assemblages thus biasing the species count.

A final consideration deserves a brief mention. In this study, the spatial resolution of the bioclimatic layers (approximately 0.64 km^2) was lower than the area characterizing most of the ponds and lakes included in the analysis (Figure 2). In other words, the bioclimate spatial grid "matched" the dimensions of the waterbodies, the latter being completely included (and described in terms of climate and vegetation cover) within the same grid cell. Additional studies including a wider size range of waterbodies as well as bioclimatic layers resolved at different spatial resolutions are necessary to provide a more complete picture of the actual relationships linking bioclimatic factors and the diversity of lentic zooplankton at multiple regional and environmental scales.

Author Contributions: Conceptualization, G.B. and G.M.; methodology, G.B., S.M. and G.M.; software, G.M.; validation, G.B.; formal analysis, G.B. and G.M.; investigation, G.B. and S.M.; resources, G.B. and S.M.; data curation, G.M.; writing—original draft preparation, G.B. S.M. and G.M.; visualization, G.M.; supervision, G.B.

Funding: This research received no external funding.

Acknowledgments: Many collaborators allowed, during 12 years, the collection of samples. Among them we are particularly grateful to Bilal Shkurtaj (Univ. of Vlore, Albania), and Giuseppe Alfonso (Univ. of Salento, Italy). The authors finally thank two anonymous reviewers whose comments greatly improved the original manuscript.

Conflicts of Interest: The authors declare no conflict of interest.

Appendix A

Table A1. List of the 25 bioclimatic variables used in the study. Layers coded MeanT- PrecColdQ were obtained from the Wordclim v2 dataset [41] available at http://www.worldclim.org/, and refer to average monthly climate data relative to the period 1970-2000 with a 30 arc-second spatial resolution (approx. 0.92×0.70 km within the study area). The vegetation layers %tr and %nontr (% tree cover and % non-tree cover) were extracted from the MODIS Vegetation Continuous Field (VCF) product (https://lpdaac.usgs.gov/products/mod44bv006/; [93]). In the table, a third VCF variable—% bare soil (%bare, in italics)—is included, as it was used together with variables %tr and %nontr only for the estimation of habitat heterogeneity (see text for further details). The R packages *raster*, *rgdal*, and *maptools* were used for the manipulation of bioclimatic layers [94–96].

ID#	Environmental Layer	Code
1	Annual Mean Temperature	MeanT
2	Mean Diurnal Range	MeanDrange
3	Isothermality	Iso
4	Temperature Seasonality	TempSea
5	Max Temperature of Warmest Month	MaxTwarm
6	Min Temperature of Coldest Month	MinTcold
7	Temperature Annual Range	TempArange
8	Mean Temperature of Wettest Quarter	MeanTwet
9	Mean Temperature of Driest Quarter	MeanTdry
10	Mean Temperature of Warmest Quarter	MeanTwarm
11	Mean Temperature of Coldest Quarter	MeanTcold
12	Annual Precipitation	Prec
13	Precipitation of Wettest Month	PrecWetM
14	Precipitation of Driest Month	PrecDryM
15	Precipitation Seasonality	PrecSea
16	Precipitation of Wettest Quarter	PrecWetQ
17	Precipitation of Driest Quarter	PrecDryQ
18	Precipitation of Warmest Quarter	PrecWarmQ
19	Precipitation of Coldest Quarter	PrecColdQ
20	Percent tree cover	%tr
21	Percent non-tree cover	%nontr
	Percent bare soil	%bare
22	Perimeter	PER
23	Surface	SUR
24	Altitude	ELE
25	Habitat heterogeneity	H

Table A2. List of the 79 crustacean species sampled in the 40 waterbodies under analysis. Linnaean classification of species using eight taxonomic levels (i.e., class, subclass, order, suborder, infraorder, family, genus and species) and the total number of occurrences are included.

Taxon	Occurrences
Branchiopoda	
Phyllopoda	
Diplostraca	
Cladocera	
Anomopoda	
Bosminidae	
Bosmina longirostris	16
Chydoridae	
Alona quadrangularis	8
Alona rustica	4
Alonella pulchella	1
Alonella sp	1
Biapertura affinis	3
Chydorus piger	2
Chydorus sp2	1
Chydorus sphaericus	15
Coronatella rectangula	1
Disparalona leei	1
Disparalona rostrata	3
Eurycercus sp	2
Paralona pigra	1
Pleuroxus sp2	4
Pleuroxus truncatus	2
Daphniidae	
Ceriodaphnia pulchella	1
Ceriodaphnia quadrangula	2
Ceriodaphnia reticulata	2
Ceriodaphnia setosa	1
Daphnia cucullata	2
Daphnia curvirostris	2
Daphnia dentifera	1
Daphnia galeata	1
Daphnia hyalina	8
Daphnia longispina	13
Daphnia pulex	2
Daphnia rosea	5
Daphnia sp	1
Scapholeberis kingii	1
Simocephalus serrulatus	1
Simocephalus vetulus	2
Ilyocryptidae	
Ilyocryptus sp	2
Macrothricidae	
Macrothrix sp	2
Moinidae	
Moina affinis	1
Moina brachiata	3
Moina macrocopa	2
Moina micrura	3
Ctenopoda	15
Sididae	
Diaphanosoma brachyurum	12
Diaphanosoma lacustris	3

Table A2. *Cont.*

Taxon	Occurrences
Haplopoda	3
Leptodoridae	
Leptodora kindtii	3
Hexanauplia	
Copepoda	
Neocopepoda	
Gymnoplea	
Calanoida	
Diaptomidae	
Arctodiaptomus salinus	4
Eudiaptomus gracilioides	1
Eudiaptomus vulgaris vulgaris	4
Eudiaptomus zachariasi	2
Mixodiaptomus sp2	2
Mixodiaptomus tatricus	12
Neodiaptomus schmackeri	7
Podoplea	
Cyclopoida	
Cyclopidae	
Acanthocyclops capillaris	4
Acanthocyclops sp. small	3
Acanthocyclops trajani	1
Acanthocyclops vernalis	4
Cyclops abyssorum	1
Cyclops bohater	1
Cyclops ricae	1
Cyclops scutifer	7
Cyclops sp8	1
Cyclops strenuous	4
Cyclops vicinus	1
Diacyclops bicuspidatus	5
Diacyclops languidoides	1
Eucyclops serrulatus	8
Halicyclops sp	1
Macrocyclops distinctus	2
Macrocyclops fuscus	2
Megacyclops brachypus	1
Mesocyclops gracilis	1
Mesocyclops leuckarti	12
Mesocyclops sp3	2
Metacyclops stammeri	2
Microcyclops sp	4
Paracyclops affinis	2
Paracyclops cf. *ectocyclops*	1
Paracyclops fimbriatus	3
Paracyclops sp. small	3
Thermocyclops sp	1
Tropocyclops sp	4
Lernaeidae	
Lernaea sp	2
Poecilostomatoida	7
Ergasilidae	
Ergasilus sp	7

Table A3. The nine bioclimatic variables characterized by a VIF factor < 10 used for MAM analysis. For predictor abbreviations see Table A1. Iso and TempSea are expressed in dimensionless units, MeanTwet in °C, Prec and PrecColdQ in mm, %tr and %nontr in percent, SUR in ha, ELE in m, and H (referring to habitat heterogeneity estimated by the Shannon's diversity index, see text) in dimensionless units. For each variable, the among-location coefficient of variation CV and the VIF value are included.

Waterbody	Iso	TempSea	MeanTwet	Prec	PrecColdQ	%tr	%nontr	SUR	H
BEL	38.82	685.85	8.62	1023	334	1	20	25.82	0.241
BIC	39.19	668.66	8.5	1023	336	7.32	72.94	3.31	0.322
DEG	38.51	669.75	8.9	1001	331	70	20	33.07	0.348
DOR	38.66	685.47	8.68	1013	333	4.51	59.3	10.85	0.355
GJE	38.83	681.62	8.62	1018	333	6.96	65.85	2.57	0.354
GRA	35.08	621.71	−3.1	1071	323	5.36	69.24	0.69	0.33
JAH	32.83	632.66	6.77	1247	475	9.64	58.52	0.21	0.392
KAT	39.35	668.39	8.48	1022	334	9.73	67.54	2.92	0.36
KOM	39.22	672.28	8.5	1019	332	3.46	68.54	1.77	0.318
KOR 1	34.34	660.73	−2.73	1077	305	9.62	74.95	0.28	0.317
KOR 2	34.34	660.73	−2.73	1077	305	9.28	73.14	0.26	0.328
KOR BL	36.85	689.46	−1.1	1069	315	45.29	36.44	0.25	0.45
KUQ	38.51	678.82	8.78	1008	331	8.21	72.68	6.53	0.327
MER	38.76	685.2	8.77	1008	333	2	50	59.9	0.337
MIL	39.19	668.66	8.5	1023	336	10.75	66.78	5.85	0.367
PER	38.71	667.18	9.03	1001	334	9	42	26.08	0.404
PIL	35.16	621.14	8.93	1281	510	8.46	70.5	0.44	0.34
PRO 1	33.33	631.13	3.73	1098	343	3.46	57.38	0.0091	0.348
PRO 2	32.91	635.5	3.83	1089	336	19.42	55.96	0.0148	0.429
RRE	29.43	609.67	4.82	1116	349	5.39	52.62	0.0136	0.373
SEF	38.84	682.17	8.65	1013	333	5	35	85.87	0.358
SHB 1	34.7	650.77	−1.27	1091	322	11.2	71.83	0.38	0.34
SHB 2	33.75	629.99	−2.13	1121	323	10.7	71.3	1.44	0.343
SHB 3a	33.53	623.2	−2.42	1126	325	9.06	68.15	1.53	0.354
SHB 3b	33.53	623.2	−2.42	1126	325	10.66	67.87	1.17	0.361
SHB1s	34.11	630.07	−1.98	1114	323	11.2	71.83	0.0145	0.34
SHL	36.98	669.26	7.45	989	277	30	50	14.96	0.447
SHR 2	32.68	643.99	−4.73	1077	283	4.85	80.93	1.35	0.259
SHR 3	33.4	666.58	−0.53	1056	277	7.26	78.48	6.02	0.286
SHR 5	32.2	615.66	−5.32	1107	290	3.96	72.99	0.45	0.302
SHR 6	33.23	631.91	−4.05	1057	275	8.53	79.14	1.74	0.284
SHR BL	32.48	635.84	−4.7	1083	282	5.43	79.98	3.26	0.268
ULC	39.35	668.39	8.48	1022	334	8.29	63.82	6.46	0.369
VAL 1	35.34	615.94	1.23	1099	324	6.96	73.44	0.72	0.318
VAL 3	35.12	613.86	−2.25	1108	328	6.32	69.34	0.38	0.335
VAL 4	35.12	614.42	−2.03	1108	327	8.09	65.35	2.38	0.362
VAL BL	35.66	636.2	3.47	1066	298	14.58	68.43	2.43	0.365
VAL BP	35.54	617.85	1.35	1103	326	9.2	75.11	1.61	0.315
VAL D	33.4	671.99	6.63	999	311	27.69	45.71	34.74	0.463
ZVE	33.16	583.76	11	903	356	10	30	0.29	0.39
CV	7.47	4.24	167.34	6.31	13.12	108.96	25.83	202.81	13.92
VIF	2.62	1.84	6.75	5.15	6.17	1.91	2.01	4.72	1.81

Table A4. Spearman rank correlation coefficients of crustacean zooplankton species with the six axes (CAP1- CAP6) extracted by canonical analysis of principal coordinates. Coefficients > 0.55 (in absolute terms) are in bold.

Taxon	CAP1	CAP2	CAP3	CAP4	CAP5	CAP6
Acanthocyclops capillaris	0.36	0.31	−0.1	−0.01	−0.27	−0.28
Acanthocyclops sp. small	−0.37	0	0.03	−0.25	0	0
Acanthocyclops trajani	−0.22	0.19	−0.17	−0.24	−0.03	−0.03
Acanthocyclops vernalis	0.36	−0.06	0.08	−0.13	0.08	0.34
Alona quadrangularis	−0.29	0.4	−0.23	−0.37	0.19	−0.02
Alona rustica	−0.03	−0.16	−0.29	0.47	−0.12	0.38
Alonella pulchella	0.08	−0.27	−0.1	0.22	0.01	0.24
Alonella sp	−0.15	0.16	−0.13	−0.09	0.03	0.16
Arctodiaptomus salinus	0.43	0.32	−0.19	−0.03	−0.25	0.03
Biapertura affinis	−0.23	−0.08	−0.11	−0.04	−0.03	0.03
Bosmina longirostris	**0.78**	−0.27	0.08	−0.18	0.17	0
Ceriodaphnia pulchella	0.16	0.22	−0.09	−0.01	−0.09	0.13
Ceriodaphnia quadrangula	0.26	0.21	0.11	0.12	−0.29	−0.17
Ceriodaphnia reticulata	−0.17	0.16	0.29	−0.13	−0.28	0.14
Ceriodaphnia setosa	0.16	0.22	−0.09	−0.01	−0.09	0.13
Ceriodaphnia sp3	0.17	−0.22	0.19	0.02	0.1	−0.24
Chydorus piger	−0.04	0.03	0.18	0.36	0.08	−0.02
Chydorus sp2	−0.09	0.15	0.08	0.19	0.22	0.23
Chydorus sphaericus	**−0.75**	−0.14	−0.44	−0.34	−0.23	−0.2
Coronatella rectangula	−0.23	−0.02	0.22	−0.13	−0.05	−0.16
Cyclops abyssorum	0.01	−0.08	−0.26	0.27	−0.2	0.19
Cyclops bohater	−0.08	0.09	0.23	0.24	0.24	0.03
Cyclops ricae	0.23	0.06	0.05	0.12	−0.19	−0.26
Cyclops scutifer	**0.57**	−0.17	0.09	−0.12	0.11	−0.24
Cyclops sp8	−0.12	0.13	0.16	0.15	0.2	−0.08
Cyclops strenuous	0.45	0.15	−0.01	−0.03	−0.17	−0.38
Cyclops vicinus	0.15	0.12	0.03	−0.22	0.09	0.01
Daphnia cucullata	0.16	0.24	−0.09	−0.09	−0.27	0.12
Daphnia curvirostris	−0.04	0.03	0.18	0.36	0.08	−0.02
Daphnia dentifera	−0.15	0.16	−0.13	−0.09	0.03	0.16
Daphnia galeata	0.16	0.22	−0.09	−0.01	−0.09	0.13
Daphnia hyalina	−0.3	0.03	−0.39	0.44	−0.19	0.01
Daphnia longispina	−0.24	0.05	0.13	**0.65**	0	−0.35
Daphnia pulex	0.01	0.13	0.38	−0.02	−0.38	0.24
Daphnia rosea	−0.52	−0.23	−0.13	−0.54	0.08	−0.11
Diacyclops bicuspidatus	−0.13	0.21	0.46	−0.14	−0.43	0.13
Diacyclops languidoides	−0.12	0.13	0.16	0.15	0.2	−0.08
Diaphanosoma brachyurum	**0.63**	−0.17	0.21	−0.28	0.1	0.36
Diaphanosoma lacustris	0.22	0.27	−0.05	−0.2	−0.17	0.1
Disparalona leei	0.19	−0.2	0.06	−0.16	0.15	0.26
Disparalona rostrata	−0.37	−0.14	0.03	−0.17	0.16	−0.08
Ergasilus sp	**0.65**	−0.2	0.03	−0.07	−0.01	−0.2
Eucyclops serrulatus	**−0.56**	−0.03	0.03	−0.3	0.25	−0.15
Eudiaptomus gracilis	0.08	−0.27	−0.1	0.22	0.01	0.24
Eudiaptomus vulgaris vulgaris	0.4	−0.22	0.29	0.16	0.01	−0.52
Eudiaptomus zachariasi	−0.25	0.15	0	−0.2	−0.06	0.17
Eurycercus sp	0.29	−0.33	0.05	−0.22	0.19	0.33
Halicyclops sp	−0.2	0.05	0.13	−0.19	−0.12	0.08
Ilyocryptus sp	−0.05	−0.03	−0.27	−0.01	−0.19	0.01
Leptodora kindtii	0.15	−0.18	−0.04	−0.23	0.18	0.37
Lernaea sp	0.34	−0.14	−0.05	−0.14	0.13	0.2

Table A4. *Cont.*

Taxon	CAP1	CAP2	CAP3	CAP4	CAP5	CAP6
Macrocyclops distinctus	−0.25	0.15	0	−0.2	−0.06	0.17
Macrocyclops fuscus	−0.22	−0.04	−0.2	−0.02	−0.15	−0.09
Macrothrix sp	−0.36	−0.27	−0.08	−0.31	0.05	−0.04
Mesocyclops gracilis	0.09	0.1	−0.23	−0.17	−0.16	0.15
Mesocyclops leuckarti	0.39	0.39	−0.52	0.26	−0.52	0
Mesocyclops sp3	−0.2	−0.06	0.01	0.1	0.05	0.2
Metacyclops stammeri	0.04	−0.06	0.25	0.12	0.22	−0.23
Microcyclops sp	0.03	−0.26	−0.32	0.51	−0.3	0.22
Mixodiaptomus sp2	−0.2	0.02	−0.02	0.15	0.03	−0.26
Mixodiaptomus tatricus	**−0.55**	−0.16	−0.35	0.17	−0.17	−0.09
Moina affinis	−0.19	−0.23	−0.06	−0.05	−0.15	0.05
Moina brachiata	0.29	−0.05	0.11	0.26	−0.23	−0.35
Moina macrocopa	0.01	0.13	0.38	−0.02	−0.38	0.24
Moina micrura	0.17	−0.02	0.28	0.01	0.41	0.25
Neodiaptomus schmackeri	**0.56**	−0.09	−0.17	−0.37	0.08	0.28
Paracyclops affinis	−0.25	0.15	0	−0.2	−0.06	0.17
Paracyclops cf. ectocyclops	−0.17	0.03	−0.01	0.08	0.06	−0.19
Paracyclops fimbriatus	−0.17	0.22	0.28	0.34	0.39	0.11
Paracyclops sp. small	−0.39	−0.23	0.07	−0.23	−0.1	−0.07
Paralona pigra	0.16	0.22	−0.09	−0.01	−0.09	0.13
Pleuroxus sp2	−0.22	0.14	−0.09	0.09	−0.04	0.34
Pleuroxus truncatus	−0.12	−0.29	−0.1	0.01	0.02	0.17
Scapholeberis kingii	−0.2	0.05	0.13	−0.19	−0.12	0.08
Simocephalus serrulatus	−0.18	−0.14	0.1	−0.08	0.2	−0.16
Simocephalus vetulus	−0.1	0.06	−0.28	0.13	−0.12	0.25
Thermocyclops sp	0.12	−0.2	0.26	0.08	0.26	−0.33
Tropocyclops sp	−0.08	0.39	0.26	0.35	0.47	0.17

References

1. Mendonça, R.; Müller, R.A.; Clow, D.; Verpoorter, C.; Raymond, P.; Tranvik, L.J.; Sobek, S. Organic carbon burial in global lakes and reservoirs. *Nat. Commun.* **2017**, *8*, 1694. [CrossRef] [PubMed]
2. Carpenter, S.R.; Stanley, E.H.; Vander Zanden, M.J. State of the world's freshwater ecosystems: Physical, chemical, and biological changes. *Annu. Rev. Environ. Res.* **2011**, *36*, 75–99. [CrossRef]
3. Garcia-Moreno, J.; Harrison, I.J.; Dudgeon, D.; Clausnitzer, V.; Darwall, W.; Farrell, T.; Savy, C.; Tockner, K.; Tubbs, N. Sustaining Freshwater Biodiversity in the Anthropocene. In *The Global Water System in the Anthropocene: Challenges for Science and Governance*; Bhaduri, A., Bogardi, J., Leentvaar, J., Marx, S., Eds.; Springer International Publishing: Cham, Switzerland, 2014; pp. 247–270. [CrossRef]
4. Gozlan, R.E.; Karimov, B.K.; Zadereev, E.; Kuznetsova, D.; Brucet, S. Status, trends, and future dynamics of freshwater ecosystems in Europe and Central Asia. *Inland Waters* **2019**, *9*, 78–94. [CrossRef]
5. Tranvik, L.J.; Downing, J.A.; Cotner, J.B.; Loiselle, S.A.; Striegl, R.G.; Ballatore, T.J.; Dillon, P.; Finlay, K.; Fortino, K.; Knoll, L.B.; et al. Lakes and reservoirs as regulators of carbon cycling and climate. *Limnol. Oceanogr.* **2009**, *54*, 2298–2314. [CrossRef]
6. Brönmark, C.; Hansson, L.A. *The Biology of Lakes and Ponds*; Oxford University Press: Oxford, UK, 2017; p. 368.
7. Downing, J.A.; Prairie, Y.T.; Cole, J.J.; Duarte, C.M.; Tranvik, L.J.; Striegl, R.G.; McDowell, W.H.; Kortelainen, P.; Caraco, N.F.; Melack, J.M.; et al. The global abundance and size distribution of lakes, ponds, and impoundments. *Limnol. Oceanogr.* **2006**, *51*, 2388–2397. [CrossRef]
8. De Meester, L.; Declerck, S.; Stoks, R.; Louette, G.; Van De Meutter, F.; De Bie, T.; Michels, E.; Brendonck, L. Ponds and pools as model systems in conservation biology, ecology and evolutionary biology. *Aquat. Conserv. Mar. Freshw. Ecosyst.* **2005**, *15*, 715–725. [CrossRef]
9. Céréghino, R.; Boix, D.; Cauchie, H.-M.; Martens, K.; Oertli, B. The ecological role of ponds in a changing world. *Hydrobiologia* **2014**, *723*, 1–6. [CrossRef]

10. Ulrich, U.; Hörmann, G.; Unger, M.; Pfannerstill, M.; Steinmann, F.; Fohrer, N. Lentic small water bodies: Variability of pesticide transport and transformation patterns. *Sci. Total Environ.* **2018**, *618*, 26–38. [CrossRef]

11. Holgerson, M.A.; Raymond, P.A. Large contribution to inland water CO_2 and CH_4 emissions from very small ponds. *Nat. Geosci.* **2016**, *9*, 222. [CrossRef]

12. Oertli, B.; Biggs, J.; Céréghino, R.; Grillas, P.; Joly, P.; Lachavanne, J.B. Conservation and monitoring of pond biodiversity: Introduction. *Aquat. Conserv. Mar. Freshw. Ecosyst.* **2005**, *15*, 535–540. [CrossRef]

13. Søndergaard, M.; Jeppesen, E.; Jensen, J.P. Pond or lake: Does it make any difference? *Arch. Hydrobiol.* **2005**, *162*, 143–165. [CrossRef]

14. Céréghino, R.; Biggs, J.; Oertli, B.; Declerck, S. The Ecology of European Ponds: Defining the Characteristics of a Neglected Freshwater Habitat. In *Pond Conservation in Europe*; Oertli, B., Céréghino, R., Biggs, J., Declerck, S., Hull, A., Miracle, M.R., Eds.; Springer: Dordrecht, The Netherlands, 2010; pp. 1–6. [CrossRef]

15. Biggs, J.; von Fumetti, S.; Kelly-Quinn, M. The importance of small waterbodies for biodiversity and ecosystem services: Implications for policy makers. *Hydrobiologia* **2017**, *793*, 3–39. [CrossRef]

16. Belmonte, G. Calanoida (Crustacea: Copepoda) of the Italian fauna: A review. *Eur. Zool. J.* **2018**, *85*, 273–289. [CrossRef]

17. Brönmark, C.; Hansson, L.-A. Environmental issues in lakes and ponds: Current state and perspectives. *Environ. Conserv.* **2002**, *29*, 290–307. [CrossRef]

18. Solimini, A.G.; Bazzanti, M.; Ruggiero, A.; Carchini, G. Developing a multimetric index of ecological integrity based on macroinvertebrates of mountain ponds in central Italy. In *Pond Conservation in Europe*; Oertli, B., Céréghino, R., Biggs, J., Declerck, S., Hull, A., Miracle, M.R., Eds.; Springer: Dordrecht, The Netherlands, 2010; pp. 109–123. [CrossRef]

19. Martínez-Sanz, C.; Cenzano, C.S.S.; Fernández-Aláez, M.; García-Criado, F. Relative contribution of small mountain ponds to regional richness of littoral macroinvertebrates and the implications for conservation. *Aquat. Conserv. Mar. Freshw. Ecosyst.* **2012**, *22*, 155–164. [CrossRef]

20. Martinez-Sanz, C.; Fernandez-Alaez, C.; Garcia-Criado, F. Richness of littoral macroinvertebrate communities in mountain ponds from NW Spain: What factors does it depend on? *J. Limnol.* **2012**, *71*, e16. [CrossRef]

21. Oertli, B.; Joye, D.A.; Castella, E.; Juge, R.; Cambin, D.; Lachavanne, J.-B. Does size matter? The relationship between pond area and biodiversity. *Biol. Conserv.* **2002**, *104*, 59–70. [CrossRef]

22. Hamerlík, L.; Svitok, M.; Novikmec, M.; Očadlík, M.; Bitušík, P. Local, among-site, and regional diversity patterns of benthic macroinvertebrates in high altitude waterbodies: Do ponds differ from lakes? *Hydrobiologia* **2014**, *723*, 41–52. [CrossRef]

23. Krasznai-K, E.Á.; Boda, P.; Borics, G.; Lukács, B.A.; Várbíró, G. Dynamics in the effects of the species–area relationship versus local environmental factors in bomb crater ponds. *Hydrobiologia* **2018**, *823*, 27–38. [CrossRef]

24. MacArthur, R.H.; Wilson, E.O. *The Theory of Island Biogeography*; Princeton University Press: Princeton, NJ, USA, 1967.

25. Bellard, C.; Bertelsmeier, C.; Leadley, P.; Thuiller, W.; Courchamp, F. Impacts of climate change on the future of biodiversity. *Ecol. Lett.* **2012**, *15*, 365–377. [CrossRef]

26. Zhang, C.; Chen, Y.; Xu, B.; Xue, Y.; Ren, Y. How to predict biodiversity in space? An evaluation of modelling approaches in marine ecosystems. *Divers. Distrib.* **2019**, in press. [CrossRef]

27. Roque, F.O.; Guimarães, E.A.; Ribeiro, M.C.; Escarpinati, S.C.; Suriano, M.T.; Siqueira, T. The taxonomic distinctness of macroinvertebrate communities of Atlantic Forest streams cannot be predicted by landscape and climate variables, but traditional biodiversity indices can. *Braz. J. Biol.* **2014**, *74*, 991–999. [CrossRef] [PubMed]

28. Li, Z.; Jiang, X.; Wang, J.; Meng, X.; Heino, J.; Xie, Z. Multiple facets of stream macroinvertebrate alpha diversity are driven by different ecological factors across an extensive altitudinal gradient. *Ecol. Evol.* **2019**, *9*, 1306–1322. [CrossRef] [PubMed]

29. Belmonte, G.; Alfonso, G.; Beadini, N.; Kotori, P.; Mali, S.; Moscatello, S.; Shkurtaj, B. An inventory of invertebrate fauna of Albania and Macedonia Lakes. *Thalassia Salentina* **2018**, *40*, 27–38.

30. Dodson, S. Predicting crustacean zooplankton species richness. *Limnol. Oceanogr.* **1992**, *37*, 848–856. [CrossRef]

31. Shurin, J.B.; Havel, J.E.; Leibold, M.A.; Pinel-Alloul, B. Local and regional zooplankton species richness: A scale-independent test for saturation. *Ecology* **2000**, *81*, 3062–3073. [CrossRef]

32. Hobæk, A.; Manca, M.; Andersen, T. Factors influencing species richness in lacustrine zooplankton. *Acta Oecol.* **2002**, *23*, 155–163. [CrossRef]

33. Boix, D.; Gascón, S.; Sala, J.; Martinoy, M.; Gifre, J.; Quintana, X.D. A new index of water quality assessment in Mediterranean wetlands based on crustacean and insect assemblages: The case of Catalunya (NE Iberian peninsula). *Aquat. Conserv. Mar. Freshw. Ecosyst.* **2005**, *15*, 635–651. [CrossRef]

34. Alfonso, G.; Beccarisi, L.; Pieri, V.; Frassanito, A.; Belmonte, G. Using crustaceans to identify different pond types. A case study from the Alta Murgia National Park, Apulia (South-eastern Italy). *Hydrobiologia* **2016**, *782*, 53–69. [CrossRef]

35. Świerk, D.; Krzyżaniak, M. Is There a pattern for occurrence of macrophytes in Polish ponds? *Water* **2019**, *11*, 1738. [CrossRef]

36. Novikmec, M.; Svitok, M.; Kočický, D.; Šporka, F.; Bitušík, P. Surface water temperature and ice cover of Tatra Mountains lakes depend on altitude, topographic shading, and bathymetry. *Arct. Antarct. Alp. Res.* **2013**, *45*, 77–87. [CrossRef]

37. Magurran, A.E. *Measuring Biological Diversity*; Blackwell Science Ltd.: Oxford, UK, 2004.

38. Warwick, R.M.; Clarke, K.R. Taxonomic distinctness and environmental assessment. *J. Appl. Ecol.* **1998**, *35*, 532–543. [CrossRef]

39. Gotelli, N.J.; Colwell, R.K. Quantifying biodiversity: Procedures and pitfalls in the measurement and comparison of species richness. *Ecol. Lett.* **2001**, *4*, 379–391. [CrossRef]

40. Warwick, R.M.; Clarke, K.R. New 'biodiversity' measures reveal a decrease in taxonomic distinctness with increasing stress. *Mar. Ecol. Prog. Ser.* **1995**, *129*, 301–305. [CrossRef]

41. Fick, S.E.; Hijmans, R.J. WorldClim 2: New 1-km spatial resolution climate surfaces for global land areas. *Int. J. Climatol.* **2017**, *37*, 4302–4315. [CrossRef]

42. Elith, J.; Leathwick, J.R. Species distribution models: Ecological explanation and prediction across space and time. *Annu. Rev. Ecol. Evol. Syst.* **2009**, *40*, 677–697. [CrossRef]

43. Hoylman, Z.H.; Jencso, K.G.; Hu, J.; Martin, J.T.; Holden, Z.A.; Seielstad, C.A.; Rowell, E.M. Hillslope topography mediates spatial patterns of ecosystem sensitivity to climate. *J. Geophys. Res. Biogeosci.* **2018**, *123*, 353–371. [CrossRef]

44. Clarke, K.R.; Warwick, R.M. A taxonomic distinctness index and its statistical properties. *J. Appl. Ecol.* **1998**, *35*, 523–531. [CrossRef]

45. Clarke, K.R.; Warwick, R.M. A further biodiversity index applicable to species lists: Variation in taxonomic distinctness. *Mar. Ecol. Prog. Ser.* **2001**, *216*, 265–278. [CrossRef]

46. Faith, D.P.; Hawksworth, D.L. Phylogenetic pattern and the quantification of organismal biodiversity. *Philos. Trans. R. Soc. Lond. B Biol. Sci.* **1994**, *345*, 45–58. [CrossRef]

47. Anderson, M.J. A new method for non-parametric multivariate analysis of variance. *Austral. Ecol.* **2001**, *26*, 32–46.

48. Dormann, C.F.; Elith, J.; Bacher, S.; Buchmann, C.; Carl, G.; Carré, G.; Marquéz, J.R.G.; Gruber, B.; Lafourcade, B.; Leitão, P.J.; et al. Collinearity: A review of methods to deal with it and a simulation study evaluating their performance. *Ecography* **2013**, *36*, 27–46. [CrossRef]

49. Belsley, D.A.; Kuh, E.; Welsch, R.E. *Regression Diagnostics: Identifying Influential Data and Sources of Collinearity*; John Wiley & Sons: Hoboken, NJ, USA, 2005; p. 292.

50. Whittingham, M.J.; Stephens, P.A.; Bradbury, R.B.; Freckleton, R.P. Why do we still use stepwise modelling in ecology and behaviour? *J. Anim. Ecol.* **2006**, *75*, 1182–1189. [CrossRef] [PubMed]

51. Lunghi, E.; Manenti, R.; Cianferoni, F.; Ceccolini, F.; Veith, M.; Corti, C.; Ficetola, G.F.; Mancinelli, G. Inter-specific and inter-population variation in individual diet specialization: Do bioclimatic factors have a role? *Ecology* **2019**, submitted.

52. Burnham, K.P.; Anderson, D.R. *Model Selection and Multimodel Inference: A Practical Information-Theoretic Approach*; Springer: Berlin, Germany, 2002.

53. Sugiura, N. Further analysis of the data by Akaike's information criterion and the finite corrections. *Commun. Stat. Theory Methods* **1978**, *A7*, 13–26. [CrossRef]

54. Akaike, H. A new look at the statistical model identification. *IEEE Trans. Autom. Control* **1974**, *19*, 716–723. [CrossRef]

55. Royall, R.M. *Statistical Evidence: A Likelihood Paradigm*; Chapman and Hall: New York, NY, USA, 1997.

56. Mancinelli, G. Intraspecific, size-dependent variation in the movement behaviour of a brackish-water isopod: A resource-free laboratory experiment. *Mar. Freshw. Behav. Physiol.* **2010**, *43*, 321–337. [CrossRef]

57. Longo, E.; Mancinelli, G. Size at the onset of maturity (SOM) revealed in length-weight relationships of brackish amphipods and isopods: An information theory approach. *Estuar. Coast. Shelf Sci.* **2014**, *136*, 119–128. [CrossRef]

58. Fox, J.; Weisberg, H.S. *An R Companion to Applied Regression, Second Edition*; Sage Publications, Inc.: Thousand Oaks, CA, USA, 2011.

59. Anderson, M.J.; Willis, T.J. Canonical analysis of principal coordinates: A useful method of constrained ordination for ecology. *Ecology* **2003**, *84*, 511–525. [CrossRef]

60. Ratkowsky, D.A. Choosing the number of principal coordinates when using CAP, the canonical analysis of principal coordinates. *Austral. Ecol.* **2016**, *41*, 842–851. [CrossRef]

61. R Development Core Team. R: A Language and Environment for Statistical Computing. R Foundation for Statistical Computing, Vienna, Austria. 2019. Available online: http://www.R-project.org/ (accessed on 18 September 2019).

62. Visconti, A.; Caroni, R.; Rawcliffe, R.; Fadda, A.; Piscia, R.; Manca, M. Defining seasonal functional traits of a freshwater zooplankton community using $\delta^{13}C$ and $\delta^{15}N$ stable isotope analysis. *Water* **2018**, *10*, 108. [CrossRef]

63. Arfè, A.; Quatto, P.; Zambon, A.; MacIsaac, H.J.; Manca, M. Long-term changes in the zooplankton community of Lake Maggiore in response to multiple stressors: A functional principal components analysis. *Water* **2019**, *11*, 962. [CrossRef]

64. Sgarzi, S.; Badosa, A.; Leiva-Presa, À.; Benejam, L.; López-Flores, R.; Brucet, S. Plankton taxonomic and size diversity of Mediterranean brackish ponds in spring: Influence of abiotic and biotic factors. *Water* **2019**, *11*, 106. [CrossRef]

65. Stamou, G.; Katsiapi, M.; Moustaka-Gouni, M.; Michaloudi, E. Grazing potential-A functional plankton food web metric for ecological water quality assessment in Mediterranean lakes. *Water* **2019**, *11*, 1274. [CrossRef]

66. Joniak, T.; Kuczyńska-Kippen, N.; Gąbka, M. Effect of agricultural landscape characteristics on the hydrobiota structure in small water bodies. *Hydrobiologia* **2017**, *793*, 121–133. [CrossRef]

67. Kuczyńska-Kippen, N.; Basińska, A. Habitat as the most important influencing factor for the rotifer community structure at landscape level. *Int. Rev. Hydrobiol.* **2014**, *99*, 58–64. [CrossRef]

68. Alfonso, G.; Belmonte, G.; Marrone, F.; Naselli-Flores, L. Does lake age affect zooplankton diversity in Mediterranean lakes and reservoirs? A case study from southern Italy. In *Fifty Years after the "Homage to Santa Rosalia": Old and New Paradigms on Biodiversity in Aquatic Ecosystems*; Naselli-Flores, L., Rossetti, G., Eds.; Springer: Dordrecht, The Netherlands, 2010; pp. 149–164. [CrossRef]

69. Belmonte, G. Species richness in isolated environments: A consideration on the effect of time. *Biodivers. J.* **2012**, *3*, 273–280.

70. Martins, K.P.; de Silva Bandeira, M.G.; Palma-Silva, C.; Albertoni, E.F. Microcrustacean metacommunities in urban temporary ponds. *Aquat. Sci.* **2019**, *81*, 56. [CrossRef]

71. Liparoto, A.; Mancinelli, G.; Belmonte, G. Spatial variation in biodiversity patterns of neuston in the Western Mediterranean and Southern Adriatic Seas. *J. Sea Res.* **2017**, *129*, 12–21. [CrossRef]

72. Wissinger, S.A.; Oertli, B.; Rosset, V. Invertebrate Communities of Alpine Ponds. In *Invertebrates in Freshwater Wetlands: An International Perspective on their Ecology*; Batzer, D., Boix, D., Eds.; Springer International Publishing: Cham, Switzerland, 2016; pp. 55–103. [CrossRef]

73. Death, R.G.; Collier, K.J. Measuring stream macroinvertebrate responses to gradients of vegetation cover: When is enough enough? *Freshw. Biol.* **2010**, *55*, 1447–1464. [CrossRef]

74. Hoffmann, M.D.; Dodson, S.I. Land use, primary productivity, and lake area as descriptors of zooplankton diversity. *Ecology* **2005**, *86*, 255–261. [CrossRef]

75. Hessen, D.O.; Faafeng, B.A.; Smith, V.H.; Bakkestuen, V.; Walseng, B. Extrinsic and intrinsic controls of zooplankton diversity in lakes. *Ecology* **2006**, *87*, 433–443. [CrossRef] [PubMed]

76. Shurin, J.B.; Winder, M.; Adrian, R.; Keller, W.; Matthews, B.; Paterson, A.M.; Paterson, M.J.; Pinel-Alloul, B.; Rusak, J.A.; Yan, N.D. Environmental stability and lake zooplankton diversity—Contrasting effects of chemical and thermal variability. *Ecol. Lett.* **2010**, *13*, 453–463. [CrossRef] [PubMed]

77. Jacobsen, D.; Schultz, R.; Encalada, A. Structure and diversity of stream invertebrate assemblages: The influence of temperature with altitude and latitude. *Freshw. Biol.* **1997**, *38*, 247–261. [CrossRef]

78. MacArthur, R.H.; MacArthur, J.W. On bird species diversity. *Ecology* **1961**, *42*, 594–598. [CrossRef]

79. Shumka, S.; Paparisto, A.; Grazhdani, S. Identification of non-native freshwater fishes in Albania and assessment of their potential threats to the national biological freshwater diversity. *Balwois 2008 Proc. Pap.* **2008**, *40*, 1–6.

80. Shurin, J.B. Interactive effects of predation and dispersal on zooplankton communities. *Ecology* **2001**, *82*, 3404–3416. [CrossRef]

81. Reissig, M.; Trochine, C.; Queimaliños, C.; Balseiro, E.; Modenutti, B. Impact of fish introduction on planktonic food webs in lakes of the Patagonian Plateau. *Biol. Conserv.* **2006**, *132*, 437–447. [CrossRef]

82. Ferrier, S.; Guisan, A. Spatial modelling of biodiversity at the community level. *J. Appl. Ecol.* **2006**, *43*, 393–404. [CrossRef]

83. Abellán, P.; Bilton, D.T.; Millán, A.; Sánchez-Fernández, D.; Ramsay, P.M. Can taxonomic distinctness assess anthropogenic impacts in inland waters? A case study from a Mediterranean river basin. *Freshw. Biol.* **2006**, *51*, 1744–1756. [CrossRef]

84. Hu, G.; Zhang, Q. Seasonal variations in macrobenthic taxonomic diversity and the application of taxonomic distinctness indices in Bohai Bay, northern China. *Ecol. Indic.* **2016**, *71*, 181–190. [CrossRef]

85. Graham, C.H.; Fine, P.V.A. Phylogenetic beta diversity: Linking ecological and evolutionary processes across space in time. *Ecol. Lett.* **2008**, *11*, 1265–1277. [CrossRef] [PubMed]

86. Jersabek, C.D.; Brancelj, A.; Stoch, F.; Schabetsberger, R. Distribution and ecology of copepods in mountainous regions of the Eastern Alps. *Hydrobiologia* **2001**, *453*, 309–324. [CrossRef]

87. Hoxha, B.; Cane, F.; Avdolli, M.; Dauti, A. Preliminary data on eutrophication of Carstic lakes. *J. Int. Environ. Appl. Sci.* **2011**, *6*, 717–723.

88. Alfonso, G.; Russo, R.; Belmonte, G. First record of the Asian diaptomid *Neodiaptomus schmackeri* (Poppe & Richard, 1892) (Crustacea: Copepoda: Calanoida) in Europe. *J. Limnol.* **2014**, *73*, 584–592. [CrossRef]

89. Stojanovski, S.; Hristovski, N.; Cakic, P.; Baker, R.A. Preliminary investigations on the parasitic Crustacea of freshwater fishes from Macedonia. *Balwois 2006 Proc. Pap.* **2006**, 1–8.

90. Solimini, A.G.; Ruggiero, A.; Anello, M.; Mutschlechner, A.; Carchini, G.M. The benthic community structure in mountain ponds affected by livestock watering in nature reserves of Central Italy. *Ver. Int. Ver. Theor. Angew. Limnol.* **2000**, *27*, 501–505. [CrossRef]

91. Uye, S.-I. Replacement of large copepods by small ones with eutrophication of embayments: Cause and consequence. In *Ecology and Morphology of Copepods*; Ferrari, F.D., Bradley, B.P., Eds.; Springer: Dordrecht, The Netherlands, 1994; Volume 102, pp. 513–519.

92. Belmonte, G.; Cavallo, A. Body size and its variability in the copepod *Acartia margalefi* (Calanoida) from Lake Acquatina (SE Italy). *Ital. J. Zool.* **1997**, *64*, 377–382. [CrossRef]

93. DiMiceli, C.M.; Carroll, M.L.; Sohlberg, R.A.; Huang, C.; Hansen, M.C.; Townshend, J.R.G. *Annual Global Automated MODIS Vegetation Continuous Fields (MOD44B) at 250 m Spatial Resolution for Data Years Beginning Day 65, 2000–2018, Percent Tree Cover, Collection 5, Version 6*; University of Maryland: College Park, MD, USA, 2019.

94. Hijmans, R.J. Raster: Geographic Data Analysis and Modeling. *R Package Version 3.0-7*. 2019. Available online: https://CRAN.R-project.org/package=raster (accessed on 10 September 2019).

95. Bivand, R.; Keitt, T.; Rowlingson, B. Rgdal: Bindings for the Geospatial Data Abstraction Library. *R Package Version 1.4-6*. 2019. Available online: https://CRAN.R-project.org/package=rgdal (accessed on 10 September 2019).

96. Bivand, R.; Lewin-Koh, N. maptools: Tools for Reading and Handling Spatial Objects. *R Package Version 0.9-5*. 2019. Available online: https://CRAN.R-project.org/package=maptools (accessed on 10 September 2019).

Article

Dynamics of Mesozooplankton Assemblage in Relation to Environmental Factors in the Maryland Coastal Bays

Efeturi U. Oghenekaro and Paulinus Chigbu *

NSF CREST-Center for the Integrated Study of Coastal Ecosystem Processes and Dynamics in the Mid-Atlantic Region, and NOAA Living Marine Resources Cooperative Science Center, Department of Natural Sciences, University of Maryland Eastern Shore, Princess Anne, MD 21853, USA; euoghenekaro@umes.edu
* Correspondence: pchigbu@umes.edu; Tel.: +14-10-621-3034

Received: 16 August 2019; Accepted: 4 October 2019; Published: 14 October 2019

Abstract: The mesozooplankton composition and dynamics in coastal lagoons of Maryland, mid-Atlantic region, USA have received little scientific attention despite the fact that the lagoons have undergone changes in water quality in the past two decades. We compared mesozooplankton abundance and community structure among sites and seasons, and between 2012, a year of higher than average salinity (33.4), and 2013 with lower than average salinity (26.6). It was observed that the composition, diversity, and abundance of mesozooplankton in 2012 differed from those of 2013. Barnacle nauplii were abundant in 2012 contributing 31% of the non-copepod mesozooplankton abundance, whereas hydromedusae were more dominant in 2013 and contributed up to 83% of non-copepod zooplankton abundance. Gastropod veliger larvae were more abundant in 2013 than in 2012 while larvae of bivalves, polychaetes, and decapods, in addition to cladocerans and ostracods had higher abundances in 2012. The abundance and diversity of mesozooplankton were explained by variations in environmental factors particularly salinity, and by the abundance of predators such as bay anchovy (*Anchoa mitchelli*). Diversity was higher in spring and summer 2012 (dry year) than in 2013 (wet year). The reduction of salinity in fall 2012, due to high freshwater discharge associated with Hurricane Sandy, was accompanied by a decrease in mesozooplankton diversity. Spatially, diversity was higher at sites with high salinity near the Ocean City Inlet than at sites near the mouth of tributaries with lower salinity, higher nutrient levels and higher phytoplankton biomass. Perhaps, the relatively low salinity and high temperature in 2013 resulted in an increase in the abundance of hydromedusae, which through predation contributed to the reduction in the abundance of bivalve larvae and other taxa.

Keywords: mesozooplankton; salinity; abundance; distribution; diversity; Maryland Coastal Bays

---

1. Introduction

Planktonic organisms ranging in size 0.2–20 mm are referred to as mesozooplankton [1]; examples are copepods, decapod zoea, rotifers, gastropod veliger, barnacle nauplii, larvaceans, cladocerans and bivalve larvae. Mesozooplankton are major grazers of primary producers, and important predators of microzooplankton, including ciliates, thereby acting as a pathway for transfer of materials and energy from microbial food web to higher trophic levels [2,3]. Copepods are the most abundant of, and contribute the most biomass to, the mesozooplankton group in marine and estuarine zooplankton [4].

Although non-copepods do not constitute the bulk of mesozooplankton in United States mid-Atlantic estuaries, they occasionally become abundant [5–7] and play an important role in the diet of zooplanktivores [8,9]. Mesozooplankton composition, abundance and diversity in estuaries and coastal lagoons vary seasonally, spatially and between years in response to abiotic and biotic factors [10–16], and have been well documented in large river-dominated estuaries of the mid-Atlantic such as the Chesapeake Bay [12,17] and Delaware Bay [18,19]. Nevertheless, we know little about mesozooplankton composition and dynamics in Maryland Coastal Bays (MCBs), which are poorly flushed, eutrophic, largely polyhaline lagoonal systems that receive relatively small amount of freshwater directly from tributary rivers and creeks. Because of differences in the hydrography, the composition, abundance and distribution of mesozooplankton in MCBs may be different than those of river-dominated estuaries, and are expected to vary between years due to variations in climatic factors.

The coastal bays support commercially and recreationally important species of shellfish such as blue crabs (*Callinectes sapidus*) and hard clams (*Mercenaria mercenaria*) which rely on the successful recruitment of their meroplanktonic larvae to sustain the adult populations. The blue crab is the most abundant shellfish in the MCBs and requires high salinity waters for spawning and larval development [20,21]. In the Chesapeake Bay, *C. sapidus* spawn near the mouth of the estuary, or in the nearby coastal ocean [6,22]. The larvae (zoeae) are then transported into adjacent coastal waters on the continental shelf [23–25] where they develop through seven or eight zoeal stages before subsequently returning as post-larvae and juveniles to the estuary [25], while the adults remain resident in the system. Due to the high salinity (>26) levels in a large portion of the MCBs, it has been hypothesized that resident adult populations of the *C. sapidus* spawn in the high saline waters of the coastal bays and the larvae complete their development within the Bays [26]. In the 2014 updated Coastal Bays Blue Crab Fishery Management Plan Implementation document, the need for studies to determine the "level of localized reproduction and entrapment of larvae" of blue crabs in the MCBs was emphasized.

Furthermore, salinity in the MCBs increased after the opening of the Ocean City Inlet in 1933, which favored the development of hard clams (*M. mercenaria*) and, by the 1960s, the fishery superseded the commercial landings and value of oysters in the bay [27]. The population of hard clams is currently dominated by large and older adults; the recruitment is low and intermittent except in areas close to the Ocean City Inlet, and densities are lower than historical levels [27] although increasing. Efforts to restore the population including stoppage of the hydraulic clam dredging activities have not resulted in significant increases in clam density, the cause of which is unknown, although intense predation by species such as blue crab, and the frequent occurrence of brown tide, *Aureococcus anophagefferens*, blooms that negatively impact bivalve larvae have been speculated to be the cause [27]. Information on the distribution and abundance of bivalve larvae, and the environmental factors influencing their abundance would be useful for designing effective management strategies to enhance bivalve populations in the MCBs.

Within the embayments in the MCBs, nutrient levels, phytoplankton biomass and fish abundance vary spatially, seasonally and inter-annually [28–30], which in addition to variations in the abundance of gelatinous zooplankton predators can influence the distribution and abundance of mesozooplankton. For example, mortality of bivalve larvae due to predation by gelatinous zooplankton can potentially slow recovery of hard clams despite the implementation of effective management plans [31,32].

The objectives of the study were:

1. To compare mesozooplankton, particularly non-copepod taxa, abundance and community structure among the MCBs embayments;
2. To determine seasonal patterns in mesozooplankton abundance and community structure;
3. To compare mesozooplankton composition, abundance and diversity between relatively dry (2012) and wet (2013) years;
4. To assess the relationships between mesozooplankton taxa abundance and abiotic and biotic factors; and
5. To use information on decapod larval occurrence to determine whether blue crabs spawn within the MCBs.

The study addressed the following questions: Are mesozooplankton abundance and diversity higher at sites closer to the Ocean City Inlet with higher salinity, than at sites farther from the Inlet with lower salinity? Are mesozooplankton abundance and diversity higher in spring and summer 2012, a year with relatively higher salinity and lower temperature, than in 2013 when salinity was lower, and temperature higher? Do blue crabs spawn within the MCBs?

2. Materials and Methods

2.1. Study Area

The MCBs located in the Mid-Atlantic region on the east coast of the United States is separated from the Atlantic Ocean by barrier islands. This lagoonal system consists of five embayments (Figure 1) that are well mixed mostly by wave and tidal actions that facilitate the exchange of water and materials between the MCBs and the nearby coastal ocean [29,33]. The system, however, is poorly flushed, has low freshwater input from groundwater, rivers and tributaries, and therefore high salinity that is close to that of the ocean except in upstream areas of rivers and creeks [28,29]. The MCBs are relatively shallow (average depth about 1.2 m) with a maximum depth of 9.3 m at the Ocean City Inlet [29]. Some characteristics of the northern bays watershed include low forest covers, non-point sources of pollution, dead-end canals, high stream nitrate, and nutrients from agricultural activities and development [34]. Increased nutrient input has resulted in eutrophic conditions in the northern bays and tributaries, although a similar condition exists in the Newport Bay located in the southern MCBs.

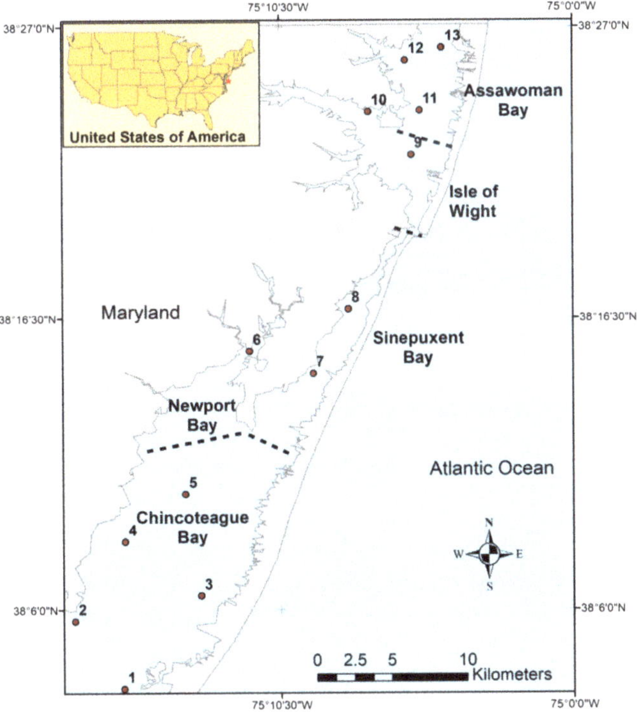

Figure 1. Map of the study area showing sampling locations within the Maryland Coastal Bays.

2.2. Sample Collection

Water quality parameters such as temperature, salinity, dissolved oxygen (DO) concentration, and pH were measured in situ using YSI 6600 sonde. Zooplankton samples were collected monthly from February 2012 to December 2013 from 13 sites (Figure 1). Sites 1–5 are located in Chincoteague Bay (CB), Site 6 in Newport Bay (NB), Sites 7 and 8 in Sinepuxent Bay (SB), Site 9 in Isle of Wight Bay (IWB), Site 10 at the mouth of St. Martin River (SMR) and Sites 11–13 in Assawoman Bay (AB). Zooplankton sampling was conducted using a plankton net, with 60-cm diameter mouth opening, 240-cm-long conical section, a mesh size of 200 μm, and a cod-end made of PVC with a volume of 1000 mL and diameter of 11 cm. Because of the shallow nature of the lagoons (average depth < 2 m), the net was towed horizontally in the upper meter of the water column similar to how zooplankton samples were collected in the shallow Barnegat Bay, NJ [35]. Sampling was for a duration of 2 min. The mouth opening of the net was equipped with a General Oceanic flow meter (Model 2030R) attached to the frame to measure the volume of water filtered by the net. Zooplankton samples were preserved in 5% buffered formalin solution immediately after collection. Biovolumes of gelatinous zooplankton (*Mnemiopsis leidyi*) were determined with a measuring cylinder.

Preserved plankton samples were processed in the lab using a dissecting microscope. A Folsom splitter was used to split samples when necessary. Samples were diluted to a known volume, and a minimum of 5 mL subsample was extracted with a Stemple pipette and 200–500 zooplankton individuals were counted. Cladocerans and crab zoeae were identified to species level when possible using a compound microscope and by reference to various publications [1,36,37]. Data on bay anchovy (*Anchoa mitchelli*) catch per unit effort (CPUE) were obtained from the Maryland Department of Natural Resources (MD DNR). Water samples for analyses of chlorophyll *a* and 19′-Butanoyloxy-Fucoxanthin (But-fuco), a diagnostic pigment for pelagophytes such as the Brown tide, *Aureococcus anophagefferens*, were collected from 0.5 m below the surface using a peristaltic pump, and placed in 2-L acid-rinsed polyethylene amber bottles. The samples were kept in ice until arrival at the lab. Water samples were then filtered using a Whatman GF/F 47 mm filter (GE Healthcare, Fisher Scientific, Hanover Park, IL, USA) and a vacuum pump for 1–4 min. A minimum of 1 L by volume of water was filtered except in situations when the filter was clogged. Chlorophyll *a* and 19′-Butanoyloxy-Fucoxanthin (But-fuco) concentrations were determined, as part of the integrated monitoring of the MCBs, using a High Performance Liquid Chromatography [38]. Freshwater discharge data for Birch branch at Showell, Maryland, USA were downloaded from the United States Geological Survey (USGS) website (http://waterdata.usgs.gov/nwis/monthly).

2.3. Data Analyses

The Mann–Whitney U test was used to compare mean abundance of mesozooplankton between years for each season. One-way ANOVA was used to compare differences in mean abundance of mesozooplankton among sites and embayments after log(x+1) transformation of data. The relationships between non-copepod taxa distribution and environmental factors were examined by canonical correspondence analysis (CCA), using CANOCO version 4.5 (Microcomputer Power, Ithaca, New York, USA.) [39]. Non-copepod mesozooplankton abundance was log(x+1) transformed; rare species were down-weighted when necessary and environmental data were standardized. Non-copepod mesozooplankton abundance matrices included only taxa that appeared in >10% of the sampling sites, representing >1% in 2012 and >0.1% in 2013 of the non-copepod community. Shannon diversity index [40] was computed using PRIMER 6 (PRIMER-e, Auckland, North Island, New Zealand. [41]. Spearman's rank correlation was applied to identify the environmental factors (water quality variables and biota) that were associated with diversity.

3. Results

3.1. Environmental Factors

Mean monthly water temperature in 2012 varied from 5°C in February to 25°C in July and then decreased to about 5 °C in December (Figure 2a). In 2013, mean temperature was 30.7 °C in July and 7 °C in December. Water temperature was on average lower in 2012 than in 2013. Lower water temperatures were observed at sites close to the inlets, while St. Martin River had the highest mean water temperature (Table 1).

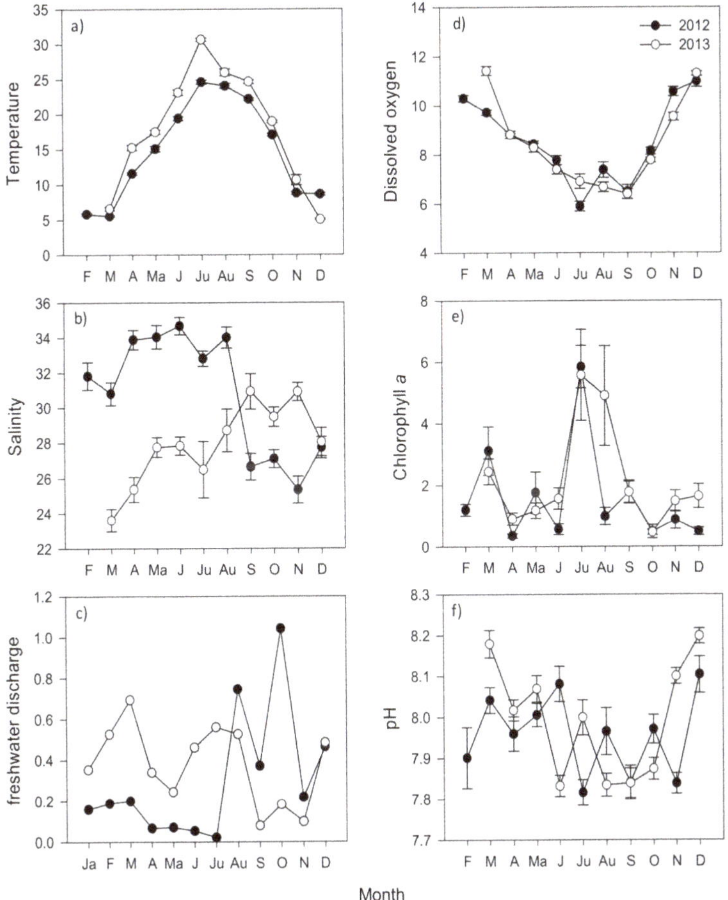

Figure 2. Seasonal patterns of environmental variables expressed as mean values ± SE: (**a**) Temperature (°C); (**b**) salinity; (**c**) Freshwater discharge (m^3 s^{-1}); (**d**) dissolved oxygen (mg L^{-1}); (**e**) Chlorophyll-*a* (μg L^{-1}); and (**f**) pH.

Mean monthly salinity was higher in 2012 (33.4) than in 2013 (26.6) during the period of March to August, but the reverse occurred from September to November (Figure 2b). This salinity pattern reflected the pattern of freshwater discharge by Birch Branch at Showell, MD (Figure 2c). From January to July, Birch Branch mean discharge was higher in 2013 (0.45 m^3 s^{-1}) than in 2012 (0.11 m^3 s^{-1}), and the reverse was the case from August to November (Figure 2c). Discharge was highest in October 2012, the month that Hurricane Sandy occurred in the area. Mean monthly salinity was between 23.6 and

34.7. Newport Bay had the lowest mean salinity level and Sinepuxent Bay had the highest salinity (Table 1). Salinity was also lower at sites in St. Martin River and Assawoman Bay relative to sites in Chincoteague and Isle of Wight Bays.

Monthly dissolved oxygen (DO) levels on the average were high in fall and winter, but relatively lower in spring and summer, although dissolved oxygen levels were > 4.6 mg L^{-1} at all sites throughout the duration of this study (Figure 2d). The lowest DO level was recorded in July (4.6 mg L^{-1}), and the highest in March (12.7 mg L^{-1}).

The seasonal patterns of chlorophyll *a* concentrations are similar in 2012 and 2013. Monthly levels of chlorophyll *a* averaged between 0.2 and 3.4 µg L^{-1} (Figure 2e). Chlorophyll *a* levels were relatively high in winter but decreased in spring, particularly in April. Chlorophyll *a* peaked in summer (July) in both years. Site 6 (Newport Bay) had the highest mean level of chlorophyll *a* in the MCBs during this study (Table 1). Monthly mean pH levels in the MCBs ranged from 7.8 to 8.2 and did not vary much within the system (Figure 2f).

Table 1. Mean physico-chemical and biotic factors measured at sampling stations within embayments of the Maryland Coastal Bays. Embayment codes are: CB, Chincoteague Bay; NB, Newport Bay; SB, Sinepuxent Bay; IWB, Isle of Wight Bay; SMR, St. Martin River; AB, Assawoman Bay. Temp, temperature; Sal, salinity; Chl *a*, chlorophyll *a*; But-Fuco, 19'-Butanoyloxy-Fucoxanthin; Cteno, Ctenophores; CPUE, Catch Per Unit Effort.

					2012				
Bay	Site	Temp (°C)	Sal	DO (mg L^{-1})	pH	Chl *a* (µg L^{1})	But-Fuco (µg L^{-1})	Cteno (mL L^{-3})	Bay Anchovy (CPUE)
CB	1–5	14.7	31.49	8.85	8.0	1.22	0.08	7.6	90
NB	6	14.7	27.07	8.68	7.9	3.93	0.18	5.2	137
SB	7–8	14.1	33.10	7.88	7.9	1.09	0.02	1.0	6
IWB	9	15.3	32.91	8.08	7.9	1.31	0.01	1.5	14
SMR	10	15.5	29.94	8.42	7.9	2.42	0.04	15.1	213
AB	11–13	14.7	29.60	8.57	7.9	1.55	0.00	13.2	103
					2013				
Bay	Site	Temp (°C)	Sal	DO (mg L^{-1})	pH	Chl *a* (µg L^{-1})	But-Fuco (µg L^{-1})	Cteno (mL L^{-3})	Bay anchovy (CPUE)
CB	1–5	17.6	29.82	8.57	8.0	1.09	0.01	9.4	59
NB	6	18.4	25.14	8.19	7.9	4.60	0.00	30.9	93
SB	7–8	17.9	29.88	7.82	7.9	1.52	0.01	8.0	25
IWB	9	17.2	28.43	8.42	8.0	1.83	0.01	16.9	497
SMR	10	18.5	25.63	8.89	10.0	4.44	0.01	23.5	152
AB	11–13	18.0	24.95	8.62	8.0	3.05	0.01	22.3	124

3.2. Temporal Pattern of Mesozooplankton Abundance

Mesozooplankton abundance expressed as individuals m^{-3} (ind. m^{-3}) varied from 5 ind. m^{-3} (Site 12) in May 2012 to 410,509 ind. m^{-3} (Site 8) in September 2013. The average monthly abundance was high (24,543 ind. m^{-3}) in February, but peaked in early March (36,878 ind. m^{-3}) during winter of 2012; it declined to a relatively low level in spring and then gradually increased from summer through fall (Figure 3). Relatively high abundance (30,032 ind. m^{-3}) of mesozooplankton was observed in December 2012. A gradual decline was also observed from winter 2013 through early spring. Densities remained low (378 ind. m^{-3}, May 2012) in spring and early summer, peaked (34,828 ind. m^{-3}) in September 2013 and then declined in fall.

Mesozooplankton abundance differed significantly between years (Table 2) in summer and fall ($p < 0.001$), but not in winter and spring ($p > 0.05$). In fall, abundance was significantly higher in 2012 than 2013 (Mann–Whitney U = 261, T = 2040.0, $p \leq 0.001$), but in summer, it was lower in 2012 than in 2013 (U = 392.0, T = 1909.0, $p \leq 0.001$).

Seasonal and spatial patterns of abundance of mesozooplankton are presented in Figure 4. For each of the seasons examined, there were no significant differences among sites in mesozooplankton abundance (ANOVA, p > 0.05) except in summer 2012 (ANOVA, p = 0.05) when the highest density was recorded (Figure 4 and Table 3) at Site 7 (Tukey test, p < 0.05). When mesozooplankton abundance was compared among bays, higher abundance was recorded in Sinepuxent Bay than in any other bay in winter and summer 2012 (ANOVA, p < 0.05), as shown in Table 3.

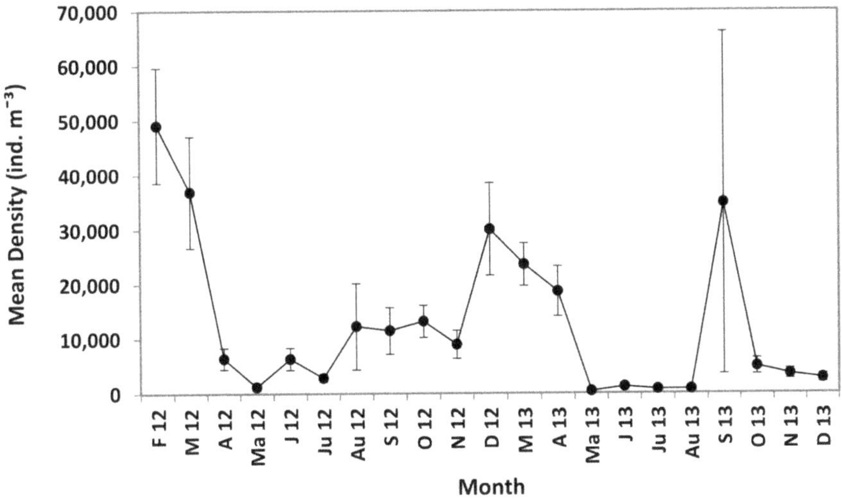

Figure 3. Temporal pattern in the abundance (±SE) of mesozooplankton in the Maryland Coastal Bays.

Table 2. Seasonal average densities (ind. m^{-3}) of mesozooplankton in the Maryland Coastal Bays. Winter, January–March; spring, April–June; summer, July–September; fall, October–December.

	Mean Density (SE)		
Season	**2012**	**2013**	**P-Value**
Winter	30,710 (5759)	23,564 (3907)	0.870
Spring	4653 (1009)	6738 (2025)	0.337
Summer	8837 (3009)	12,109 (10,489)	≤0.001
Fall	17,368 (3364)	3,697 (634)	≤0.001

Table 3. One-way ANOVA result of analyses comparing mesozooplankton density among sampling sites and embayments of the Maryland Coastal Bays in the different seasons in 2012 and 2013. d.f, degree of freedom; Fs, F-value; P, probability value; * 0.05 > p > 0.01; n.s, p> 0.05. There were insufficient data to perform ANOVA in winter 2012.

Year	Season	Sites				Embayments			
		d.f	F	P		d.f	F	P	
2012	Winter	12	1.59	0.210	n.s	4	3.22	0.027	*
	Spring	12	0.93	0.532	n.s	4	1.81	0.137	n.s
	Summer	12	2.14	0.050	*	4	2.62	0.042	*
	Fall	12	1.16	0.360	n.s	4	1.51	0.215	n.s
2013	Winter	12	-	-		4	3.01	0.092	n.s
	Spring	12	0.32	0.979	n.s	4	0.83	0.535	n.s
	Summer	12	1.00	0.476	n.s	4	1.10	0.379	n.s
	Fall	12	0.70	0.737	n.s	4	1.62	0.183	n.s

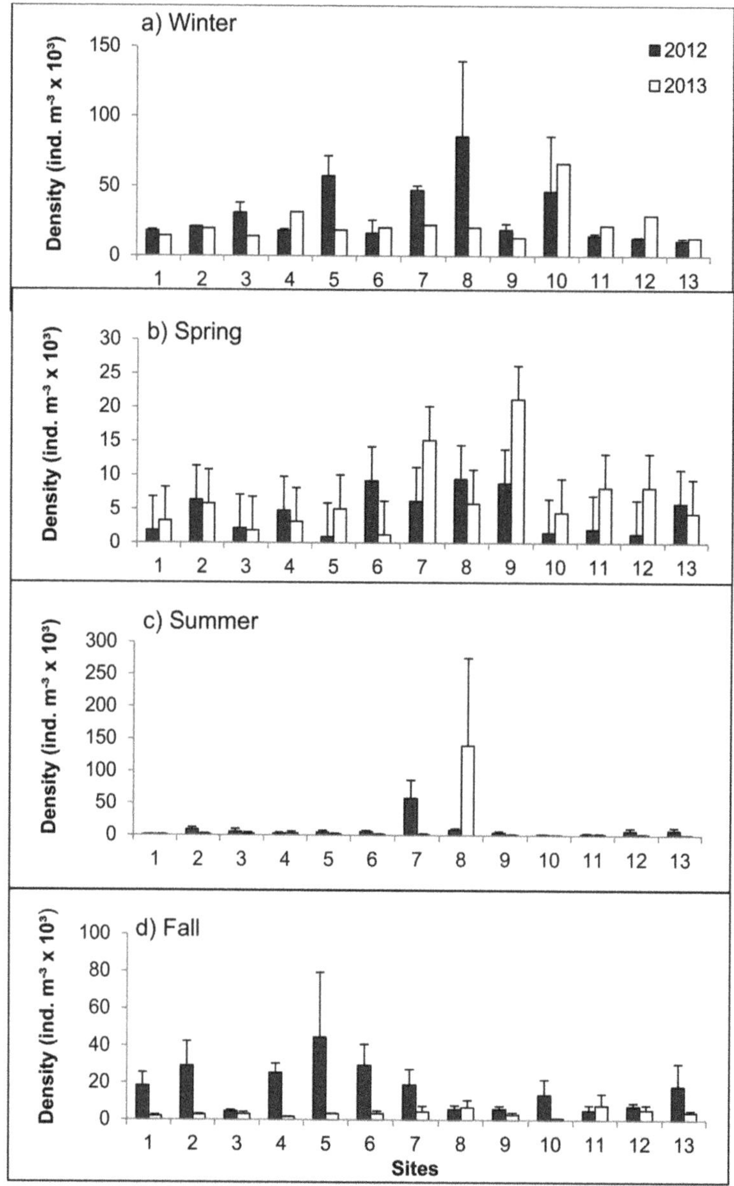

Figure 4. Spatial patterns in the densities (+SE) of mesozooplankton in the Maryland Coastal Bays in 2012 (year of below average freshwater discharge) and 2013 (year of above average freshwater discharge): (**a**) winter; (**b**) spring; (**c**) summer; and (**d**) fall.

3.3. Community Composition, Relative Abundance and Distribution of Non-Copepod Mesozooplankton in MCBs

Mesozooplankton community in the MCBs was represented by 14 major taxa dominated by copepods, but only groups that are relatively abundant are presented in Figure 5. Non-copepods contributed about 5% of the total mesozooplankton abundance. A detailed description of copepod species composition and abundance in the MCBs has been made [42].

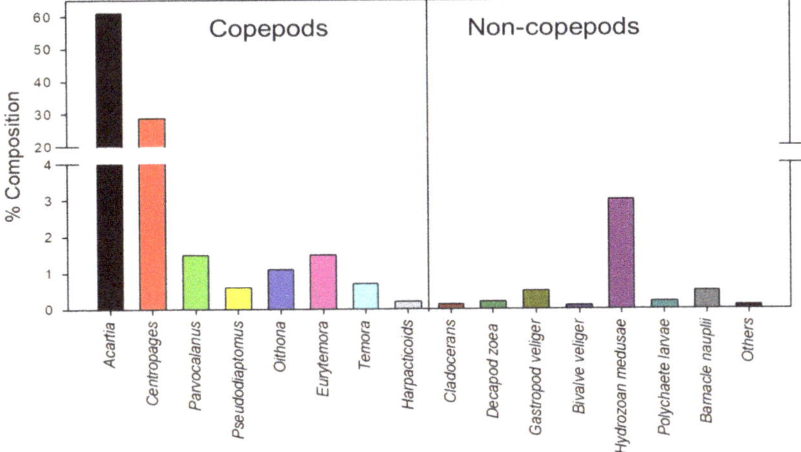

Figure 5. Mesozooplankton community composition in the Maryland Coastal Bays based on samples collected in 2012 and 2013.

The non-copepod community on the average was dominated by hydromedusae (3%), followed by gastropod veliger (0.5%), and barnacle nauplii (0.5%) (Figure 5). Cladocerans (0.1%) and larval stages of decapods (0.2%), polychaetes (0.2%), bivalves (0.1%), and fish (0.1%) were also relatively abundant within the non-copepod community. Larvaceans, nematodes, amphipods, and isopods made up <1% of the zooplankton community.

The percentage contribution of non-copepods to total zooplankton abundance was as high as 98% at Site 12 (AWB) in the summer when copepod relative abundance was low. Monthly averaged contribution varied between 0–31.9% in 2012 and 0.4–76.5% in 2013 (Figure 6). Non-copepods were most represented in May 2012 and June 2013.

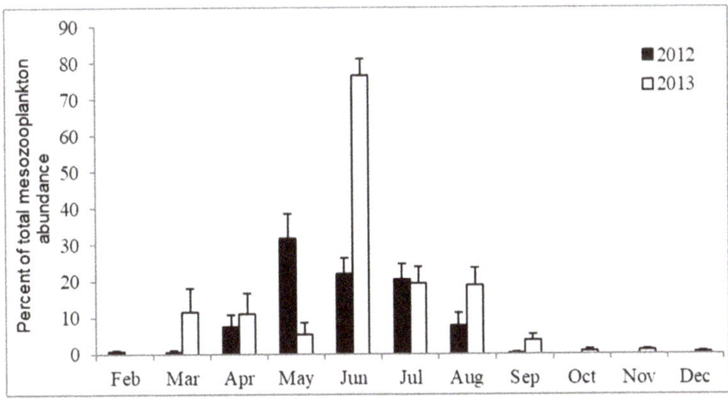

Figure 6. Percentage contribution by number of non-copepod taxa to total mesozooplankton abundance (+SE) in the Maryland Coastal Bays.

In 2012, barnacle nauplii were the most abundant non-copepods, with mean density of 810 ind. m^{-3} in Newport Bay (Figure 7a,b). Annual peaks were observed in April 2012 (368 ind. m^{-3}) and

September 2013 (102 ind. m^{-3}). The maximum barnacle nauplii density (4645 ind. m^{-3}) was attained in April 2012 in Newport Bay and mean abundance of barnacle nauplii was significantly higher in Newport Bay than in other embayments (ANOVA, p = 0.02). Although, barnacle nauplii were absent in fall 2012, they were present in all seasons in 2013 (Figure 8a–d). Maximum density in 2013 was 1119 ind. m^{-3} in September at Site 3 (Chincoteague Bay). Barnacle nauplii were more abundant in spring and summer than in the fall and winter (ANOVA, p < 0.001).

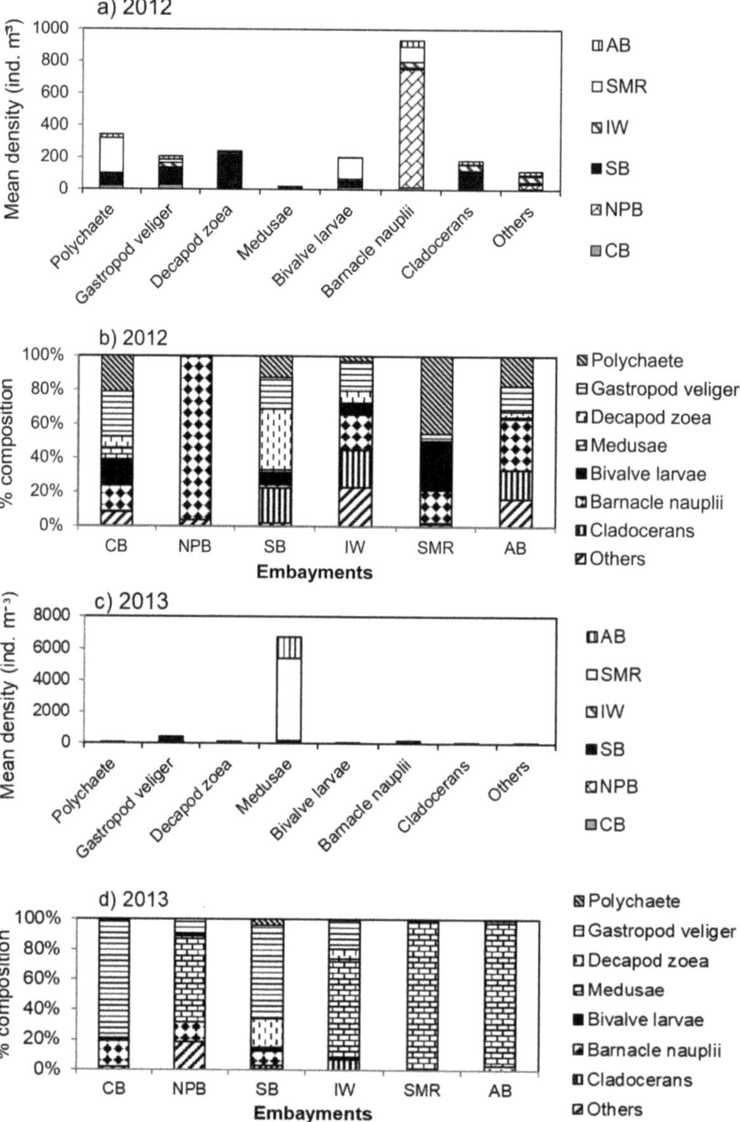

Figure 7. Spatial distribution, abundance (**a,c**) and percentage composition (**b,d**) of non-copepod mesozooplankton in the MCBs during 2012 (below average) and 2013 (above average) freshwater discharge from winter to summer. CB, Chincoteague Bay; NB, Newport Bay; SB, Sinepuxent Bay; IWB, Isle of Wight Bay; SMR, St. Martin River; AB, Assawoman Bay.

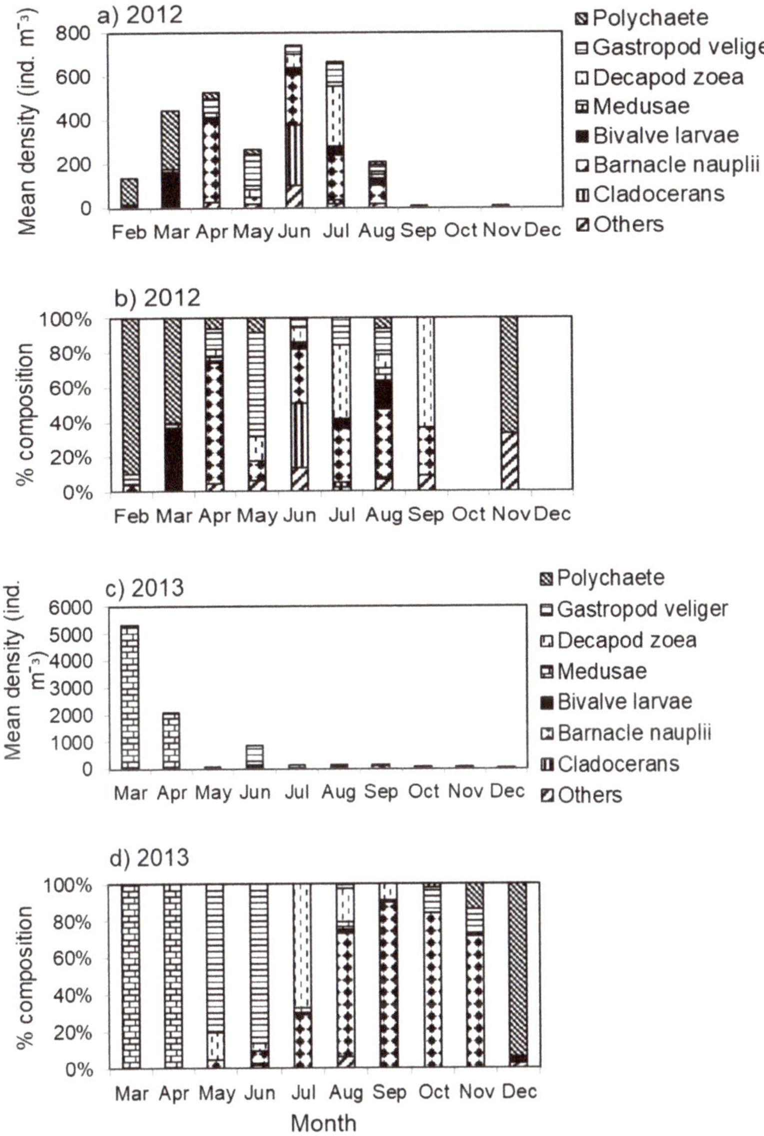

Figure 8. Temporal patterns in abundance (**a,c**) and percent contribution (**b,d**) of non-copepod mesozooplankton in the Maryland Coastal Bays during 2012 (year of below average freshwater discharge from winter to summer) and 2013 (year of above average freshwater discharge from winter to summer).

In 2013, hydromedusae were the most abundant non-copepod mesozooplankton (Figure 7c,d). Mean densities as high as 1367 ind. m^{-3} and 5156 ind. m^{-3} were recorded in Assawoman Bay and St. Martin River, respectively, in 2013. Hydromedusae were not observed at St. Martin River and Newport Bay sites in 2012 and also were not observed in Sinepuxent Bay in 2013 (Figure 7b,d). They were present from March to August in 2012 and from March to September in 2013, but were not found in May 2013 (Figure 8a–d). Average hydromedusae abundance was less than 20 ind. m^{-3} each month and at all sampling sites in 2012. Monthly mean densities peaked in April, 2012 (16 ind. m^{-3}) and in March 2013 (5252 ind. m^{-3}). The highest density recorded in 2013 (50,399 ind. m^{-3}, March, Site 10) was 298 times greater than the highest density (181 ind. m^{-3}, March, Site 5) in 2012.

Gastropod larvae were intermittently present and their monthly mean density reached a peak in May 2012 (158 no m^{-3}, Figure 8a). A secondary peak (103 ind. m^{-3}) occurred in July of the same year. Mean density in 2013 peaked in June (721 ind. m^{-3}). Gastropod larvae were abundant (107 ind. m^{-3}) in Sinepuxent Bay, but were not observed in Newport Bay in 2012 (Figure 7a–b). A maximum density (1299 ind. m^{-3}) of the larvae was observed in May 2012 at Site 7 (Sinepuxent). In 2013, maximum density (1852 ind. m^{-3}) occurred at Site 4 (Chincoteague Bay) in June. Gastropod larvae were least abundant in Assawoman Bay (2 ind. m^{-3}) in 2013.

Polychaete larvae occurred from February through August in 2012 and were most abundant (266 ind. m^{-3}) in March (Figure 8a). Mean polychaete larval densities never exceeded 10 ind. m^{-3} in 2013, occurred sporadically, and were most abundant in March (9 ind. m^{-3}). Polychaete larvae were more abundant in 2012 than 2013 (Mann–Whitney U-test, p = 0.0002). Peak numbers were 2384 ind. m^{-3} in March 2012 at Site 10 located in St. Martin River and 60 ind. m^{-3} in March 2013 at Site 5 in Chincoteague Bay (Figure 7a–d).

Bivalve larvae were present from February through August in 2012 (Figure 8a), but their occurrence was irregular in 2013 (Figure 8c) with mean monthly densities < 5 ind. m^{-3}. Bivalve larvae were better represented in plankton samples from 2012 than 2013 (Mann–Whitney U-test, p = 0.001) and relatively abundant in March during both years (Figure 7). Average density was 161 ind. m^{-3} in March 2012, and 4 ind. m^{-3} in March 2013. Maximum densities of 1490 ind. m^{-3} at Site 10 (St. Martin River) in 2012 and 58 ind. m^{-3} at Site 8 (Sinepuxent Bay) in 2013 were recorded in March. Bivalve larvae were only found in Sinepuxent Bay and Isle of Wight Bay (Site 9) in 2013, but their distribution in 2012 was widespread except at Sites 3 and 6 where they were not observed. Densities did not differ significantly amongst embayments (ANOVA, p = 0.12).

Cladocerans (*Evadne nordmanni*, *E. spinifera*, *Pseudoevadne tergestina*, *Pleopis polyphemoides*, *Podon intermedius*) were most abundant in Sinepuxent and Isle of Wight Bays (ANOVA, p = 0.002) during summer and were absent from Newport Bay (Figure 7b).

Decapod larvae (shrimps and crabs) were prevalent in Sinepuxent Bay where they occurred from March through November in 2012 and 2013 (Figure 7; Figure 8). In July, they comprised 42% and 67% of non-copepod zooplankton abundance, and 10% and 11% of total mesozooplankton abundance in 2012 and 2013, respectively. Larval stages of crabs *Panopeus herbstii*, *Neopanope texana*, *Uca* spp. and *Pinnixia* spp. were observed in zooplankton samples collected in 2012. Zoeae of *Callinectes sapidus*, *Rhithropanopeus harrisii*, *Ovalipes ocellatus*, *Cancer irroratus*, *Emerita* spp., *Hemigrapus* sp., *Libinia dubia*, *Ocydopode spp.*, *Petrolisthes armatus*, and *Lepidopa websteri* were also present.

Fish larvae were first observed in April 2012 (mean: 16 ind. m^{-3}) when they were most abundant (Figure 8). They were observed in December 2013 (0.1 ind. m^{-3}), but mean densities were very low. The highest density of fish larvae (202 ind. m^{-3}) was observed at Site 6 in Newport Bay (April 2012).

Larvaceans occurred from April to June 2012 and in August 2013. Average densities ranged from 1 to 5 ind. m^{-3}. The maximum density of larvaceans was 23 ind. m^{-3}, at Site 8 (Sinepuxent Bay) in August, 2013. Other non-copepods irregularly found in samples included amphipods, isopods, nematodes, and ostracods.

3.4. Relationships Between Non-Copepod Mesozooplankton and Environmental Factors

In 2012 (Figure 9a), the first two CCA axes explained 47.1% of the cumulative percentage variance of taxa-environment relationship and the correlations were 0.89 and 0.89 for the first and second axes, respectively. The biplot in 2012 showed that the main variability (29.6% of variance) was due to the positive association of larval shrimp and bivalves with salinity (p = 0.046). In contrast, barnacle nauplii and ostracods were negatively related to salinity (p < 0.05). Gastropod and polychaete larvae showed a positive relationship with pH (p = 0.022), while crab zoeae, *Evadne* spp. and *Pleopis* sp. related inversely to pH. Hydromedusae neither associated closely with salinity nor pH in 2012.

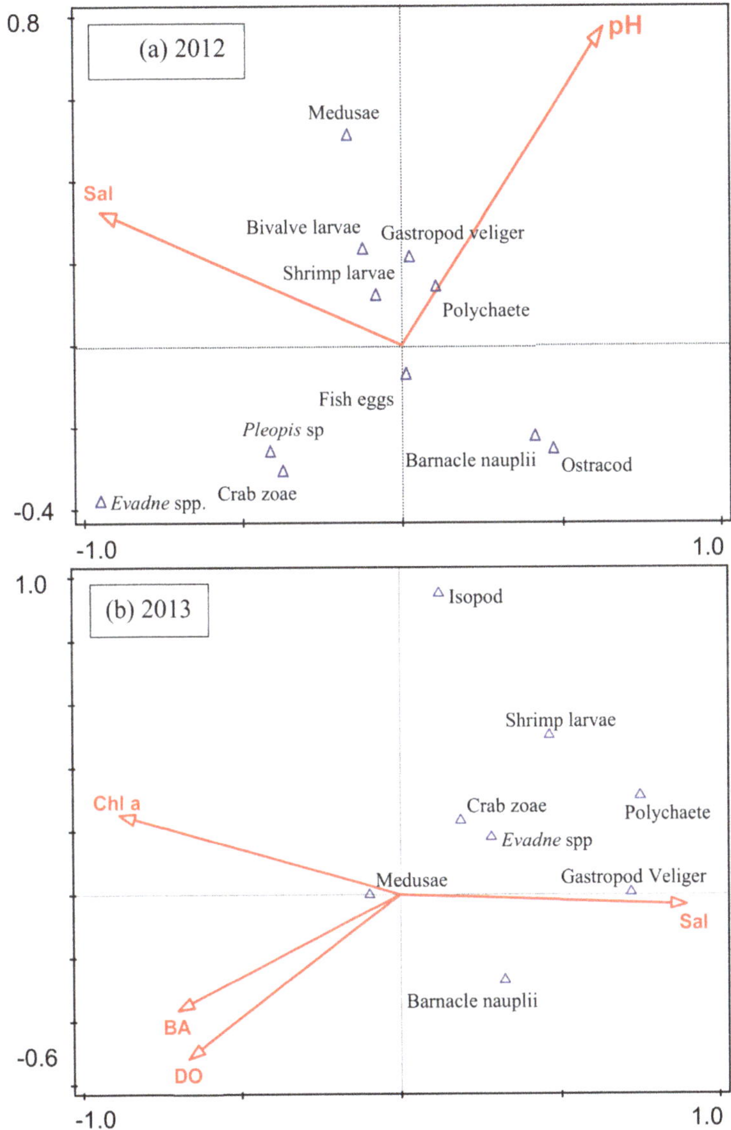

Figure 9. Results of CCA analysis showing environmental variables and taxa relationships in: 2012 (**a**); and 2013 (**b**) in MCBs. BA, bay anchovy; DO, Dissolved oxygen; Sal, Salinity; Chl *a*, Chlorophyll *a*.

In 2013 (Figure 9b), the first two axes of the CCA explained 70% of the cumulative variation in taxa-environment relationship, and the correlation was 0.98 for the first axis and 0.91 for the second axis. The taxa-environment biplot (Figure 9b.) showed that the variability (63.7% of variance) was accounted for by the positive association of gastropod veliger with salinity (p = 0.026), and inverse relationship of hydromedusae to salinity (p < 0.05). Barnacle larvae were negatively aligned to Chl. *a* (p = 0.034); bay anchovy larvae were positively correlated to DO (p = 0.04), which was also plotted inversely with cladocerans, decapod zoae and polychaete larvae. Temperature, pH, and ctenophores correlated strongly with DO and were not selected by the model.

3.5. Mesozooplankton Diversity

The diversity of mesozooplankton varied temporally and spatially and was affected by salinity such that diversity was higher at higher salinity than at lower salinity (Figure 10). The Shannon–Wiener diversity index was high (≥2.5) in 2012 at all sites except at Site 6 (H′ = 2.2) in Newport Bay with the lowest diversity (Figure 10). Diversity index value in Newport Bay (Site 6) in 2012 (H′ = 2.2) was slightly lower than in 2013 (H′ = 2.3). On the average, diversity was significantly higher (t = 6.2; p ≤ 0.001) in 2012 than in 2013. Sinepuxent Bay supported the highest diversity (H′ = 2.8 in 2012; H′ = 2.6 in 2013) of mesozooplankton in both years. Diversity decreased from Sinepuxent Bay (Sites 7 and 8) close to the Ocean City Inlet to Assawoman Bay (Site 13) in the northern MCBs in both 2012 and 2013. Likewise, diversity in Chincoteague Bay decreased from Site 5 to Sites 1 and 2 in both years.

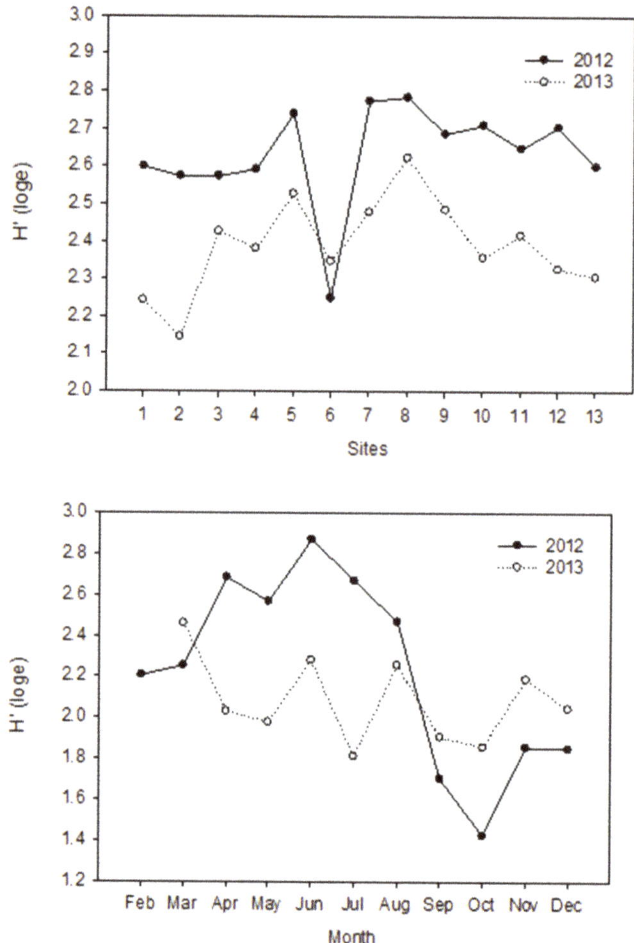

Figure 10. Spatial (**top**) and seasonal (**bottom**) patterns in Shannon–Wiener diversity index of mesozooplankton in the MCBs in 2012 and 2013 from winter to summer.

Diversity was higher in spring and summer (H′ > 2.4) than in fall (H′ < 1.9) of 2012. In 2013, diversity in March and from September to December was higher (t = −2.3; p ≤ 0.06) than diversity in 2012. Periods of maximum and minimum diversity index values were different in 2012 and 2013. Mesozooplankton was most diverse in June (H′ = 2.9) and March (H′ = 2.5), but least diverse in October (H′ = 1.4) and July (H′ = 1.8) in 2012 and 2013, respectively.

When examined spatially, diversity indices values in 2012 were significantly correlated negatively with Chl. *a* (r = −0.7, p = 0.01) and positively with salinity (r = 0.7, p = 0.02), but not with But-fuco (r = −0.5, p = 0.08). In 2013, diversity also correlated positively with salinity (r = 0.3, p < 0.050). All other environmental factors were weakly correlated with mesozooplankton spatial diversity in 2013, and no significant relationship was observed (p > 0.05). With respect to monthly diversity, H′ was significantly correlated with ctenophores (r = 0.7, p = 0.02) in 2012, but not in 2013 (r = − 0.1, p = 0.9). Other variables such as temperature (r = − 0.4, p = 0.2) and salinity (r = − 0.4, p = 0.3) showed weak negative correlations with diversity in 2013.

4. Discussion

This is the first study to describe mesozooplankton assemblage dynamics in the MCBs. The mesozooplankton composition, abundance and diversity varied spatially and temporally in the MCBs, driven in part by fluctuations in environmental factors, particularly salinity. Lower salinity due to increased freshwater input coupled with higher ambient temperature in 2013 (Figure 2) relative to that of 2012, perhaps promoted high densities of hydromedusae in winter and early spring. Gelatinous zooplankton can exert high predatory pressure on crustacean zooplankton species as well as on gastropod veligers, barnacle nauplii, polychaete larvae, and ichthyoplankton (fish larvae) [32,43–47]. The feeding rate of ctenophores, in particular, increases with water temperature [48], possibly contributing to the decline in annual peak abundance observed for some non-copepod taxa (polychaete larvae, ostracods, bivalve and fish larvae). Nevertheless, the high freshwater inflow associated with Hurricane Sandy in fall 2012 accompanied by the relatively low salinity that occurred through spring 2013 might have contributed to the observed low abundances of the taxa in the MCBs.

4.1. Mesozooplankton Community Composition and Taxa-Environment Relationships

The percent contribution of non-copepods (e.g., decapod larvae, cladocerans) to mesozooplankton abundance in this study was highest in late spring and early summer when copepod density was relatively low [42]. Similar observations were made in the lower Narragansett Bay [16] and in other mid-Atlantic estuaries [6,19,35,49]. The increase in non-copepod taxa relative abundance that occurred when copepod density was low is important since they serve as alternate prey for zooplanktivorous fishes. Bay anchovy is abundant in summer in MCBs [50], and the juveniles usually make up the majority of estuarine bay anchovy population biomass in summer and fall [51]. Although copepods contribute the bulk of bay anchovy's diet, non-copepod taxa, especially larvae of barnacles, bivalves, and polychaetes, as well as amphipods, crab zoeae, mysids, ostracods, cladocerans, and ichthyoplankton seasonally dominate the diet in mid-Atlantic estuaries such as Long Island Sound, Cheseapeake Bay and MCBs [8,9,52–55].

Temporal variations in freshwater discharge and salinity influenced the relative abundance of mesozooplankton taxa. For example, Barnacle nauplii dominated the non-copepod community, especially in Newport Bay from January to August 2012. During this period, freshwater discharge into MCBs was relatively low and salinity was high. Naupliar stages of barnacles were absent in fall of 2012, which might have been due to the dramatic reduction in salinity caused by excessive rainfall that was associated with Hurricane Sandy that increased freshwater inputs into the system possibly increasing mortality of the larvae and perhaps the adult stages. Mortality of barnacle larvae is higher at low salinity (<6) than at higher (>14) salinity [56], and high rainfall with associated high freshwater discharge has been reported to depress the abundance of some species of barnacles. Higher densities of barnacles were collected during dry season than wet season in coastal lagoons in Panama [56] and in Lagos Harbor, Nigeria [57]. The salinity observed in the MCBs in spring of 2013 (wet year) was much higher (23–26) than the levels observed during the rainy season in these other lagoons, though the salinity in MCBs during the period Hurricane Sandy occurred might have been much lower than the values we recorded after the event. Spring and summer peaks of barnacle naupliar observed in the MCBs were also reported for the Delaware River estuary [19]. In 2013, a year of above average freshwater discharge into MCBs and lower salinity, hydromedusae dominated non-copepod community particularly in Assawoman Bay and at the mouth of St. Martin River. Larvae of polychaetes and bivalves, in addition to ostracods, and larvaceans, were better represented in the plankton in 2012 when salinity was higher, than in 2013 when salinity was lower. In estuaries that experience much horizontal gradient in salinity, seawater penetration can bring in zooplankton species from the coastal ocean during periods of low freshwater inputs [11].

There were seasonal changes in the relative abundance of mesozooplankton taxa in the MCBs. Hydromedusae were present from winter through summer, but scarce in fall (October–December). Hydromedusae were abundant in Delaware Bay River estuary during winter [18] and in Long Island estuary in April [49]. Polychaete and bivalve larvae were also abundant in winter to spring, with a shift in seasonal dominance from polychaetes to barnacles, gastropod larvae, crab zoea, barnacles, shrimps, and finally polychaete larvae in 2012. This seasonal pattern was different in 2013 when a shift in dominance of the major taxa occurred from hydromedusae to gastropod larvae, crab zoea, barnacles, and finally polychaetes.

The spatial distributions of mesozooplankton taxa were related to variations in salinity. Larvaceans were positively correlated to salinity, and found only in Sinepuxent Bay close to the Ocean City inlet in 2013. Bivalve larvae also showed a similar preference for high salinity bay areas in 2013, and gastropod larval densities decreased at sites near the mouths of MCBs tributaries. The relatively high abundance of barnacle nauplii in Newport Bay and St. Martin River in both years as well as the occurrence of polychaete and bivalve larvae in high abundance in St. Martin River in the dry year (2012) and within the open waters of Sinepuxent Bay in the wet year (2013), suggests that salinity is an important factor that influenced their distributions and abundance. As salinity increases towards the ocean, clams, crustaceans and polychaete worms dominate the benthos of the coastal bays [27], thus accounting for the high abundance of meroplanktonic larvae of these organisms around Sinepuxent Bay. The occurrence of gastropod and polychaete larvae in high abundance in embayments closest to the coastal ocean (downstream) is contrary to the findings in Pagan River, a sub-estuary of Chesapeake Bay [17] and in Mondego estuary, Western Portugal [58] where the larvae were abundant upstream, although this might be due to differences in the species reported in the systems. In the Delaware River estuary, polychaete larvae occurred from the entrance of the river to the middle of the bay and peaked in abundance from winter to spring [19]. The lowest salinity at which they were observed was 24.1. In this study, annual peaks in the abundance of polychaete larvae were recorded at salinity levels >30.

4.2. Occurrence of Blue Crab Larvae in the MCBs

Blue crab, *C. sapidus* zoeae, were common in July in this study which is similar to the month during which they were observed by other investigators in nearby estuaries [18,37]. The study confirms what was previously suspected based on the capture of gravid females by the MD DNR that blue crabs spawn in the MCBs. Blue crab zoeae were most prevalent in Sinepuxent Bay and the northern bays and rarely occurred in southern bay areas with adequate salinity for larval development, although no sampling occurred near the Chincoteague Inlet in Virginia. This supports results of previous studies that indicated that blue crabs hatch their eggs near the mouth of estuaries [6,22]. The occurrence of zoeae at sites located in Assawoman Bay close to tributary creeks, however, suggests that spawning occurs not only close to the Ocean City Inlet, but also in areas that are some distance away from the inlet. It is also possible that crab larvae that hatched near the inlets were dispersed into Assawoman Bay by waves and tidal action. The proximity of spawning locations to the inlet enhances easy transportation of zoeae from the MCBs into the nearby continental shelf waters where further development takes place [22,25]. Additional studies are needed, however, to determine if blue crab larvae complete their development from zoeae to megalopae within the MCBs.

4.3. Diversity of Mesozooplankton

Mesozooplankton assemblage was most diverse in spring and summer of 2012 when salinity was relatively high, especially in areas close to the Ocean City Inlet, connecting the bays to the Atlantic Ocean. Intrusion of seawater in estuaries can act as a major driver for zooplankton diversity [59]. In Pearl River estuary China, zooplankton diversity and species richness were enhanced in the middle and lower portions with high diversity at salinity level >25 [11]. The higher zooplankton diversity in spring and summer months in 2012 might have been contributed by the increase in the abundance of cladocerans, decapod larvae, and other meroplanktonic forms in the system during that period, some of which were likely transported into the bays from coastal ocean by tidal action.

The most diverse assemblage was observed in areas with high salinity, but low chlorophyll *a*. The observed negative correlation between chlorophyll *a* and diversity was because chlorophyll *a* levels were higher in areas close to MCBs tributaries (Newport Bay and St. Martin River) with lower salinity and high nutrient levels due to freshwater inflow [38,60], which also had lower mesozooplankton diversity than in areas close to the inlets. Howson et al. [35] observed higher zooplankton diversity in the southern bay of the Barnegat Bay, NJ, with more oceanic influence and less anthropogenic impact, than in the more nutrient enriched northern bay.

4.4. Implications of Temporal Variability in Environmental Factors on Mesozooplankton

Variability in freshwater discharge altered salinity and influenced the abundance and distribution of mesozooplankton in the MCBs. High freshwater discharge that began in fall 2012 and continued through spring 2013 lowered salinity that likely favored hydromedusae whose abundance was highest in March and April, but perhaps disfavored bivalve and barnacle larvae. During this period, salinity was about 23 (March) and 26 (April) compared with about 30.5 (March) and 34 (April) in 2012. Salinity has been reported to influence the abundance and distribution of hydromedusae in other estuaries [61,62]. In Mississippi Sound, the incidence of hydromedusae was highest at 25.1 to 30 below and above which the abundance decreased [61]. With the occurrence and high abundance of hydromedusae, ctenophores [42] and scyphozoan jellyfish in the MCBs, and their documented negative impacts on the standing stock of planktonic communities [31,45,47,48], the slow rate of recovery of the hard clam population abundance in the MCBs is not surprising. This is because, peak abundance of bivalve larvae (March) coincided with the increase in the abundance of hydromedusae (March and April), which was followed by ctenophores (April–June) especially in 2013 when salinity was relatively low [42]. Hydromedusae are known predators of zooplankton [44] and when numerically abundant, are able to exert their potential maximum clearance impact [63]. Ctenophores as predators of bivalve larvae [31,43,64] cause high mortality, with a potential of consuming up to 94.1% of the larvae during spawning periods [32].

Non-copepod zooplankton are important prey items for carnivorous zooplankton, larval fish, and adult planktivorous and forage fish such as bay anchovy. These organisms are relatively abundant when copepod abundance is low in the MCBs. During years with high freshwater input, the MCBs salinity levels in some areas may not be favorable for the development of some decapod crab zoeae and bivalve larvae (*Mercenaria mercenaria*) that prefer high salinity. In addition, the temperature range (~17–30 °C) tolerated by *M. mercenaria* larvae is reduced when salinity decreases from the optimum range of 26–27 [65]. Thus, increased temperature, reduced salinity as well as high densities of gelatinous zooplankton (hydromedusae and ctenophores) can affect the breeding success of finfish and shellfish species in the MCBs. Gelatinous zooplankton can change food web dynamics and decrease recruitment of fish and shellfish populations by preying on the larval forms or via competition for zooplankton prey [45,46].

5. Conclusions

Results of this study indicate that variations in freshwater discharge affect the abundance, distribution and diversity of mesozooplankton in MCBs. These changes in mesozooplankton structure likely influence trophic relationships and recruitment of finfish and shellfish larvae to the adult populations. The percent contribution of non-copepods to mesozooplankton abundance in MCBs was highest in late spring and early summer when copepod density was relatively low. The composition and abundance of mesozooplankton differed between 2012 and 2013 such that barnacle larvae were the most abundant of non-copepod mesozooplankton, whereas hydromedusae were the most abundant group in 2013. Mesozooplankton diversity varied temporally and spatially, and was higher at higher salinity than at lower salinity. Seasonally, diversity was higher in spring and summer 2012 than in fall. Additionally, in 2012 when salinity was comparatively high because of low precipitation, diversity was higher than in 2013 when salinity was relatively low. Blue crab zoeae were common in the MCBs in the summer, especially in Sinepuxent Bay and the northern bays confirming that the adults spawn in the MCBs, although further studies are needed to determine if they complete their life cycle in the bays. Additional studies are also needed to investigate the predation rates of ctenophores, jellyfish and zooplanktivores such as bay anchovy on various zooplankton taxa. This will be useful for understanding the role of non-copepod community as alternate prey for fish in the MCBs during spring and summer minimum of copepods, and will also provide more information on the impact of predation on the survival of bivalve larvae and their recruitment to the adults.

Author Contributions: This study was conducted as a joint effort of all authors. Conceptualization, P.C. and E.U.O.; Methodology, E.U.O. and P.C.; Software, E.U.O.; Validation, E.U.O. and P.C.; Formal Analysis, E.U.O.; Investigation, E.U.O. and P.C; Resources, P.C.; Data Curation, E.U.O.; Writing—Original Draft Preparation, E.U.O.; Writing—Review and Editing, E.U.O. and P.C.; Visualization, E.U.O.; Supervision, P.C.; Project Administration, P.C.; and Funding Acquisition, P.C.

Funding: This publication was made possible by the National Science Foundation Center for Research Excellence in Science and Technology (NSF-CREST) award number (1036586) to the Center for the Integrated Study of Ecosystem Processes and Dynamics, University of Maryland Eastern Shore, and in part by the National Oceanic and Atmospheric Administration, Office of Education, Educational Partnership Program award number (NA16SEC4810007) to the Living Marine Resources Cooperative Science Center at the University of Maryland Eastern Shore. Its contents are solely the responsibility of the award recipient and do not necessarily represent the official views of the U.S. Department of Commerce, National Oceanic and Atmospheric Administration.

Acknowledgments: We would like to thank Kam Tang and Jamie Pierson for their valuable comments on the earlier draft of this manuscript, and Capt. Christopher Daniels (Boat Captain), Ozuem Oseji, Andres Morales-Nunez, Kingsley Nkeng, Chinwe Otuya, Abena Okyere Acheampong and Jessica Mazile for help with collection and enumeration of plankton samples. Data for bay anchovy were provided by Steve Doctor, Maryland Department of Natural Resources (MD DNR). Young Tsuei (DOEE) and Tracy Bishop (UMES) assisted with creation of map of the sampling sites.

Conflicts of Interest: The authors declare no conflict of interest. The funders had no role in the design of the study; in the collection, analyses or interpretation of data; in the writing of the manuscript or in the decision to publish the results.

References

1. Johnson, W.S.; Allen, D.M. *Zooplankton of the Atlantic and Gulf Coasts: A Guide to Their Identification and Ecology,* 2nd ed.; The Johns Hopkins University Press: Baltimore, MD, USA, 2012; 452p.
2. Kleppel, G.S. On the diets of calanoid copepods: Review. *Mar. Ecol. Prog. Ser.* **1993**, *99*, 183–195. [CrossRef]
3. Durbin, E.; Kane, J. Seasonal and spatial dynamics of *Centropages typicus* and *C. hamatus* in the western North Atlantic. *Prog. Oceanogr.* **2007**, *72*, 249–258. [CrossRef]
4. Verity, P.G.; Smetacek, V. Organism life cycles, predation, and the structure of marine pelagic ecosystems. *Mar. Ecol. Prog. Ser.* **1996**, *130*, 277–293. [CrossRef]
5. Bosch, H.F.; Taylor, W.R. Marine cladocerans in the Chesapeake Bay estuary. *Crustaceana* **1968**, *15*, 161–164. [CrossRef]
6. Sandifer, P.A. Distribution and abundance of decapod crustacean larvae in the York River Estuary and Adjacent Lower Chesapeake Bay, Virginia, 1968–1969. *Chesap. Sci.* **1973**, *14*, 235–257. [CrossRef]

7. Steinberg, D.K.; Condon, R.H. Zooplankton of the York River. *J. Coast. Res.* **2009**, *57*, 66–79. [CrossRef]

8. Hartman, K.J.; Howell, J.; Sweka, J.A. Diet and daily ration of bay anchovy in the Hudson River, New York. *Am. Fish. Soc.* **2004**, *133*, 762–771. [CrossRef]

9. Klebasko, M.J. Feeding Ecology and Daily Ration of Bay Anchovy (*Anchoa mitchilli*) in the Mid-Chesapeake Bay. Master' Thesis, University of Maryland, College Park, MD, USA, 1991.

10. Mallin, M.A. Zooplankton abundance and community structure in a mesohaline North Carolina estuary. *Estuaries* **1991**, *14*, 481–488. [CrossRef]

11. Li, K.Z.; Yin, J.Q.; Huang, L.M.; Tan, Y.H. Spatial and temporal variations of mesozooplankton in the Pearl River estuary, China. *Estuar. Coast. Shelf Sci.* **2006**, *67*, 543–552. [CrossRef]

12. Kimmel, D.G.; Roman, M.R.; Zhang, X. Spatial and temporal variability in factors affecting mesozooplankton dynamics in Chesapeake Bay: Evidence from biomass size spectra. *Limnol. Oceanogr.* **2006**, *51*, 131–141. [CrossRef]

13. Elliot, D.T.; Kaufman, R.S. Spatial and temporal variability of mesozooplankton and tintinnid ciliates in a seasonally hypersaline estuary. *Estuaries Coasts* **2007**, *30*, 418–430. [CrossRef]

14. Shiganova, T.A. Invasion of the Black Sea by the ctenophore *Mnemiopsis leidyi* and recent changes in pelagic community structure. *Fish. Oceanogr.* **1998**, *7*, 305–310. [CrossRef]

15. Sage, L.E.; Herman, S.S. Zooplankton of the Sandy Hook Bay area, N.J. *Chesap. Sci.* **1972**, *13*, 29–39. [CrossRef]

16. Hulsizer, E. Zooplankton of the lower Narragansett Bay, 1972–1973. *Chesap. Sci.* **1976**, *17*, 260–270. [CrossRef]

17. Davis, L.N.; Marshall, H.G. Mesozooplankton distribution and abundance in the Pagan River: A nutrient enriched sub-estuary of the James River, Virginia. *Va. J. Sci.* **1998**, *49*, 151–162.

18. Deevey, G.B. The zooplankton of the surface waters of the Delaware Bay region. *Bull. Bingham Oceanogr. Collect.* **1960**, *17*, 5–53.

19. Cronin, L.E.; Daiber, J.D.; Hulbert, E.M. Quantitative seasonal aspects of zooplankton in the Delaware River Estuary. *Chesap. Sci.* **1962**, *3*, 63–93. [CrossRef]

20. Sandoz, M.; Rogers, R. The effect of environmental factors on hatching, moulting, and survival of zoea larvae of the blue crab *Callinectes sapidus* Rathbun. *Ecology* **1944**, *25*, 216–228. [CrossRef]

21. Costlow, J.D., Jr. The effect of salinity and temperature on survival and metamorphosis of megalops of the blue crab *Callinectes sapidus*. *Helgol Wiss Meeresunters* **1967**, *15*, 84–97. [CrossRef]

22. McConaugha, J.R.; Johnson, D.F.; Provenzano, A.J.; Maris, R.C. Seasonal distribution of larvae of *Callinectes sapidus* (Crustacea; Decapoda) in the waters adjacent to Chesapeake Bay. *J. Crustacean Biol.* **1983**, *3*, 582–591. [CrossRef]

23. Sulkin, S.D.; Van Heukelem, W.; Kelly, P.; Van Heukelem, L. The behavioral basis of larval recruitment in the crab *Callinectes sapidus* Rathbun: A laboratory investigation of ontogenetic changes in geotaxis and barokinesis. *Biol. Bull.* **1980**, *159*, 402–417. [CrossRef]

24. Provenzano, A.J.; McConaugha, J.R.; Philips, K.; Johnson, D.F.; Clark, J. Vertical distribution of first stage larvae of the blue crab, *Callinectes sapidus* at the mouth of Chesapeake Bay. *Estuar. Coast. Shelf Sci.* **1983**, *16*, 489–500. [CrossRef]

25. Epifanio, C.E. Transport of blue crab *Callinectes sapidus* larvae in the waters off Mid-Atlantic States. *Bull. Mar. Sci.* **1995**, *57*, 713–725.

26. MD DNR, Maryland Department of Natural Resources Coastal Bays Fisheries Advisory Committee. *Coastal Bays Blue Crab Fisheries Management Plan*; MD DNR: Annapolis, MD, USA, 2001; 44p.

27. Jesien, R.V.; Bollinger, A.; Brinker, D.F.; Casey, J.F.; Doctor, S.B.; Etgen, C.P.; Eyler, T.B.; Hall, M.R.; Hoffman, M.L.; Kilian, J.V.; et al. Diversity of Life in the Coastal Bays. In *Shifting Sands: Environmental and Cultural Changes in Maryland's Coastal Bays*; Dennison, W.C., Thomas, J.E., Cain, C.J., Carruthers, T.J.B., Hall, M.R., Jesien, R.V., Wazniak, C.E., Wilson, D.E., Eds.; IAN Press; University of Maryland Center for Environmental Science (UMCES): Cambridge, MD, USA, 2009; pp. 293–344.

28. Boynton, W.R.; Hagy, J.D.; Murray, L.; Stokes, C.; Kemp, W.M. A comparative analysis of eutrophication patterns in a temperate coastal lagoon. *Estuaries* **1996**, *19*, 408–421. [CrossRef]

29. Wazniak, C.; Wells, D.; Hall, M. The Maryland coastal bays ecosystem. In *Maryland Coastal Bays Ecosystem Health Assessment*; Wazniak, C., Hall, M., Eds.; DNR-12-1202-0009; Maryland Department of Natural Resources: Annapolis, MD, USA, 2004.

30. Glibert, P.M.; Hinkle, D.C.; Sturgis, B.; Jesien, R.V. Eutrophication of a Maryland/Virginia Coastal Lagoon: A tipping point, ecosystem changes, and potential Causes. *Estuaries Coasts* **2014**, *37*, S128–S146. [CrossRef]

31. Quaglietta, C.E. Predation by *Mnemiopsis leidyi* on Hard Clam Larvae and Other Natural Zooplankton in Great South Bay, NY. Master's Thesis, State University of New York, Stony Brook, NY, USA, 1987.

32. McNamara, M.E.; Lonsdale, D.J.; Cerrato, R.M. Shifting abundance of the ctenophore *Mnemiopsis leidyi* and its implications for larval bivalve mortality. *Mar. Biol.* **2010**, *157*, 401–412. [CrossRef]

33. Kang, X.; Xia, M.; Pitula, J.S.; Chigbu, P. Dynamics of water and salt exchange in Maryland Coastal Bays. *Estuar. Coast. Shelf Sci.* **2017**, *189*, 1–16. [CrossRef]

34. Thomas, J.E.; Woerner, J.L.; Abele, R.W.; Blazer, D.P.; Blazer, G.P.; Cain, C.J.; Dawson, S.L.; McGinty, M.K.; Schupp, C.A.; Spaur, C.C.; et al. Isle of Wight Bay. In *Shifting Sands: Environmental and Cultural Changes in Maryland's Coastal Bays*; Dennison, W.C., Thomas, J.E., Cain, C.J., Carruthers, T.J.B., Hall, M.R., Jesien, R.V., Wazniak, C.E., Wilson, D.E., Eds.; IAN Press; University of Maryland Center for Environmental Science (UMCES): Cambridge, MD, USA, 2009; pp. 101–116.

35. Howson, U.A.; Buchanan, G.A.; Nickels, J.A. Zooplankton community dynamics in a western mid-Atlantic lagoonal estuary. *J. Coast. Res.* **2017**, *78*, 141–168. [CrossRef]

36. Egloff, D.A.; Fofonoff, P.W.; Onbé, T. Reproductive biology of marine cladocerans. *Adv. Mar. Biol.* **1997**, *31*, 79–168.

37. Sandifer, P.A. Morphology and Ecology of Chesapeake Bay Decapod Crustacean Larvae. Ph.D. Thesis, University of Virginia, Charlottesville, VA, USA, 1972.

38. Oseji, O.F.; Fan, C.; Chigbu, P. Composition and dynamics of phytoplankton in the coastal bays of Maryland, USA, revealed by microscopic counts and diagnostic pigments analyses. *Water* **2019**, *11*, 368. [CrossRef]

39. Ter Braak, C.J.F.; Šmilauer, P. *Canoco Reference Manual and User's Guide: Software for Ordination, Version 5.0*; Microcomputer Power: Ithaca, NY, USA, 2012.

40. Shannon, C.E. A mathematical theory of communication. *Bell Syst. Technol. J.* **1948**, *27*, 379–423. [CrossRef]

41. Clarke, K.R.; Gorley, R.N. *PRIMER v6: User Manual/Tutorial (Plymouth Routines in Multivariate Ecological Research*; PRIMER-E Limited: Plymouth, UK, 2006.

42. Oghenekaro, E.U.; Chigbu, P.; Oseji, O.F.; Tang, K.W. Seasonal factors influencing the abundance of copepods in the Maryland Coastal Bays. *Estuaries and Coasts* **2018**, *41*, 495–506. [CrossRef]

43. Burrell, V.G., Jr.; Van Engel, W.A. Predation by and distribution of a ctenophore, *Mnemiopsis leidyi* A. Agassiz, in the York River estuary. *Estuar. Coast. Mar. Sci.* **1976**, *4*, 235–242. [CrossRef]

44. Mills, C.E. Medusae, siphonophores and ctenophores as planktonorous predators in changing global ecosystems. *ICES J. Mar. Sci.* **1995**, *52*, 575–581. [CrossRef]

45. Purcell, J.E.; Shiganova, T.A.; Decker, M.B.; Houde, E.D. The ctenophore *Mnemiopsis* in native and exotic habitats: U.S. estuaries versus the Black Sea basin. *Hydrobiologia* **2001**, *451*, 145–176. [CrossRef]

46. Purcell, J.E.; Arai, M.N. Interactions of pelagic cnidarians and ctenophores with fishes: A review. *Hydrobiologia* **2001**, *451*, 24–27.

47. Purcell, J.E.; Decker, M.B. Effects of climate on relative predation by scyphomedusae and ctenophores on copepods in Chesapeake Bay during 1987–2000. *Limnol. Oceanogr.* **2005**, *50*, 376–387. [CrossRef]

48. Kremer, P. Predation by the ctenophore *Mnemiopsis leidyi* in Narragansett Bay, Rhode Island. *Estuaries* **1979**, *2*, 97–105. [CrossRef]

49. Turner, J.T. The annual cycle of zooplankton in a Long Island Estuary. *Estuaries* **1982**, *5*, 261–274. [CrossRef]

50. Murphy, R.F.; Secor, D.H. Fish and Blue crab assemblage structure in a U.S. mid-Atlantic Coastal lagoon complex. *Estuaries Coasts* **2005**, *29*, 1121–1131. [CrossRef]

51. Vouglitois, J.J.; Able, K.W.; Kurtz, R.J.; Tighe, K.A. Life history and population dynamics of the bay anchovy in New Jersey. *Trans. Am. Fish. Soc.* **1987**, *116*, 141–153. [CrossRef]

52. Kinch, J.C. Trophic habits of the juvenile fishes within artificial waterways: Marco Island, Florida. *Contrib. Mar. Sci.* **1979**, *22*, 77–90.

53. Vasquez, A.V. Energetics, Trophic Relationships and Chemical Composition of Bay Anchovy, *Anchoa mitchilli*, in Chesapeake Bay. Master's Thesis, University of Maryland, College Park, MD, USA, 1989.

54. O'Brien, M. Marine Influence on Juvenile Fish Trophic Ecology and Community Dynamics in Maryland's Northern Coastal Lagoons. Master's Thesis, Marine Estuarine Environmental Science, University of Maryland, College Park, MD, USA, 2013.

55. Mayor, E.D. Population Dynamics of Mysids and Their Role in the Trophic Ecology of Fishes in Maryland Coastal Bays. Ph.D. Thesis, University of Maryland Eastern Shore, Princess Anne, MD, USA, 2015.

56. Starczak, V.; Pérez-Brunius, P.; Levine, H.E.; Gyory, J.; Pineda, J. The role of season and salinity in influencing barnacle distributions in two adjacent coastal mangrove lagoons. *Bull. Mar. Sci.* **2011**, *87*, 275–299. [CrossRef]

57. Sandison, E.E. The effect of salinity fluctuations on the life cycle of *Balanus pallidus stutsburi* Darwin in Lagos Harbour, Nigeria. *J. Anim. Ecol.* **1966**, *35*, 363–378. [CrossRef]

58. Vieira, L.; Azeiteiro, U.; Ré, P.; Pastorinho, R.; Marques, J.C.; Morgado, F. Zooplankton distribution in a temperate estuary (Mondego estuary southern arm: Western Portugal). *Acta Oecol.* **2003**, *24*, S163–S173. [CrossRef]

59. Hwang, J.S.; Kumar, R.; Hsieh, C.W.; Kuo, A.Y.; Souissi, S.; Hsu, M.H.; Wu, J.T.; Liu, W.C.; Wang, C.F.; Chen, Q.C. Patterns of zooplankton distribution along the estuarine, marine and riverine portions of Danshuei ecosystem in northern Taiwan. *Zool. Stud.* **2010**, *49*, 335–352.

60. Oseji, O.F.; Chigbu, P.; Oghenekaro, E.; Waguespack, Y.; Chen, N. Spatial and temporal patterns of phytoplankton composition and abundance in the Maryland Coastal Bays: The influence of freshwater discharge and anthropogenic activities. *Estuar. Coast. Shelf Sci.* **2018**, *207*, 119–131. [CrossRef]

61. Gunter, G.; Ballard, B.S.; Venkataramiah, A. A review of salinity problems of organisms in United States coastal areas subject to the effects of engineering works. *Gulf Res. Rep.* **1974**, *4*, 380–475. [CrossRef]

62. Santhakumari, V.; Tiwari, L.R.; Nair, V.R. Species composition, abundance and distribution of hydromedusae from Dharamtar estuarine system, adjoining Bombay harbor. *Indian J. Mar. Sci.* **1999**, *28*, 158–162.

63. Hansson, L.J.; Moeslund, O.; Kiørboe, T.; Riisgard, H.U. Clearance rates of jellyfish and their potential predation impact on zooplankton and fish larvae in a neritic ecosystem (Limfijorden, Denmark). *Mar. Ecol. Prog. Ser.* **2005**, *304*, 117–131. [CrossRef]

64. Nelson, T.C. On the occurrence and food habits of ctenophores in New Jersey inland coastal waters. *Biol. Bull.* **1925**, *48*, 92–111. [CrossRef]

65. Eversole, A.G. *Species Profile: Life History and Environmental Requirements of Coastal Fishes and Invertebrates (South Atlantic)—Hard Clam*; TR EL-82-4; U.S. Army Corps of Engineers: Washington, DC, USA, 1987; 33p.

Article

Dynamics of *Calanus* Copepodite Structure during Little Auks' Breeding Seasons in Two Different Svalbard Locations

Kaja Balazy *, Emilia Trudnowska and Katarzyna Błachowiak-Samołyk

Institute of Oceanology, Polish Academy of Sciences, Powstańców Warszawy 55, 81-712 Sopot, Poland
* Correspondence: kaja@iopan.pl; Tel.: +48-58-7311696

Received: 15 May 2019; Accepted: 3 July 2019; Published: 9 July 2019

Abstract: Populations dynamics of key zooplankton species in the European Arctic, *Calanus finmarchicus* and *Calanus glacialis* (hereafter defined as *Calanus*) may be sensitive to climate changes, which in turn is of great importance for higher trophic levels. The aim of this study was to investigate the complete copepodite structure and dynamics of *Calanus* populations in terms of body size, phenology and their relative role in the zooplankton community over time in different hydrographic conditions (two fjords on the West Spitsbergen Shelf, cold Hornsund vs. warm Kongsfjorden), from the perspective of their planktivorous predator, the little auk. High-resolution zooplankton measurements (taken by nets and a laser optical plankton counter) were adapted to the timing of bird's breeding in the 2015 and 2016 summer seasons, and to their maximal diving depth ($\leq$50 m). In Hornsund, the share of the *Calanus* in zooplankton community was greater and the copepodite structure was progressively older over time, matching the little auks timing. The importance of *Calanus* was much lower in Kongsfjorden, as represented mainly by younger copepodites, presumably due to the Atlantic water advections, thus making this area a less favourable feeding ground. Our results highlight the need for further studies on the match/mismatch between *Calanus* and little auks, because the observed trend of altered age structure towards a domination of young copepodites and the body size reduction of *Calanus* associated with higher seawater temperatures may result in insufficient food availability for these seabirds in the future.

Keywords: population dynamics; size; match-mismatch; Spitsbergen; laser optical plankton counter

1. Introduction

Calanus finmarchicus and *Calanus glacialis* (hereafter *Calanus*) co-exist and dominate the mesozooplankton biomass in the shelf waters of the Arctic Ocean [1–3]. Originally, both species have different centers of distribution and thus are adapted to different environmental conditions and consequently adopt different life-history strategies [4–7]. As dynamic changes in the environmental conditions and in timing of primary production in the Arctic [8] cause high variability in the development rate and length of life cycles of the *Calanus* species [9], comprehensive phenological studies of these copepods are essential, especially in the context of their availability to planktivores in the warming Arctic. The important role of *Calanus* spp. in marine ecosystem functioning is based on the transfer of omega-3 fatty acids (long-chain PUFAs produced by marine algae), which are crucial for the growth and reproduction of all marine organisms [10]. *Calanus* spp. are full of lipids (up to 50–70% in dry mass) [11–13], and this makes them extremely nutritious for planktivores. Until recently, the amount of lipids was suggested to be species-specific, with a higher lipid content attributed to *C. glacialis* compared to the smaller *C. finmarchicus* [4,5,13]. This assumption emphasized the different role of these species in the Arctic ecosystem. However, in most of the studies the identification between these two morphologically very similar species was based on their size [13–15], while according to recent molecular studies the size ranges of *C. glacialis* and *C. finmarchicus* overlap and are generally much broader than previously assumed [16–20]. Therefore, the problem with a correct separation of these species could lead to substantial confusion dealing with their life history, population dynamics and ecological roles. Because the lipid content was found to depend on body size rather than species, the analysis of the *Calanus* spp. size seems to be more appropriate to understanding the process of energy transfer from *Calanus* to higher trophic levels [20].

The individual size of *Calanus* spp. is tightly coupled with the ambient seawater temperature, with smaller individuals observed in warmer seawater conditions [16,21,22]. Temperature increase affects not only intra-species size variability but leads also to alteration of the zooplankton community size structure, which is considered to be more important than shifts in biomass [23,24]. The scenarios of pelagic food web modifications due to the increased seawater temperature indicate that smaller boreal species, due to faster reproduction, will have increasingly important roles in the higher latitudes [15,20,25]. A northward expansion of organisms, better adapted to warm water conditions, has already been observed [26–28]. Fluctuations in planktonic production may lead to a disturbance in interactions between predators and prey (match/mismatch) [29,30] and as a consequence may severely disrupt the functioning of the whole ecosystem [31–34]. However, new hypotheses have also emerged, suggesting that energy transfer to higher trophic levels may be more efficient than previously assumed, because of the accelerated zooplankton development [20]. Unfortunately, such promising scenarios may not apply to strictly specialized visual predators, actively selecting larger *Calanus* spp. individuals, such as the little auk [35,36], the keystone planktivorous seabird in the Arctic [37]. The little auk requires their prey to occur in high proportion in relation to other zooplankters in the seawater [14,37], because high abundances of small zooplankters may hinder the detection of their preferred large, energy-rich prey. The reproduction period of little auks is tightly matched in time with particular phase of *Calanus* development, because they actively select mainly the lipid-full fifth copepodite stage (CV) [38–43]. A recent study of *Calanus* spp. phenology in Greenland waters [44] showed that little auks select foraging grounds where availability of their main prey is matched in time with their high food demands. This might be especially important in Svalbard, where some of the world's largest colonies of these birds are located [45] and which is now threatened by the new climate state due to progressing Atlantification. Little auks are the main fertilizers of ornithogenic tundra and thus play an important role in the Arctic ecosystem [46]. Therefore, studies of the phenology of their main prey are important for better understanding threats for both marine and terrestrial ecosystems and their interactions.

Consequently, two main goals emerged in this study: (1) to test the size of zooplankton both on individual (*Calanus* copepodite stages) and community (abundance-weighted mean size) levels; and (2) to investigate dynamics of complete copepodite structure of *Calanus* in the context of food demand of the little auk during the summer chick rearing period in two different regions and summer seasons on the West Spitsbergen Shelf. We hypothesized that both the population dynamics of *Calanus* and the taxonomic composition of zooplankton vary significantly between more Atlantic and Arctic water dominated regions, which in turn provide different feeding conditions for planktivores.

2. Materials and Methods

2.1. Study Area

Research was conducted on the west coast of Spitsbergen in two fjords, representing different hydrographic regimes: Kongsfjorden, located in the north and Hornsund in the south (Figure 1). Hornsund is situated on the south-western tip of Spitsbergen and is influenced by the coastal Sørkapp Current, which carries less-saline, cold Arctic water [47,48]. Kongsfjorden is exposed to advection from the warmer Atlantic water of the West Spitsbergen Current (WSC) [48–50]. The two different currents are separated by a density gradient that forms the large frontal system along the West Spitsbergen Shelf (WSS) [51]. Thus, Kongsfjorden receives twice as much Atlantic water than Hornsund, and this results in 1 °C higher water temperatures and 0.5 higher salinities compared to Hornsund [48]. However, in recent years, gradual warming has been observed in both fjords [48,52].

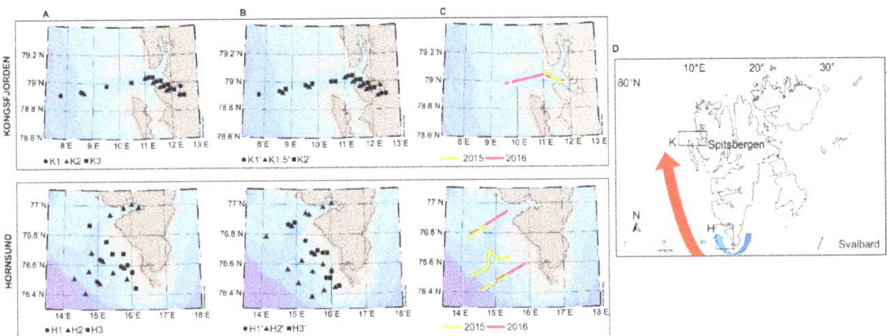

Figure 1. Zooplankton sampling stations in 2015 (**A**) and 2016 (**B**) from two fjords: Kongsfjorden (upper panel) and Hornsund (lower panel) located along the coast of Spitsbergen and sampled during three study periods each year (marked by various shapes of points). The sections of laser optical plankton counter (LOPC) surveys (**C**) are marked with yellow and pink lines. Map of the study area (**D**) with current patterns in the Spitsbergen region (simplified from Sakshaug et al. 2009).

2.2. Sampling Protocol

Zooplankton samples from the little auk foraging grounds were collected during two consecutive summer seasons (hereafter study season) of 2015 and 2016 (Figure 1 and Figure S1) in two study locations of cold Hornsund, and warm Kongsfjorden, both inside the fjords and in the adjacent areas located on the WSS (Figure 1). In Hornsund, samples were collected three times during each season, (hereafter study periods): three, 13, and seven samples in 2015 (hereafter respectively H1, H2, H3) and five, 15, and seven in 2016 (respectively H1', H2', H3'). In Kongsfjorden nine, five, and six samples were collected in 2015 (hereafter respectively K1, K2, K3) and nine, eight, and nine in 2016 (K1', K1.5', K2'). For exact sampling dates see Figure S1. In total, during 12 zooplankton sampling campaigns, 96 zooplankton samples were collected: 50 in the Hornsund area, and 46 in the Kongsfjorden area. Most sampling was performed directly from the R/V Oceania (IO PAN). Two sampling campaigns in the Kongsfjorden area (K1 and K2') were performed on board of the R/V Lance (Norwegian Polar Institute), one on board of a Zodiac boat (K3 in 2015) and one in Hornsund on board the S/Y Magnus Zaremba (H3' in 2016). Most of the samples (91) were collected using a WP2-type net (0.25 m^2 opening area) fitted with filtering gauze of 180 μm mesh size, which is best suited to catch all copepodite stages of *Calanus*. Five samples (K2) were collected using a WP2-type net with 60 μm mesh size. All zooplankton samples were collected from the mid-surface water layer (i.e., upper 50 m), which was arbitrarily chosen taking into consideration the little auk maximal diving depth of 50 m [53]. After collection, all samples were preserved in 4% formaldehyde solution in borax-buffered seawater and transported to laboratory for analysis.

2.3. Laboratory Analyses

Detailed laboratory analyses of each sample were performed according to a standard procedure, following the instructions given by Kwasniewski et al. [14]. First, zooplankters larger than 0.5 cm were removed from the sample, identified, and counted. Then, 2 mL subsamples were taken from each sample with a macropipette according to the sub-sampling method described by Harris et al. [54]. Subsamples were taken until at least 400 individuals were counted from a single sample. All organisms were identified to the lowest possible taxonomic level, typically species. Special focus was put on *Calanus* and their copepodite stages, which were identified according to criteria described by Kwasniewski et al. [55]. The abbreviations (CI-AF) refer to six successive copepodite stages of

Calanus, i.e., CI, CII, CIII, CIV and CV as the first five copepodite stages, and AF to adult females. After microscope analysis, the number of individuals in each sample was converted into abundance (ind. m^{-3}), basing on the volume of filtered water.

Due to the similar function of both species according to the trait-based approach [20] and difficulties in distinguishing them properly, for the purposes of this work *C. glacialis* and *C. finmarchicus* have been merged into one group, hereafter *Calanus*. However, the measurements of the prosome length of all copepodite stages of *Calanus* were performed for at least 30 individuals from each copepodite stage in the subsamples. To compare the prosome length of all the *Calanus* copepodite stages, 12,800 individuals were measured.

2.4. Temperature Measurements

Temperature data for comparison with the measurements of body size of *Calanus* and zooplankton community from net samples were obtained from 52 stations, (35 in Hornsund and 17 in Kongsfjorden) in two years of study by vertical profiles using a conductivity–temperature–depth sensor (CTD)). Measurements were first all binned into 1 m depth intervals, and then averaged over the upper 50 m water column for each station.

2.5. LOPC Measurements

To measure the size and concentrations of *Calanus*-type particles in different temperatures, the continuous oscillatory profiles, from the surface to 50 m depth, were performed with a conductivity–temperature–depth sensor (CTD; SBE 911plus, Seabird Electronics Inc., Washington, DC, USA) and a laser optical plankton counter (LOPC; Brooke Ocean Technology Ltd., Dartmouth, NS, Canada). The LOPC is an in-situ sensor that provides data on the abundance and size structure of a plankton community by measuring each particle passing through a sampling tunnel of a 49 cm^2 cross section. As the particle passes the sensor, the portion of blocked light is measured and recorded as a digital size and converted to the equivalent spherical diameter (ESD). The ESD is the diameter of a sphere that would represent the same cross-sectional area as the particle being measured with the use of a semi-empirical formula based on calibration with spheres of known diameters [56–58]. A *Calanus*-type group of particles, that involved only older life stages (app. CV), was selected basing on size (0.9–2.5 mm ESD) and transparency (attenuance index > 0.4) [59]. Then, the measurements (size + concentration of *Calanus*-type) were grouped according to a few seawater temperature ranges (<4 °C, 4–6 °C, and >6 °C) to calculate the 1d kernel density (R function geom_density) estimates of the dominating size in particular water temperatures.

2.6. Data Analyses

Multivariate nonparametric permutational ANOVA (PERMANOVA) [60] was used to test differences in: *Calanus* prosome length of individual copepodite stages; *Calanus* copepodite structure (CI-AF, stage index, and zooplankton species composition among fixed factors of regions and study periods. Prior to the analyses, abundance data were square-root-transformed [61]. The distribution of centroids representing particular samples was illustrated with a non-metric multi-dimensional scaling (nMDS) using Bray–Curtis similarities ordinations. The calculation of the Pseudo-F and *p* values was based on 999 permutations of the residuals under a reduced model [62]. To assess the magnitude of the spatial variation at each gradient, the estimated components of variation (ECV) as a percentage of the total variation were used. The relationship between seawater temperature and the *Calanus* prosome length and the mean size of zooplankton organisms was tested with the use of Pearson linear correlation. The *Calanus* stage index was calculated as a weighted mean on the basis of relative abundance of particular life stages, with each stage given values from 1 (CI) to 6 (AF) [63], where AF stage was represented by adult females. The mean size of zooplankton organisms in the sample was calculated as a weighted mean based on a local database with detailed measurements made by the Svalbard fiords in the scope of the Dwarf project (Polish-Norwegian Research project no. Pol-Nor/201992/93/2014). To determine zooplankton community structure, relative abundance of species/taxa was used. For each study period, a median number of individuals of a given taxa was calculated. Taxa constituting more than 5% of total zooplankton abundance were distinguished; the rest were grouped into an "others" category. To compare the prosome length frequency distribution of a copepodite stage CV of *C. finmarchicus* and *C. glacialis*, the number of all CVs in the sample (ind. m^{-3}) was divided by the total number of measured individuals in the sample. To determine the relative abundance of measured individuals, each measurement of a given individual was multiplied by a factor: the abundance of CV (ind. m^{-3})/number of measured individuals. Measured individuals were classified into size classes every 0.05 mm.

3. Results

3.1. Size Response of Calanus and Mesozooplankton to Different Seawater Temperatures

The prosome length of *Calanus* copepodite stages based on comprehensive morphometrical analysis differed significantly between the two fjords in the case of all life stages (CI–CV), except for adults (AF) (Table 1). In general, the median prosome length of *Calanus* individuals was larger in Hornsund than in Kongsfjorden (Figure 2).

Table 1. Results of one-factor multivariate PERMANOVA for the prosome length of *Calanus* copepodite stages in Hornsund and Kongsfjorden. Bold means $p < 0.05$, df are degrees of freedom, MS represents means of squares, $\sqrt{ECV}$ are square root of the estimated components of variance.

Copepodite Stage	Factor	df	MS	Pseudo-F	p	$\sqrt{ECV}$
CI	Fjord	1	1274.8	212.98	**0.001**	1.40
CII	Fjord	1	5180.9	560.51	**0.001**	2.43
CIII	Fjord	1	8748.4	1086.40	**0.001**	2.74
CIV	Fjord	1	5728.8	720.28	**0.001**	1.98
CV	Fjord	1	2277.5	237.43	**0.001**	1.43
AF	Fjord	1	1.2	0.22	0.630	−0.18

The prosome length of individual copepodite stages of *Calanus* significantly correlated with temperature in both fjords for CI-CV (Pearson correlation coefficient, R = −0.37, $p < 0.001$ (CI); R = −0.48, $p < 0.001$ (CII); R = −0.39, $p < 0.001$ (CIII); R = −0.38, $p < 0.001$ (CIV); R = −0.37, $p < 0.001$ (CV), Figure 3), and did not correlate for AF (Pearson correlation coefficient, R = 0.01, $p = 0.88$, Figure 3). Generally, a relatively large variation in the body size of each *Calanus* copepodite stage was observed for each temperature recorded, except for the highest values (>7 °C), at which only relatively small individuals were observed in Kongsfjorden.

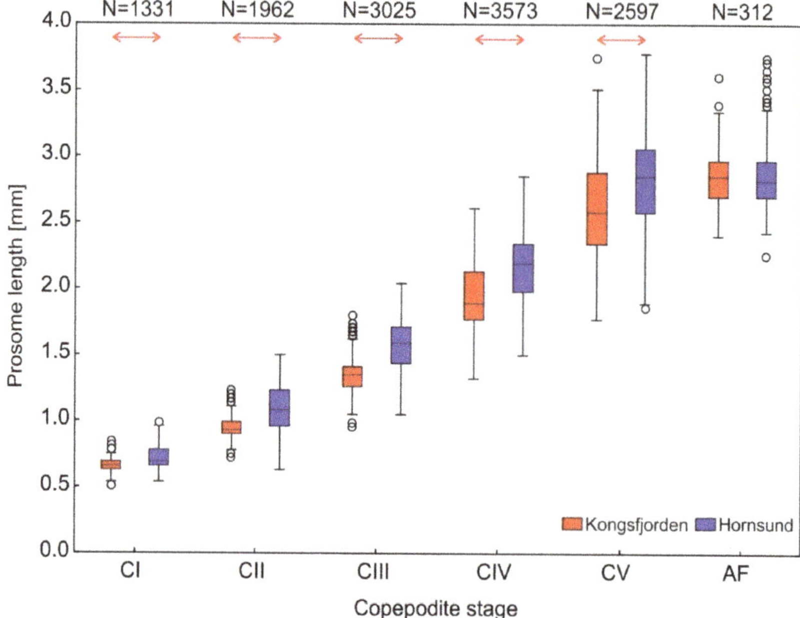

Figure 2. Prosome length of *Calanus* copepodite stages: CI–adult females (AF) in Kongsfjorden (red) and Hornsund (blue). Horizontal black lines in the boxes show the median, box represents percentiles, whiskers indicate ranges, dots represent values outside the range, red arrows show statistically significant size differences between two investigated fjords, N indicates number of measured individuals.

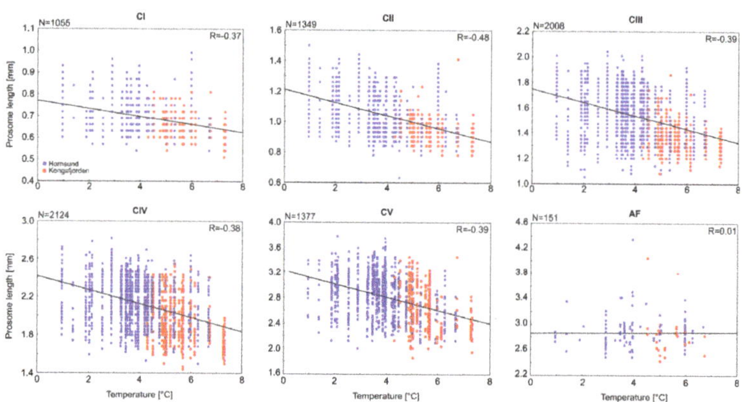

Figure 3. The relation between seawater temperature and prosome length of *Calanus* copepodite stages CI–AF. Dots represent individuals measured, blue in Hornsund and red in Kongsfjorden. Trendline is marked in black, N indicates the number of measured individuals, R indicates Pearson correlation coefficient.

Additionally, a significant negative correlation between seawater temperature and mean zooplankton body size was observed (Pearson's correlation R = −0.46; p = 0.001; Figure S2). Larger animals (>1200 μm) were observed at lower seawater temperatures (i.e., below 4 °C) in Hornsund, while smaller organisms were dominating mainly in warmer (4–7.5 °C) Kongsfjorden.

Moreover, according to the measurements of the automatic LOPC method of the *Calanus*-type particles, individual size of older life stages of *Calanus* (CIV and CV) also differed in relation to the seawater temperature (Figure 4). In Hornsund in both years the largest *Calanus*-type particles (c.a. 1.5 mm ESD) were observed in low temperatures (<4 °C), while the highest densities of smallest individuals (1.0–1.2 mm ESD) were noted at higher temperatures (>6 °C). In Kongsfjorden, the size range of older *Calanus* differed between two years, with larger size fraction dominating in 2015 (Figure 4).

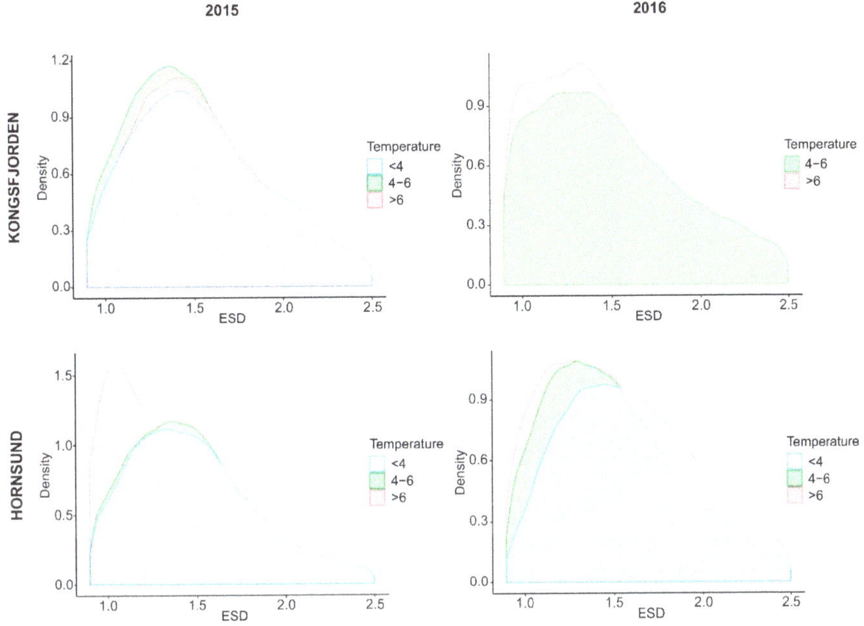

Figure 4. Size distribution of Calanus-type particles expressed as equivalent spherical diameter (ESD, mm) recorded by laser optical plankton counter (LOPC) in different temperature ranges in Kongsfjorden and Hornsund in 2015 and 2016.

3.2. Zooplankton Community Structure

The zooplankton community structure in Kongsfjorden was significantly different between the two years (PERMANOVA, MS = 6127.4, Pseudo-F = 16.13, $p = 0.001$) and among study periods in 2015 (PERMANOVA, MS = 1776.3, Pseudo-F = 5.22, $p = 0.003$) and in 2016 (PERMANOVA, MS = 2133.0, Pseudo-F = 4.49, $p = 0.001$). Because samples in K2 period of 2015 were collected with different net mesh size, they were excluded from the current analysis (see M and M for details). In 2015, *Calanus* was the most abundant taxon in K1, when it reached 47% of the total zooplankton abundance. Its percentage decreased over time reaching 25% in K3 (Figure 5). In 2016, the *Calanus* percentage was relatively low, and decreased from 12% in K1′ and K1.5′ to 5% in K2′ (Figure 5). The second numerous taxon, *Oithona similis* was more significant in 2015 (30% in K1 and 46% in K3) than in 2016 (20% throughout three study periods). The share of the Copepoda nauplii was higher (maximum in K1′) in 2016 than in 2015. Similarly as in Hornsund, the percentage of Bivalvia veligers were especially high in 2016 (17–26%) and very low in 2015 (≤5% in K3). Another important zooplankton component in Kongsfjorden was *Limacina helicina* with 16% to 31% contribution to the overall zooplankton abundances in 2016. The percentage of *Pseudocalanus* spp. was rather constant over time.

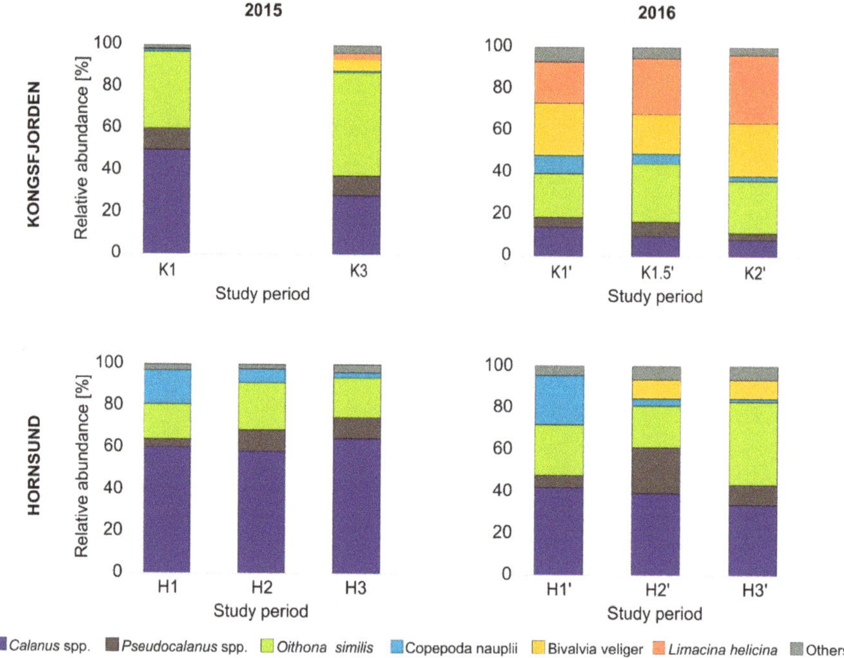

Figure 5. Zooplankton community structure in Kongsfjorden and Hornsund.

The zooplankton community structure in Hornsund also differed between the two years (PERMANOVA, MS = 3245.0, Pseudo-F = 5.83, p = 0.001) and between study periods (PERMANOVA, MS = 5222.6, Pseudo-F = 9.39, p = 0.001) of both years (PERMANOVA, MS = 1661.3, Pseudo-F = 2.98, p = 0.003). In both years, *Calanus* dominated the zooplankton community, constituting approx. 60% of total zooplankton abundance in all three study periods of 2015 and approximately 40% in 2016. The second most numerous taxon was *O. similis* with at least 20% of total zooplankton abundance throughout all the periods of both study seasons (Figure 5). Copepoda nauplii composed a significant percentage of total zooplankton abundance, especially in H1 and H1' (20%). The contribution of *Pseudocalanus* spp. was in general the most important in H2 and H2' study periods, while Bivalvia veligers comprised about 10% of all taxa in H2' and H3' in summer 2016.

3.3. Calanus Copepodite Structure

The copepodite structure of *Calanus* in Kongsfjorden clearly differed between the years (PERMANOVA, MS = 6079.7, Pseudo-F = 15.19, p = 0.001) and between the study periods only in 2015 (PERMANOVA post hoc, MS = 1782.3, Pseudo-F = 4.45, p = 0.013). *Calanus* copepodite stage structures were much more similar among study periods within one year than among corresponding study periods of the two years (Figure 6), which was indicated by higher estimated components of variance (ECV) for years other than for study periods (19.5% vs. 10.8%). The *Calanus* copepodite stage index in Kongsfjorden differed between years (PERMANOVA, MS = 698.1, Pseudo-F = 46.40, p = 0.001), but not between study periods (PERMANOVA, MS = 21.9, Pseudo-F = 1.46, p = 0.257). In general, in 2015 the median values of *Calanus* copepodite stage index were higher than in 2016, when they remained relatively low (Figure 7).

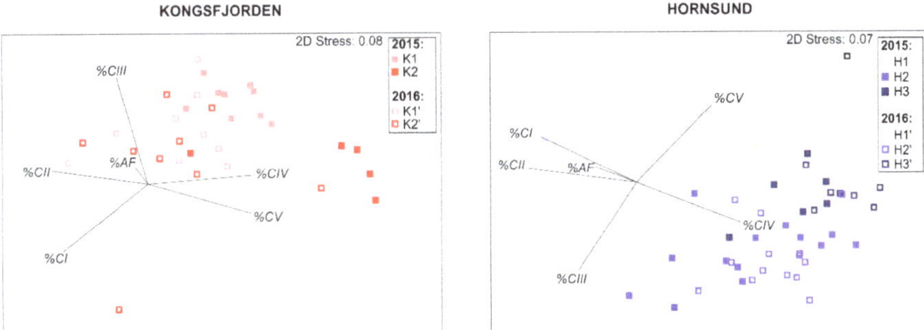

Figure 6. The non-metric multi-dimensional scaling (nMDS) of the *Calanus* copepodite structure in two Spitsbergen fjords: Kongsfjorden and Hornsund. Vectors indicate the direction of best correlating variables determined as a percentage of each copepodite stage. Their lengths correspond with the strength of the correlation.

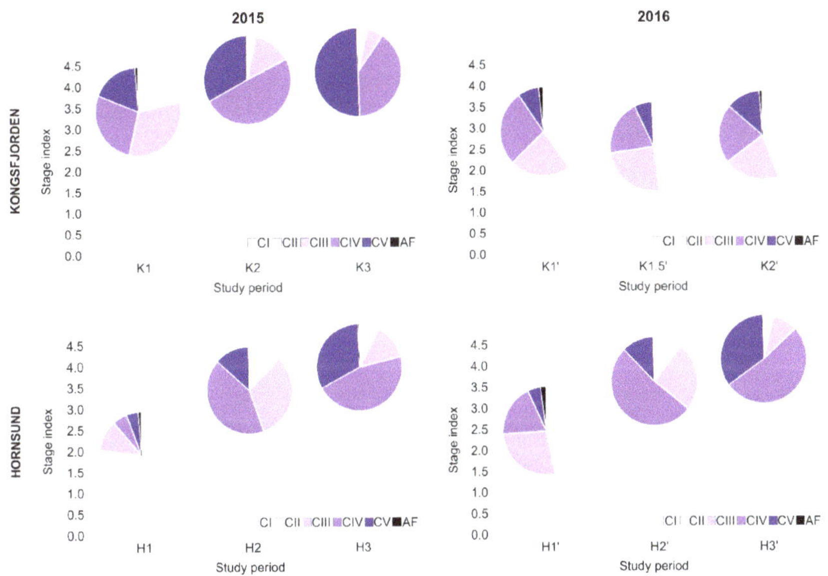

Figure 7. *Calanus* stage index (Y axis) together with copepodite stage structure (different colours) in Kongsfjorden and Hornsund in particular study periods (X axis) in 2015 and 2016.

The copepodite structure of *Calanus* in Hornsund differed first of all between study periods (PERMANOVA, MS = 3724.2, Pseudo-F = 31.16, p = 0.001), but also slightly between years (PERMANOVA, MS = 475.1, Pseudo-F = 3.98, p = 0.026; Figure 6). The higher share of ECV was obtained for the factor of periods than for a factor of year (15.9% vs. 4.4%). The *Calanus* copepodite stage index in Hornsund also differed between years (PERMANOVA, MS = 49.5, Pseudo-F = 7.62, p = 0.014) and study periods (PERMANOVA, MS = 746.9, Pseudo-F = 57.56, p = 0.001). Differences in the copepodite stage index had a stepwise character in both years, with a slightly higher median values in all investigated periods observed in 2016 (Figure 7).

Large differences in *Calanus* copepodite structure in Kongsfjorden were only observed in 2015, when the proportion of early copepodite stages (CI–CIII) decreased from over 50% in K1 to about 10% in K3 (Figure 7). The CIV was the dominating life stage in K2, then it was outnumbered by CV in K3. The proportion of AF was very low with a slightly higher value, but not exceeding 2% in K1. In 2016, the composition of *Calanus* community was dominated by young copepodite stages and was very similar among study periods (Figure 7). The CI–CIII constituted approximately 60% of all copepodite stages with about 20% share for each stage. *Calanus* population in Hornsund in both years changed over time (Figure 7). Although the highest importance of the youngest copepodites of *Calanus* (CI and CII) were observed early in both years, (H1 and H1′), their contribution was almost two times higher in 2015 than in 2016. Thus, the correlation of these stages with the ordination coordinates was stronger concerning samples from 2015 (Figure 6). The highest proportion of CIII was observed in 2015 in H2 (31%) and in 2016 in both H1′ and H2′ (25%). The highest proportion of CIV was observed in the second and the third study periods in both years and was slightly higher in 2016. Peak occurrence of CV took place in the third period of both years (H3 and H3′). The proportion of AF was relatively low throughout the study, but slightly higher in the first study periods of both years and correlated positively with the youngest life stages.

The *Calanus* copepodite structure differed significantly between Hornsund and Kongsfjorden in corresponding periods only in 2016 (PERMANOVA, MS = 5558.3, Pseudo-F = 18.95, p = 0.001). In 2015, the copepodite structure in both fjords was relatively similar in corresponding periods H3 and K3 (PERMANOVA, MS = 642.8, Pseudo-F = 3.35, p = 0.054) with the predominance of late stages (CIV and CV) in both regions. In turn, in 2016 in the comparable H2′ and K1.5′ periods, the copepodite structure differed significantly, because of the high percentage of early stages CI–CIII recorded in Kongsfjorden, and the domination of CIV in Hornsund. Also the copepodite stage index for *Calanus* differed between two fjords only in 2016 (PERMANOVA, MS = 258.5, Pseudo-F = 34.61, p = 0.001), while in 2015 in corresponding study periods the stage index in H3 and K3 was similar (Figure 7). In analogous periods of 2016 (H2′ and K1.5′) a higher median value of stage index was observed in Hornsund (Figure 7).

4. Discussion

The phenology of *Calanus* is a critical factor for little auks reproduction success during their breeding season, as was shown recently in a spatial perspective study in Greenland [44] and in time perspective study in Svalbard waters [64]. Therefore it is now of vital importance to study the match in time and space between the availability of older life stages of *Calanus* and little auks, because as was shown by this study, the development rate and the age structure of *Calanus* may differ significantly depending on the region, water temperature and time in the season. The issue is alarming not only because temperature warming has been shown to accelerate development of *Calanus* [65], but also because the altered phenology of many species is becoming an increasingly important problem for trophic interactions [31,64,66] and thus entire food webs. To date, disturbance in interactions between predators and prey (match/mismatch) have been observed in many groups of organisms, e.g., between fish and plankton [67,68], insects and plants [69], birds and insects [70–72] shorebirds and arthropods [73] or seabirds and zooplankton [33,74,75]. The high variability in *Calanus* development, smaller body size and lower proportion in their concentration in relation to smaller zooplankton taxa in warmer Kongsfjorden, observed in our study, are in line with predicted scenarios of pelagic system modifications in the future Arctic towards faster development and prevailing role of smaller organism size [15,20,25]. Even though shortening life cycles and body size reduction of *Calanus* are expected not to have negative consequences for top predators [20], it will probably be important to the little auk, which is dependent on the availability of large, energy-rich prey [35].

Little auk preferentially search for large, lipid-rich copepods to cover the high energy costs incurred during foraging trips and feeding underwater [76–78]. Because the lipid content is strongly related to the body size, in this work *Calanus* size rather than species affiliation was utilized as the main qualitative trait. This approach is suggested as more appropriate for understanding the process of energy transfer to higher trophic levels on a larger scale [20]. Combining two species into one *Calanus* category was also supported by the recently discussed and clearly demonstrated problem of misidentification of *C. glacialis* and *C. finmarchicus* [17,19,20]. In order to gain a broader view on the *Calanus* population characteristics, in this study at the first step individuals of both species were identified and separated in accordance with traditional morphological classification [55], by measuring the prosome length. Results have shown that neither the specimens classified as *C. glacialis* nor *C. finmarchicus* were following a normal size distribution (Figure 8), which was in opposition to what was demonstrated for this stage for both species by molecular methods [20] (Figure 1). A right-skewed size distribution of *C. glacialis* both in Hornsund and in Kongsfjorden observed in our study is most probably caused by the fact that larger individuals classified as *C. finmarchicus* are in fact smaller individuals of *C. glacialis* [20]. This confirms that the size criterion is no more a reliable tool to accurately classify an individual for a given species [79], thus justifying our approach to combine them into one group.

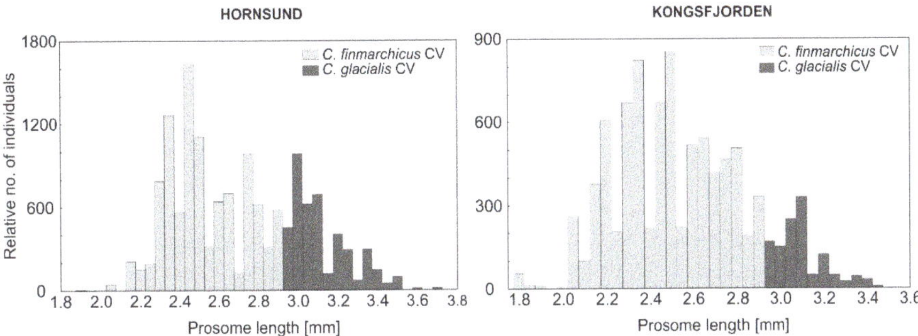

Figure 8. Prosome length frequency distribution of copepodite stage CV of *Calanus finmarchicus* and *C. glacialis* classified according to morphological identification after Kwasniewski et al. [55] in Hornsund (n = 1798) and Kongsfjorden (n = 799).

The body size of individual copepodite stages (CI–CV) of *Calanus* differed clearly between the two investigated fjords, with smaller prosome length observed in warmer Kongsfjorden than in colder Hornsund. First of all, such a difference may be explained by expected differences in proportions of *Calanus* species (*C. glacialis* vs. *C. finmarchicus*), which is also of importance for little auks [66]. But the change in individual size within particular species has also to be taken into consideration. Such observations are especially important in the light of recent studies demonstrating a great range of size plasticity of *Calanus* [17,19,20] and considering the fact that temperature is a key factor determining body size in copepods [21,80]. In this work, the smallest prosome length of *Calanus* CI-CV was observed at highest temperatures (>7 °C), whereas the largest individuals were observed in seawater temperature <4 °C. A similar trend was observed by LOPC measurements of *Calanus*-type particles selected for the older life stages, with the predominance of the largest individuals observed at temperatures below 4 °C and the smallest ones observed at temperatures above 6 °C in Hornsund. Such a clear differentiation in size modes can be explained by the co-occurrence of two water masses carrying two different *Calanus* species [14] and/or that the younger (CIV) copepodite stage was dominating in warmer, close to surface waters, while older (CV) life stage prevailed in colder conditions. Although the fraction of older *Calanus*-type particles was characterized by similar sizes in all temperature ranges in Kongsfjorden in 2015, the dominating size of this fraction was evidently shifted towards smaller sizes in the warmer season in 2016. This could be caused by different relative roles between CIV and CV life stages, or by the methodological bias, because the location of transects in both years slightly differed. In 2015 transects were performed closer to the interior part of the fjord, while in 2016 transects were mainly located in open shelf waters. Smaller individuals representing mainly *C. finmarchicus* can dominate in open waters with prevailing Atlantic water masses [19], while inside the fjord usually both species co-exist [17]. Hence the larger size of *Calanus*-type particles observed in 2015 could have resulted from a higher share of the larger species (*C. glacialis*). Despite differences in species composition, the significant relation between the body size of *Calanus* and seawater temperature observed in this study agrees with the assumptions of the temperature-size-rule (TSR) [81,82], which states that ectotherms grow slower, but mature at a larger body size in colder environments. The smaller size of *Calanus* in warmer temperatures observed in this study may be explained by the fact that organisms tend to be smaller in response to warming [83–85] and progressive reduction of *Calanus* body size is predicted with increasing seawater temperatures [86]. The body size of *C. glacialis* was found to vary considerably along its geographical range [19,20,83,87]. In experimental studies on *C. finmarchicus*, its prosome length also significantly decreased with increasing sea temperatures. Moreover, the largest individuals of *C. glacialis* were recorded after development in waters with temperatures not exceeding 3 °C [16].

In addition to individual size of *Calanus*, the proportion of *Calanus* in the overall community is very important for visual planktivores such as little auks, which need their prey to occur not only in a high concentration but also as easily visible [37]. In general, *Calanus* species are the key element of zooplankton communities in Svalbard waters, especially in terms of biomass [1,88,89], however their proportion in total zooplankton abundance is highly variable in time, space and under different hydrographic conditions [63]. In this study the proportion of *Calanus* in total zooplankton abundance was higher in Hornsund than in Kongsfjorden in both studied years, which confirmed more favourable foraging conditions for little auks in Hornsund than Kongsfjorden [77,90]. A similar predominance of *Calanus* in Hornsund was also recorded in 2007 [91]. Likewise, Trudnowska et al. [92] found higher proportions of *Calanus* in zooplankton communities in the colder Hornsund than in the Atlantic-influenced Magdalenefjorden. The lower proportion of *Calanus* in Kongsfjorden could result from the strong advection of the Atlantic Water carrying high concentrations of small copepods (e.g., *Oithona similis*) [27,93,94]. Therefore, the Atlantic water masses are typically avoided by little auks, due to high proportions of small copepods, which hinder detection of preferred prey [14,37]. The abundance of *O. similis*, which was really high in Kongsfjorden in this study, has been increasing gradually in Spitsbergen fjords since 2006 [91] as a consequence of the progressive Atlantification of these waters [27]. The increasing importance of small copepods in the zooplankton composition [89] is one of the most spectacular examples of the progressing warming that have already been documented [23].

Studies of *Calanus* development are challenging because its reproductive strategies are highly variable in time and space due to corresponding changes in environmental conditions and food supply [6,20,86,95]. In this study the development of the *Calanus* population and in consequence also its stage index, followed similar trends in Hornsund in both years. A similar gradual development of *Calanus* population, reflected by a dominance of young stages CI–CIII during the first study period and older CIV–CV stages during the third period was observed in Rijpfjorden [6]. Such similarity in the *Calanus* age structure observed in Rijpfjorden and Hornsund indicates a coincident timing of reproductive events and its synchrony with ice algae bloom in April and phytoplankton bloom in July [96]. In addition, the presence of early copepodite stages in all the studied periods (this study) might suggest continuous reproduction, or at least, the presence of more than one generation of *Calanus*, which is likely in high latitudes according to several new studies [20,65,97]. In Kongsfjorden, the trend of a gradual population development was observed in this work only in 2015. This observation coincided with the seasonal dynamic of the *Calanus* population structure emphasized for this fjord by a year-round investigation of Lischka and Hagen [94] with a higher contribution of early copepodite stages in July and a more advanced population in August. In turn in 2016, *Calanus* age structure was very similar in all three studied periods in Kongsfjorden and persisted relatively young according to low stage index. To some extent this might be caused by shorter time intervals between sampling periods in this year, but most probably it was caused by different advection impacts, according to different sea surface temperatures in the two years investigated (Figure 9). The events of advection are often associated with a transport of younger populations [55,93,95], which could explain the higher contribution of early stages in 2016 in Kongsfjorden in K1′ (7–8 °C SST, Figure 9). This fact was also confirmed by a multiyear study conducted in the WSC region, where the copepodite structure of *C. finmarchicus* was younger during 'warmer' than 'colder' summers [27].

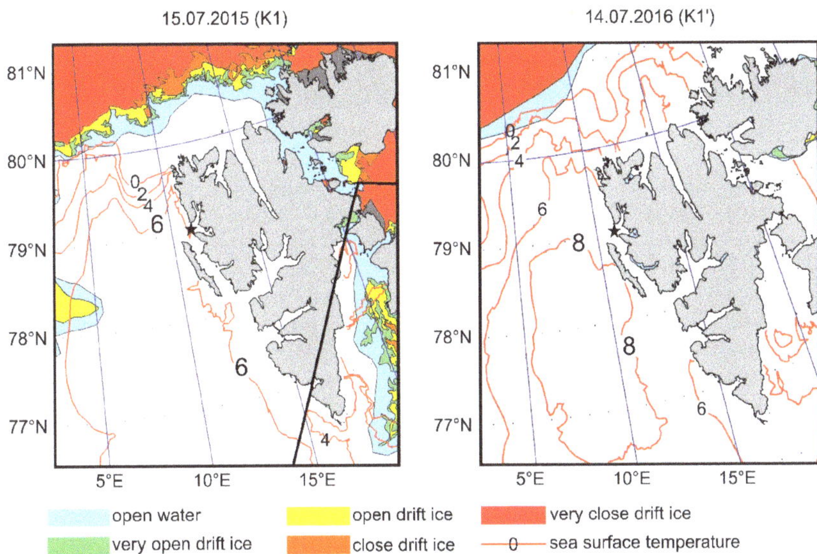

Figure 9. Sea surface temperature isoline and sea ice conditions (colour surfaces) in Svalbard waters in 2015 during K1 (**a**) and in 2016 during K1' (**b**) study periods. Star icon represents Kongsfjorden study area. Data source: Norwegian Meteorological Institute.

5. Conclusions

This study confirms the difficulty in proper *Calanus finmarchicus* and *Calanus glacialis* recognition, which, together with the similar functions of these species proved in recent studies, was an important argument for aggregating them into a single *Calanus* group. This study also provides evidence that the development rate and size structure of *Calanus* is highly variable in time, space and in relation to seawater temperature. Furthermore, according to the observations of this research, the copepodite structure is much more dynamically affected by Atlantification in Kongsfjorden than in Hornsund. Additionally, seawater temperature was confirmed to correlate negatively with both the mean size of mesozooplankton organisms and the body length of *Calanus* copepodites (CI–CV). These results support the hypothesis about shortening the life span and associated reduction of the body size of *Calanus* along with climate warming. The accelerated development of *Calanus* can cause a significant shift in time in availability of its fifth copepodite stage in the foraging grounds of their key predator, the little auks. These findings confirm the hypothesis of the possible mismatch in timing between the availability of *Calanus* CV and the little auks highest food demands and therefore highlight the necessity to continue further seasonal studies of *Calanus* phenology in Svalbard waters.

Supplementary Materials: The following are available online at http://www.mdpi.com/2073-4441/11/7/1405/s1, Figure S1: Zooplankton sampling dates in two fjords (Kongsfjorden and Hornsund) along the coast of Spitsbergen. Codes for sampling in Kongsfjorden: K1, K2, K3 in 2015 and K1', K1.5', K2' in 2016. Codes for sampling in Hornsund: H1, H2, H3 in 2015 and H1', H2', H3' in 2016, Figure S2: Relationship between seawater temperature and the weighted mean size of zooplankton in communities sampled in Hornsund (blue) and Kongsfjorden (red). Each dot represents mean individual zooplankton size in a particular sample.

Author Contributions: K.B.-S. and K.B. were responsible for the research design. K.B. sampled zooplankton material and analysed the data. E.T. analysed LOPC data and prepared related figure. K.B. prepared drafted the text and figures and performed statistical analyses. All authors participated in discussions and editing.

Funding: This study received support from Polish Ministry of Science and Higher Education decision No. 3605/SEAPOP/2016/2 for international, co-funding project SEAPOP II for years 2016–2020. K.B. has been supported as a PhD student by the Centre for Polar Studies, Leading National Research Centre, Poland. E.T. was financed by Polish National Science Centre (ecoPlast No. 2017/27/B/NZ8/00631).

Acknowledgments: We thank Agnieszka Promińska for CTD data. We acknowledge all people who were involved in extensive zooplankton sampling campaigns: Anette Wold, Anna Maria Kubiszyn, Justyna Wawrzynek, Maciej Jan Ejsmond, Wojciech Moskal, Michał Procajło and Mateusz Ormańczyk.

Conflicts of Interest: The authors declare no conflict of interest.

References

1. Błachowiak-Samołyk, K.; Søreide, J.E.; Kwasniewski, S.; Sundfjord, A.; Hop, H.; Falk-Petersen, S.; Hegseth, E.N. Hydrodynamic control of mesozooplankton abundance and biomass in northern Svalbard waters (79–81°N). *Deep Sea Res. Part II Top. Stud. Oceanogr.* **2008**, *55*, 2310–2324. [CrossRef]

2. Carstensen, J.; Weydmann, A.; Olszewska, A.; Kwasniewski, S. Effects of environmental conditions on the biomass of *Calanus* spp. in the Nordic Seas. *J. Plankton Res.* **2012**, *34*, 951–966. [CrossRef]

3. Aarflot, J.M.; Skjodal, H.R.; Dalpadado, P.; Skern-Mauritzen, M. Contribution of *Calanus* species to the mesozooplankton biomass in the Barents Sea. *ICES J. Mar. Sci.* **2017**, *75*, 2342–2354. [CrossRef]

4. Scott, C.L.; Kwasniewski, S.; Falk-Petersen, S.; Sargent, J.R. Lipids and life strategies of *Calanus finmarchicus*, *Calanus glacialis* and *Calanus hyperboreus* in late autumn, Kongsfjorden, Svalbard. *Polar Biol.* **2000**, *23*, 510–516. [CrossRef]

5. Falk-Petersen, S.; Mayzaud, P.; Kattner, G.; Sargent, J. Lipids and life strategy of Arctic *Calanus*. *Mar. Biol. Res.* **2009**, *5*, 18–39. [CrossRef]

6. Daase, M.; Falk-Petersen, S.; Varpe, Ø.; Darnis, G.; Søreide, J.E.; Wold, A.; Leu, E.; Berge, J.; Philippe, B.; Fortier, L. Timing of reproductive events in the marine copepod *Calanus glacialis*: A pan-Arctic perspective. *Can. J. Fish. Aquat. Sci.* **2013**, *70*, 871–884. [CrossRef]

7. Melle, W.; Runge, J.; Head, E.; Plourde, S.; Castellani, C.; Licandro, P.; Pierson, J.; Jónasdóttir, S.; Johnson, C.; Broms, C.; et al. The North Atlantic Ocean as habitat for *Calanus finmarchicus*: Environmental factors and life history traits. *Prog. Oceanogr.* **2014**, *129*, 244–284. [CrossRef]

8. Kubiszyn, A.M.; Piwosz, K.; Wiktor, J.M., Jr.; Wiktor, J.M. The effect of inter-annual Atlantic water inflow variability on the planktonic protist community structure in the West Spitsbergen waters during the summer. *J. Plankton Res.* **2014**, *36*, 1190–1203. [CrossRef]

9. Ji, R.; Ashjian, C.J.; Campbell, R.G.; Chen, C.; Gao, G.; Davis, C.S.; Cowles, G.W.; Beardsley, R.C. Life history and biogeography of *Calanus* copepods in the Arctic Ocean: An individual-based modeling study. *Prog. Oceanogr.* **2012**, *96*, 40–56. [CrossRef]

10. Ackman, R.G. *Marine Biogenic Lipids, Fats and Oils*; CRC Press: Boca Raton, FL, USA, 1989.

11. Miller, C.B.; Crain, J.A.; Morgan, C.A. Oil storage variability in *Calanus finmarchicus*. *ICES J. Mar. Sci.* **2000**, *57*, 1786–1799. [CrossRef]

12. Lee, R.F.; Hagen, W.; Kattner, G. Lipid storage in marine zooplankton. *Mar. Ecol. Prog. Ser.* **2006**, *307*, 273–306. [CrossRef]

13. Mayzaud, P.; Falk-Petersen, S.; Noyon, M.; Wold, A.; Boutoute, M. Lipid composition of the three co-existing *Calanus* species in the Arctic: Impact of season, location and environment. *Polar Biol.* **2016**, *39*, 1819–1839. [CrossRef]

14. Kwasniewski, S.; Gluchowska, M.; Jakubas, D.; Wojczulanis-Jakubas, K.; Walkusz, W.; Karnovsky, N.; Blachowiak-Samolyk, K.; Cisek, M.; Stempniewicz, L. The impact of different hydrographic conditions and zooplankton communities on provisioning little auks along the west coast of Spitsbergen. *Prog. Oceanogr.* **2010**, *87*, 72–82. [CrossRef]

15. Wassmann, P.; Duarte, C.M.; Agustí, S.; Sejr, M.K. Footprints of climate change in the Arctic marine ecosystem. *Glob. Chang. Biol.* **2011**, *17*, 1235–1249. [CrossRef]

16. Parent, G.J.; Plourde, S.; Turgeon, J. Overlapping size ranges of *Calanus* spp. off the Canadian Arctic and Atlantic coasts: Impact on species' abundances. *J. Plankton Res.* **2011**, *33*, 1654–1665. [CrossRef]

17. Gabrielsen, T.M.; Merkel, B.; Søreide, J.E.; Johansson-Karlsson, E.; Bailey, A.; Vogedes, D.; Nygård, H.; Varpe, Ø.; Berge, J. Potential misidentifications of two climate indicator species of the marine arctic ecosystem: *Calanus glacialis* and *C. finmarchicus*. *Polar Biol.* **2012**, *35*, 1621–1628. [CrossRef]

18. Nielsen, T.G.; Kjellerup, S.; Smolina, I.; Hoarau, G.; Lindeque, P. Live discrimination of *Calanus glacialis* and *C. finmarchicus* females: Can we trust phenological differences? *Mar. Biol.* **2014**, *161*, 1299–1306. [CrossRef]

19. Choquet, M.; Kosobokova, K.; Kwaśniewski, S.; Hatlebakk, M.; Dhanasiri, A.K.S.; Melle, W.; Daase, M.; Svensen, C.; Søreide, J.E.; Hoarau, G. Can morphology reliably distinguish between the copepods *Calanus finmarchicus* and *C. glacialis*, or is DNA the only way? *Limnol. Oceanogr. Methods* **2018**, *16*, 237–252. [CrossRef]

20. Renaud, P.E.; Daase, M.; Banas, N.S.; Gabrielsen, T.M.; Søreide, J.E.; Varpe, Ø.; Cottier, F.; Falk-Petersen, S.; Halsband, C.; Vogedes, D.; et al. Pelagic food-webs in a changing Arctic: A trait-based perspective suggests a mode of resilience. *ICES J. Mar. Sci.* **2018**, *75*, 1871–1881. [CrossRef]

21. Campbell, R.; Wagner, M.M.; Teegarden, G.J.; Boudreau, C.A.; Durbin, E.G. Growth and development rates of the copepod *Calanus finmarchicus* reared in the laboratory. *Mar. Ecol. Prog. Ser.* **2001**, *221*, 161–183. [CrossRef]

22. Wilson, R.J.; Speirs, D.C.; Heath, M.R. On the surprising lack of differences between two congeneric calanoid copepod species, *Calanus finmarchicus* and *C. helgolandicus*. *Prog. Oceanogr.* **2015**, *134*, 413–431. [CrossRef]

23. Richardson, A.J.; Schoeman, D.S. Climate impact on plankton ecosystems in the Northeast Atlantic. *Science* **2004**, *305*, 1609–1612. [CrossRef] [PubMed]

24. Lane, P.V.Z.; Llinás, L.; Smith, S.L.; Pilz, D. Zooplankton distribution in the western Arctic during summer 2002: Hydrographic habitats and implications for food chain dynamics. *J. Mar. Res.* **2008**, *70*, 97–133. [CrossRef]

25. Beaugrand, G.; Ibanez, F.; Lindley, J.A.; Reid, P.C. Diversity of calanoid copepods in the North Atlantic and adjacent seas: Species associations and biogeography. *Mar. Ecol. Prog. Ser.* **2002**, *232*, 179–195. [CrossRef]

26. Woodworth-Jefcoats, P.A.; Polovina, J.J.; Drazen, J.C. Climate change is projected to reduce carrying capacity and redistribute species richness in North Pacific pelagic marine ecosystems. *Glob. Chang. Biol.* **2016**, *23*, 1000–1008. [CrossRef]

27. Gluchowska, M.; Dalpadado, P.; Beszczynska-Möller, A.; Olszewska, A.; Ingvaldsen, R.B.; Kwasniewski, S. Interannual zooplankton variability in the main pathways of the Atlantic water flow into the Arctic Ocean (Fram Strait and Barents Sea branches). *ICES J. Mar. Sci.* **2017**, *74*, 1921–1936. [CrossRef]

28. Haug, T.; Bogstad, B.; Chierici, M.; Gjøsæter, H.; Hallfredsson, E.H.; Høines, Å.S.; Hoel, A.H.; Ingvaldsen, R.B.; Jørgensen, L.L.; Knutsen, T.; et al. Future harvest of living resources in the Arctic Ocean north of the Nordic and Barents seas: A review of possibilities and constraints. *Fish. Res.* **2017**, *188*, 38–57. [CrossRef]

29. Hjort, J. Fluctuations in the great fisheries of Northern Europe viewed in the light of biological research. *Reun. Cons. Int. Explor. Mer.* **1914**, *20*, 1–228.

30. Cushing, D.H. Plankton production and year-class strength in fish populations: An update of the match/mismatch hypothesis. *Adv. Mar. Biol.* **1990**, *26*, 249–293.

31. Stenseth, N.C.; Mysterud, A.; Ottersen, G.; Hurrell, J.W.; Chan, K.S.; Lima, M. Ecological effects of climate fluctuations. *Science* **2002**, *297*, 1292–1296. [CrossRef]

32. Both, C.; Bouwhuis, S.; Lessells, C.M.; Visser, M.E. Climate change and population declines in a long-distance migratory bird. *Nature* **2006**, *441*, 81–83. [CrossRef] [PubMed]

33. Hipfner, J.M. Matches and mismatches: Ocean climate, prey phenology and breeding success in a zooplanktivorous seabird. *Mar. Ecol. Prog. Ser.* **2008**, *368*, 295–304. [CrossRef]

34. Nakazawa, T.; Doi, H. A perspective on match/mismatch of phenology in community contexts. *Oikos* **2012**, *121*, 489–495. [CrossRef]

35. Kidawa, D.; Jakubas, D.; Wojczulanis-Jakubas, K.; Stempniewicz, L.; Trudnowska, E.; Boehnke, R.; Keslinka-Nawrot, L.; Błachowiak-Samołyk, K. Parental efforts of an Arctic seabird, the little auk *Alle alle*, under variable foraging conditions. *Mar. Biol. Res.* **2015**, *4*, 349–360. [CrossRef]

36. Enstipp, R.; Descamps, S.; Fort, J.; Grémillet, D. Almost like a whale-first evidence of suction feeding in a seabird. *J. Exp. Biol.* **2018**, *221*, jeb182170. [CrossRef] [PubMed]

37. Stempniewicz, L.; Darecki, M.; Trudnowska, E.; Błachowiak-Samołyk, K.; Boehnke, R.; Jakubas, D.; Keslinka-Nawrot, L.; Kidawa, D.; Sagan, S.; Wojczulanis-Jakubas, K. Visual prey availability and distribution of foraging little auks (*Alle alle*) in the shelf waters of West Spitsbergen. *Polar Biol.* **2013**, *36*, 949–955. [CrossRef]

38. Mehlum, F.; Gabrielsen, G.W. The diet of high arctic seabirds in coastal and ice-covered, pelagic areas near the Svalbard archipelago. *Polar Res.* **1993**, *11*, 1–20. [CrossRef]

39. Wojczulanis, K.; Jakubas, D.; Walkusz, W.; Wenneberg, L. Differences in food delivered to chicks by males and females of little auks (*Alle alle*) on South Spitsbergen. *J. Ornithol.* **2006**, *147*, 543–548. [CrossRef]

40. Jakubas, D.; Wojczulanis-Jakubas, K. Subcolony variation in phenology in breeding parameters in little auk *Alle alle*. *Polar Biol.* **2011**, *34*, 31–39. [CrossRef]

41. Hovinen, J.E.H.; Wojczulanis-Jakubas, K.; Jakubas, D.; Hop, H.; Berge, J.; Kidawa, D.; Karnovsky, N.J.; Steen, H. Fledging success of little auks in the high Arctic: Do provisioning rates and the quality of foraging grounds matter? *Polar Biol.* **2014**, *37*, 665–674. [CrossRef]

42. Boehnke, R.; Gluchowska, M.; Wojczulanis-Jakubas, K.; Jakubas, D.; Karnovsky, N.J.; Walkusz, W.; Kwasniewski, S.; Błachowiak-Samołyk, K. Supplementary diet components of little auk chicks in two contrasting regions on the west Spitsbergen coast. *Polar Biol.* **2015**, *38*, 261–267. [CrossRef] [PubMed]

43. Boehnke, R.; Balazy, K.; Jakubas, D.; Wojczulanis-Jakubas, K.; Błachowiak-Samołyk, K. Meso-scale variations in little auk chicks diet composition in NW Spitsbergen. *Polar Res.* **2017**, *36*, 1409585. [CrossRef]

44. Møller, E.F.; Johansen, K.L.; Agersted, M.D.; Rigét, F.; Clausen, D.S.; Larsen, J.; Lyngs, P.; Middelbo, A.; Mosbech, A. Zooplankton phenology may explain the North Water polynya's importance as a breeding area for little auks. *Mar. Ecol. Prog. Ser.* **2018**, *605*, 207–223. [CrossRef]

45. Keslinka, L.K.; Wojczulanis-Jakubas, K.; Jakubas, D.; Neubauer, G. Determinants of the little auk (*Alle alle*) breeding colony location and size in W and NW coast of Spitsbergen. *PLoS ONE* **2019**, *14*, e0212668. [CrossRef] [PubMed]

46. González-Bergonzoni, I.; Johansen, K.L.; Mosbech, A.; Landkildehus, F.; Jeppesen, E.; Davidson, T.A. Small birds, big effects: The little auk (*Alle alle*) transforms high Arctic ecosystems. *Proc. R. Soc. B Biol. Sci.* **2017**, *284*, 20162572. [CrossRef]

47. Cottier, F.; Tverberg, V.; Inall, M.; Svendsen, H.; Nilsen, F.; Griffiths, C. Water mass modification in an Arctic fjord through cross-shelf exchange: The seasonal hydrography of Kongsfjorden, Svalbard. *J. Geophys. Res.* **2005**, *110*, C12005. [CrossRef]

48. Prominska, A.; Cisek, M.; Walczowski, W. Kongsfjorden and Hornsund hydrography-comparative study based on a multiyear survey in fjords of west Spitsbergen. *Oceanologia* **2017**, *59*, 397–412. [CrossRef]

49. Saloranta, E.; Svendsen, H. Across the Arctic Front west of Spitsbergen: High-resolution CTD sections from 1998–2000. *Polar Res.* **2001**, *20*, 174–184. [CrossRef]

50. Nilsen, F.; Cottier, F.; Skogseth, R.; Mattson, S. Fjord-shelf exchanges controlled by ice and brine production: The interannual variation of Atlantic Water in Isfjorden, Svalbard. *Cont. Shelf Res.* **2008**, *28*, 1838–1853. [CrossRef]

51. Sakshaug, E.; Johnsen, G.; Kovacs, K. *Ecosystem Barents Sea*; Tapir Academic Press: Trondheim, Norway, 2009.

52. Pavlov, A.K.; Tverberg, V.; Ivanov, B.V.; Nilsen, F.; Falk-Petersen, S.; Granskog, M.A. Warming of Atlantic Water in two Spitsbergen fjords over the last century (1912–2009). *Polar Res.* **2013**, *32*, 11206. [CrossRef]

53. Amélineau, F.; Bonnet, D.; Heitz, O.; Mortreux, V.; Harding, A.M.A.; Karnovsky, N.; Walkusz, W.; Fort, J.; Grémillet, D. Microplastic pollution in the Greenland Sea: Background levels and selective contamination of planktivorous diving seabirds. *Environ. Pollut.* **2016**, *219*, 1131–1139. [CrossRef] [PubMed]

54. Harris, R.; Wiebe, L.; Lenz, J.; Skjoldal, H.R.; Huntley, M. *ICES Zooplankton Methodology Manual*; Academic Press: London, UK, 2000.

55. Kwasniewski, S.; Hop, H.; Falk-Petersen, S.; Pedersen, G. Distribution of *Calanus* species in Kongsfjorden, a glacial fjord in Svalbard. *J. Plankton Res.* **2003**, *25*, 1–20. [CrossRef]

56. Herman, A.W. Design and calibration of a new optical plankton counter capable of sizing small zooplankton. *Deep-Sea Res. A* **1992**, *39*, 395–415. [CrossRef]

57. Herman, A.W.; Beanlands, B.; Phillips, E.F. The next generation of Optical Plankton Counter: The Laser-OPC. *J. Plankton Res.* **2004**, *26*, 1135–1145. [CrossRef]

58. Herman, A.W.; Harvey, M. Application of normalized biomass size spectra to laser optical plankton counter net intercomparisons of zooplankton distributions. *J. Geophys. Res.* **2006**, *111*, C5. [CrossRef]

59. Basedow, S.L.; Tande, K.S.; Norrbin, M.F.; Kristiansen, S.A. Capturing quantitative zooplankton information in the sea: Performance test of laser optical plankton counter and video plankton recorder in a *Calanus finmarchicus* dominated summer situation. *Prog. Oceanogr.* **2013**, *108*, 72–80. [CrossRef]

60. Anderson, M.J.; Gorley, R.N.; Clarke, K.R. *PERMANOVA+ for PRIMER: Guide to Software and Statistical Methods*; Primer-E Ltd.: Plymouth, UK, 2008.

61. Clarke, K.R.; Gorley, R.N. *Primer*; Primer-E Ltd.: Plymouth, UK, 2001.

62. Anderson, M.J.; Braak, C.J.F. Permutation tests for multi-factorial analysis of variance. *J. Stat. Comput. Simul.* **2003**, *73*, 85–113. [CrossRef]

63. Kwasniewski, S.; Gluchowska, M.; Walkusz, W.; Karnovsky, N.J.; Jakubas, D.; Wojczulanis-Jakubas, K.; Harding, A.M.A.; Goszczko, I.; Cisek, M.; Beszczynska-Möller, A.; et al. Interannual changes in zooplankton on the West Spitsbergen Shelf in relation to hydrography and their consequences for the diet of planktivorous seabirds. *J. Mar. Sci.* **2012**, *69*, 890–901. [CrossRef]

64. Jakubas, D.; Wojczulanis-Jakubas, K.; Boehnke, R.; Kidawa, D.; Błachowiak-Samołyk, K.; Stempniewicz, L. Intra-seasonal variation in zooplankton avialability, chick diet and breeding performance of a high Arctic planktivorous seaird. *Polar Biol.* **2016**, *39*, 1547–1561. [CrossRef]

65. Weydmann, A.; Walczowski, W.; Carstensen, J.; Kwaśniewski, S. Warming of subarctic waters accelerates development of a key marine zooplankton *Calanus finmarchicus*. *Glob. Chang. Biol.* **2018**, *24*, 172–183. [CrossRef]

66. Jakubas, D.; Wojczulanis-Jakubas, K.; Iliszko, L.M.; Strøm, H.; Stempniewicz, L. Habitat foraging niche of a High Arctic zooplanktivorous seabird in a changing environment. *Sci. Rep.* **2017**, *7*, 16203. [CrossRef]

67. Ottersen, G.; Planque, B.; Belgrano, A.; Post, E.; Reid, P.C.; Stenseth, N.C.H. Ecological effects of the North Atlantic Oscillation. *Oecologia* **2001**, *128*, 1–14. [CrossRef] [PubMed]

68. Beaugrand, G.; Brander, K.M.; Lindley, J.A.; Souissi, S.; Reid, P.C. Plankton effect of cod recruitment in the North Sea. *Nature* **2003**, *426*, 661–664. [CrossRef] [PubMed]

69. Visser, M.; Holleman, L. Warmer springs disrupt the synchrony of oak and winter moth phenology. *Proc. R. Soc. B Biol. Sci.* **2001**, *268*, 289–294. [CrossRef] [PubMed]

70. Sanz, J.J.; Potti, J.; Moreno, J.; Merino, S.; Frias, O. Climate change and fitness components of a migratory bird breeding in the Mediterranean region. *Glob. Chang. Biol.* **2003**, *9*, 112. [CrossRef]

71. Thomas, D.W.; Blondel, J.; Perret, P.; Lambrechts, M.M.; Speakman, J.R. Energetic and fitness costs of mismatching resource supply and demand in seasonally breeding birds. *Science* **2001**, *291*, 2598–2600. [CrossRef]

72. Visser, E.M.; Adriaensen, F.; Van Balen, J.H.; Blondel, J.; Dhondt, A.A.; Van Dongen, S.; Du Feu, C.; Ivankina, E.V.; Kerimov, A.B.; De Laet, J.; et al. Variable responses to large-scale climate change in European Parus populations. *Proc. R. Soc. B Boil. Sci.* **2003**, *270*, 367–372.

73. McKinnon, L.; Picotin, M.; Bolduc, E.; Juillet, C.; Bêty, J. Timing of breeding, peak food availability, and effects of mismatch on chick growth in birds nesting in the High Arctic. *Can. J. Zool.* **2012**, *90*, 961–971. [CrossRef]

74. Mackas, D.L.; Batten, S.; Trudel, M. Effects on zooplankton of a warmer ocean: Recent evidence from the Northeast Pacific. *Prog. Oceanogr.* **2007**, *75*, 223–252. [CrossRef]

75. Bertram, D.F.; Mackas, D.L.; Welch, D.W.; Boyd, W.S.; Ryder, J.L.; Galbraith, M.; Hedd, A.; Morgan, K.; O'Hara, P.D. Variation in zooplankton prey distribution determines marine foraging distributions of breeding Cassin's Auklet. *Deep-Sea Res. Pap.* **2017**, *129*, 32–40. [CrossRef]

76. Konarzewski, M.; Taylor, J.R.E.; Gabrielsen, G.W. Chick energy requirements and adult energy expenditures of Dovekies (*Alle alle*). *Auk* **1993**, *110*, 603–609. [CrossRef]

77. Karnovsky, N.J.; Kwasniewski, S.; Weslawski, J.M.; Walkusz, W.; Beszczynska-Møller, A. The foraging behavior of little auks in a heterogeneous environment. *Mar. Ecol. Prog. Ser.* **2003**, *253*, 289–303. [CrossRef]

78. Jakubas, D.; Wojczulanis, K.; Walkusz, W. Response of dovekie to changes in food availability. *Waterbirds* **2007**, *30*, 421–428. [CrossRef]

79. Ragonese, S.; Bianchini, M.L. Growth, mortality and yield-per-recruit of the deep-water shrimp Aristeus antennatus (Crustacea-Aristeidae) of the Strait of Sicily (Mediterranean Sea). *Fish. Res.* **1996**, *26*, 125–137. [CrossRef]

80. Huntley, M.E.; Lopez, M.D.G. Temperature-dependent production of marine copepods: A global synthesis. *Am. Nat.* **1992**, *140*, 201–242. [CrossRef] [PubMed]

81. Atkinson, D. Temperature and organism size—A biological law for ectotherms? *Adv. Ecol. Res.* **1994**, *25*, 1–58. [CrossRef]

82. Angilletta, M.J.; Steury, T.D.; Sears, M.W. Temperature, growth rate, and body size in ectotherms: Fitting pieces of a life-history puzzle. *Integr. Comp. Biol.* **2004**, *44*, 498–509. [CrossRef]

83. Gillooly, J.F.; Brown, J.H.; West, G.B.; Savage, V.M.; Charnov, E.L. Effects of size and temperature on metabolic rate. *Science* **2001**, *293*, 2248–2251. [CrossRef]

84. Daufresne, M.; Lengfellner, K.; Sommer, U. Global warming benefits the small in aquatic ecosystems. *Proc. Natl. Acad. Sci. USA* **2009**, *106*, 12788–12793. [CrossRef]

85. Gardner, J.L.; Peters, A.; Kearney, M.R.; Joseph, L.; Heinsohn, R. Declining body size: A third universal response to warming? *Trends Ecol. Evol.* **2011**, *26*, 285–291. [CrossRef]

86. Banas, N.S.; Møller, E.F.; Nielsen, T.G.; Lisa, B.E. Copepod life strategy and population viability in response to prey timing and temperature: Testing a new model across latitude, time, and the size spectrum. *Front. Mar. Sci.* **2016**, *3*, 255. [CrossRef]

87. Leinaas, H.P.; Jalal, M.; Gabrielsen, T.M.; Hessen, D.O. Inter- and intraspecific variation in body- and genome size in calanoid copepods from temperate and arctic waters. *Ecol. Evol.* **2016**, *6*, 5585–5595. [CrossRef] [PubMed]

88. Daase, M.; Eiane, K. Mesozooplankton distribution in northern Svalbard waters in relation to hydrography. *Polar Biol.* **2007**, *30*, 969–981. [CrossRef]

89. Weydmann, A.; Carstensen, J.; Goszczko, I.; Dmoch, K.; Olszewska, A.; Kwasniewski, S. Shift towards the dominance of boreal species in the Arctic: Inter-annual and spatial zooplankton variability in the West Spitsbergen Current. *Mar. Ecol. Prog. Ser.* **2014**, *501*, 41–52. [CrossRef]

90. Brown, Z.W.; Welcker, J.; Harding, A.M.A.; Walkusz, W.; Karnovsky, N.J. Divergent diving behaviour during short and long trips of a bimodal forager, the little auk *Alle alle*. *J. Avian Biol.* **2012**, *43*, 215–226. [CrossRef]

91. Gluchowska, M.; Kwasniewski, S.; Prominska, A.; Olszewska, A.; Goszczko, I.; Falk-Petersen, S.; Hop, H.; Weslawski, J.M. Zooplankton in Svalbard fjords on the Atlantic-Arctic boundary. *Polar Biol.* **2016**, *39*, 1785–1802. [CrossRef]

92. Trudnowska, E.; Sagan, S.; Kwasniewski, S.; Darecki, M.; Blachowiak-Samaolyk, K. Fine-scale zooplankton vertical distribution in relation to hydrographic and optical characteristics of the surface waters on the Arctic shelf. *J. Plankton Res.* **2015**, *37*, 120–133. [CrossRef]

93. Willis, K.; Cottier, F.; Kwasniewski, S.; Wold, A.; Falk-Petersen, S. The influence of advection on zooplankton community composition in an Arctic fjord (Kongsfjorden, Svalbard). *J. Mar. Syst.* **2006**, *61*, 39–54. [CrossRef]

94. Lischka, S.; Hagen, W. Seasonal dynamics of mesozooplankton in the Arctic Kongsfjord (Svalbard) during year-round observations from August 1998 to July 1999. *Polar Biol.* **2016**, *39*, 1859–1878. [CrossRef]

95. Espinasse, M.; Halsband, C.; Varpe, Ø.; Gislason, A.; Gudmundsson, K.; Falk-Petersen, S.; Eiane, K. Interannual phenological variability in two North-East Atlantic populations of *Calanus finmarchicus*. *Mar. Biol. Res.* **2018**, *14*, 752–767. [CrossRef]

96. Søreide, J.E.; Leu, E.; Berge, J.; Graeve, M.; Falk-Petersen, S. Timing of blooms, algal food quality and *Calanus glacialis* reproduction and growth in a changing Arctic. *Glob. Chang. Biol.* **2010**, *16*, 3154–3163. [CrossRef]

97. Gluchowska, M.; Trudnowska, E.; Goszczko, I.; Kubiszyn, A.M.; Blachowiak-Samolyk, K.; Walczowski, W.; Kwasniewski, S. Variations in the structural and functional diversity of zooplankton over vertical and horizontal environmental gradients en route to the Arctic Ocean through the Fram Strait. *PLoS ONE* **2017**, *12*, e0171715. [CrossRef] [PubMed]

Article

The Copepod *Acartia tonsa* Dana in a Microtidal Mediterranean Lagoon: History of a Successful Invasion

Elisa Camatti *, Marco Pansera and Alessandro Bergamasco

Consiglio Nazionale delle Ricerche, Istituto di Scienze Marine (CNR ISMAR), Arsenale Tesa 104, Castello 2737/F, 30122 Venezia, Italy; marco.pansera@ve.ismar.cnr.it (M.P.); alessandro.bergamasco@ismar.cnr.it (A.B.)
* Correspondence: elisa.camatti@ismar.cnr.it; Tel.: +39-041-2407-978

Received: 13 May 2019; Accepted: 5 June 2019; Published: 8 June 2019

Abstract: The Lagoon of Venice has been recognized as a hot spot for the introduction of nonindigenous species. Several anthropogenic factors as well as environmental stressors concurred to make this ecosystem ideal for invasion. Given the zooplankton ecological relevance related to the role in the marine trophic network, changes in the community have implications for environmental management and ecosystem services. This work aims to depict the relevant steps of the history of invasion of the copepod *Acartia tonsa* in the Venice lagoon, providing a recent picture of its distribution, mainly compared to congeneric residents. In this work, four datasets of mesozooplankton were examined. The four datasets covered a period from 1975 to 2017 and were used to investigate temporal trends as well as the changes in coexistence patterns among the *Acartia* species before and after *A. tonsa* settlement. Spatial distribution of *A. tonsa* was found to be significantly associated with temperature, phytoplankton, particulate organic carbon (POC), chlorophyll *a*, and counter gradient of salinity, confirming that *A. tonsa* is an opportunistic tolerant species. As for previously dominant species, *Paracartia latisetosa* almost disappeared, and *Acartia margalefi* was not completely excluded. In 2014–2017, *A. tonsa* was found to be the dominant *Acartia* species in the lagoon.

Keywords: *Acartia tonsa*; Lagoon of Venice; nonindigenous species; zooplankton distribution; coexistence patterns; niche overlaps; long-term ecological research

---

1. Introduction

Alien invasive species, also called non-native or nonindigenous species (NIS), together with marine pollution, overexploitation of living resources, and physical alteration of habitats, represent the main threats to the world's oceans on local, regional, and global scales [1]. The planktonic crustacean *Acartia tonsa* Dana (1849) (calanoid copepod) is an NIS recently introduced in the Mediterranean Sea [2]. Several studies have shown that *Acartia* NIS are colonizing coastal areas and estuaries by propagation or introduction (e.g., [3,4]). The ability of Acartiidae to cross geographic barriers relies mainly in their capability of producing resting stages [5]. These NIS are modifying the status of native species, which are subject to competitive pressure [6]. *A. tonsa* is widely distributed in estuarine environments along the Atlantic coasts of North and South America [7] and the Pacific coast of North America [8] where it is the most abundant species. It appeared in the European coasts in the first half of the 20th century, possibly transferred by ship ballast waters [9,10]. In the Mediterranean Sea, *A. tonsa* was first reported in 1985 in the Etang de Berre, a eutrophic lagoon near Marseilles, France [2]. Only after 1985 was the presence of *A. tonsa* confirmed in several Italian transitional waters such as in a lagoon of the Po River Delta (northern Adriatic Sea) [11], in the Lagoon of Lesina [12], and in the Lagoon of Venice [13]. In estuarine ecosystems, *A. tonsa* generally reaches relatively high abundances and becomes often the dominant zooplankton species during summer [14], provided that high concentrations of particulate organic carbon and particulate organic matter are available. In fact, the life cycle of this copepod is strictly dependent on the quantity of the available food—the larger the trophic supplies are, the more accelerated its growth rate is [15–17].

The Lagoon of Venice, a large Mediterranean lagoon located in the northwest coast of the Adriatic Sea, presents marked habitat heterogeneity, and the classification of its habitats is still a matter of debate [18,19]. The lagoon is considered the main hotspot for invasive species in the whole Italian coast with the presence of more than 60 NIS, including 29 invertebrates and 34 macrophyte species among which *A. tonsa* also appears [20–25]. The lagoon is also part of the Long-Term Ecological Research (LTER) network (http://www.lteritalia.it), a global network of research sites located in a wide array of ecosystems. LTER research is a fundamental tool for monitoring environmental changes over time. Zooplankton communities exhibit an intrinsically high variability in transitional environments, and the elucidation of coexistence patterns is a critical question in ecology as well as in accounting for differences in abundances among species. In addition to zooplankton ecological relevance related to the role in the marine trophic network, changes in the community have implications for environmental management and ecosystem services.

This work describes relevant steps in the history of invasion and establishment of *A. tonsa* in the Lagoon of Venice with respect to local stressors. This work provides a recent picture of its distribution along gradients of environmental parameters and identifying the relevant biogeochemical quantities assisting or limiting the successful colonization of the species, mainly compared to congeneric residents in the lagoon (*Acartia margalefi* Alcaraz (1976), *Paracartia latisetosa* Krichagin (1873), and *Acartia clausi* Giesbrecht (1889)). The study will focus on abundance, distribution, and niche interactions of the Acartiidae community to address the question of niche separation and to investigate multidimensional niche breadths under a temporal and trophic gradient.

The work aims to identify which ecological factors are most important for *A. tonsa* population structure and organization and to provide a possible key to disentangle the roles of *Acartia* lagoon dominant species based on their niche characteristics. Identification or exclusion of possible overlaps among realized niches in the space of considered environmental variables will aim to highlight the specialization of each species and the relevance of coexistence mechanisms in the structuration of the community. Combining spatial and temporal frames in the three different periods taken into consideration, along a gradient from the mainland to the inlets, determining species' relative abundance distributions (RADs) in a given habitat and time will support the comparison of the reciprocal position of the species of interest within the mesozooplankton community and also how they eventually interacted in a competitive way.

2. Materials and Methods

2.1. Study Area

The Lagoon of Venice is a large Mediterranean lagoon with an area of approximately 500 km^2 stretching along the northwest coast of the Adriatic Sea (Figure 1). Three inlets connect the lagoon to the sea, and its general circulation results from the superposition of tide, wind, and topographic control [26,27]. The effective renewal rate of water is on the order of a few days for the areas closest to the inlets and up to a month for the innermost areas [26,28]. Its depth is very shallow, some 1.3 m on average and more than 10 m in the deepest channels connecting the three inlets with the city of Venice and with the industrial area of Marghera. The Lagoon of Venice is surrounded by urban and industrial areas. Moreover, fishery, mariculture, and tourist activities are very developed. Because of the complex interplay of the variety of simultaneously occurring stressors, the lagoon experiences high variability in most of the environmental parameters, showing high habitat heterogeneity [29,30].

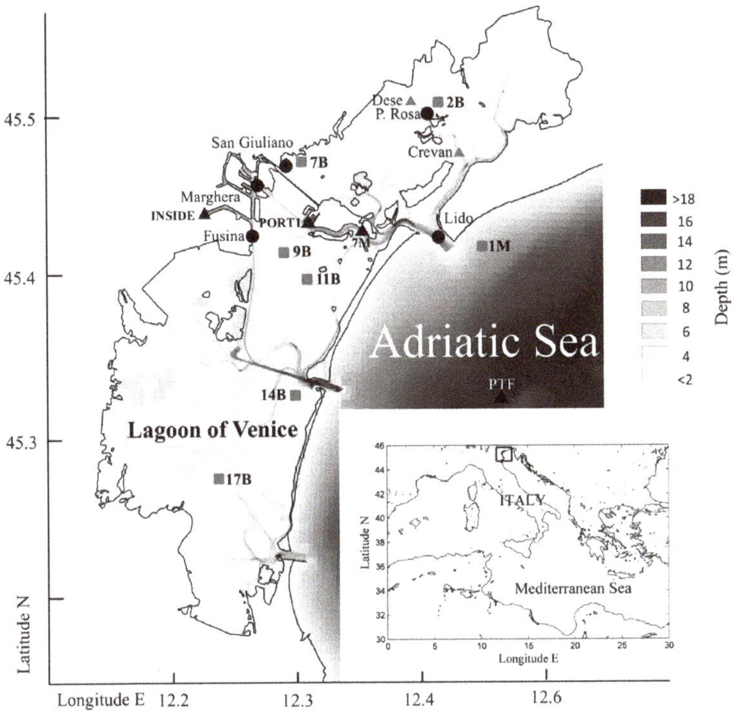

Figure 1. Study area (Lagoon of Venice, Italy) and location of sampling stations during 1975–1980 (grey triangles, including Lido), 1997–2002 (black circles), 2003–2004 (grey squares), and 2014–2017 (black triangles, including San Giuliano, Marghera, Fusina, and Lido).

2.2. Zooplankton and Environmental Variables Datasets

Four zooplankton datasets (DSs) from surveys carried out in the Lagoon of Venice were considered in this study: the first dataset refers to the situation before the settlement of *A. tonsa* (DS1: period 1975–1980, monthly sampling) and the three after the first record *A. tonsa* in the lagoon (DS2: period 1997–2002, monthly sampling; DS3: period 2003–2004, monthly sampling; and DS4: period 2014–2017, seasonal sampling).

The first dataset (DS1), concerning data prior to the establishment of *A. tonsa*, was used for RAD analysis to compare the different relative species' abundance distributions as a function of the presence of *A. tonsa* along an expected environmental gradient from the sea to the mainland (Lido inlet, Crevan, Dese; Figure 1). The second dataset (DS2), the most evenly distributed in terms of spatial and temporal coverage, was used to investigate patterns and trends of *A. tonsa* almost immediately after its first record in an intermediate area of the lagoon, called Palude della Rosa. In this case, zooplankton samplings were collected from May 1997 to April 2002 at five stations located in the northern part of the lagoon (San Giuliano, Marghera, Fusina, Lido, and Palude della Rosa; Figure 1) characterized by a complex interplay of freshwater and marine inputs [26,28] and by anthropogenic pressure [31]. San Giuliano collects urban waste water from the town of Mestre, where phytoplankton blooms often develop [32,33]. Marghera is influenced by industrial pollution. Fusina is affected by thermal pollution from a thermo-electric power plant. Lido, in the northernmost inlet of the lagoon, is characterized by substantial lagoon shelf exchanges. Palude della Rosa is a typical lagoon environment, influenced both by freshwater and, to a lesser extent, by shelf water transported this far by tides.

The recent spatial distribution of *A. tonsa* populations in the lagoon, and its potential diverse response in areas subjected to different natural and anthropogenic influences, were investigated through the analysis of samples derived from monthly samplings, from April 2003 to March 2004, at six stations inside the lagoon and one station in the Adriatic Sea site just outside the Lido inlet (third data set DS3, Figure 1). Thus, with respect to the 1997–2002 stations, this dataset provided more coverage and representativeness of different areas in the lagoon with different natural and anthropogenic stressors, including a station on the outer shelf. Station 1M (Figure 1) was chosen to verify the presence of the species in the coastal sea, whereas the other stations were selected based on the following rationale: station 2B was influenced by freshwaters; station 7B collected urban waste; station 9B was affected by thermal pollution; station 11B was located in an intermediate area between northern and central lagoon basins; station 14B was in a site with seagrass meadow; and station 17B was located in the southern lagoon basin (Figure 1). Station 1M was not sampled in September 2003 and February 2004, and station 17B was not sampled in April 2003. All samples reported abundances of phytoplankton and mesozooplankton organisms as well as physical and chemical parameters related to water quality, in particular: temperature, salinity, dissolved oxygen, chlorophyll *a*, and particulate organic carbon (POC).

Samples related to the most recent period 2014–2017 (DS4) were used to confirm the presence and distribution of *A. tonsa* ten years after the previous survey along a gradient from the inner lagoon to the outer shelf. In this case, eight stations were sampled. The Inside station (Figure 1) was located in the inner channel inside the industrial harbour of Marghera, Port1 was in the lee of the cruise ships docks, 7M was at the yachting mooring docks, and PTF was the CNR Acqua Alta Oceanographic tower (http://www.ismar.cnr.it/infrastructures/piattaforma-acqua-alta), situated few miles offshore the lagoon in the Northern Adriatic. The remaining four stations corresponded to Marghera, Fusina, San Giuliano, and Lido sites of the period 1997–2002. All the considered 65 samples reported abundances of mesozooplankton organisms (66 present taxa) as well as physical and chemical parameters related to water quality, in particular: temperature, salinity, dissolved oxygen, chlorophyll *a*, nutrients, and Secchi disk.

Sampling procedures and analyses were performed in the same way for all datasets; thus, data from different years and projects were homogeneous and comparable with each other. Zooplankton samples were collected by a plankton horizontal sampler (200 μm mesh net size) and preserved with borax buffered formaldehyde for microscopic analysis. Taxonomic and quantitative zooplankton determinations were performed using a Zeiss stereomicroscope at the lowest possible taxonomic level (species for copepods and cladocerans). Each sample was poured into a beaker with 200 cm^3 of filtered seawater to allow a thorough mixing for random distribution of the organisms. At least four aliquots of the samples were analysed while the entire sample was checked against the identification of rare species [34,35]. Phytoplankton samples resulting from the DS3 were collected in 250 cm^3 dark glass bottles and fixed with 10 cm^3 of 20% hexamethylentetramine buffered formaldehyde [36]. Counts were taken according to Utermöhl's (1958) [37] method, using an inverted microscope (Zeiss Axiovert 35) after 2 to 10 cm^3 subsamples had settled for 24 h [38]. Temperature, salinity, and dissolved oxygen were measured in situ using a CTD (Conductivity Temperature Depth) Idronaut Ocean Seven 316 Multiprobe. Particulate organic carbon (POC) was determined using a CHN (Carbon Hydrogen Nitrogen) analyser [39], whereas surface chlorophyll *a* (Chl *a*) was analysed with spectrophotometric methods [40].

2.3. Statistical Analyses

In order to characterize and explain the *A. tonsa* population dynamics and trends after its arrival in the lagoon, a seasonal Kendall test (SK) [41] was performed over the data set 1997–2002 (DS2). The test performed the Mann–Kendall (MK) trend test for individual seasons of the year, where season was defined by the user (here, corresponding to monthly values). It then combined the individual results into an overall test to depict whether the dependent variable changed in a consistent direction (monotonic trend) over time or not.

To ascertain if species abundances and environmental parameters were associated, the nonparametric Kendall rank correlation test was carried out [42] on the 2003–2004 dataset (DS3). A correspondence analysis (CA) [43] was carried out on this dataset to highlight the possible relationships and groupings between stations (representative of different habitats) and zooplankton species (proxy of different water masses). CA is the most suitable statistical technique to analyse enumerative data [44]. As it is based on the chi squared metric, this algorithm automatically weights both low and high frequencies. The dataset was organized in a species/station matrix. Statistical analyses were performed using MATLAB® software.

Species RADs were used to support the comparison of the whole zooplankton community throughout the four decades and the different habitats covered by the available datasets. In particular, three types of habitats were identified along a gradient from the mainland to the sea: the innermost shallow zones characterized by low water circulation (inner), the inlet areas at the interface between the lagoon and the Adriatic Sea with strong marine characters (inlet), and the transition areas with less direct links to the sea (intermediate). For these three habitats, all available samples of each of the three decades (decade D1: 1975–1985, from data set DS1; decade D3: 1995–2005, from data set DS3; recent period D4: 2005–2017, from DS4) were integrated to obtain 3 × 3 snapshots (before invasion, early settlement, and more mature condition in each habitat) of the RADs of the overall zooplankton community.

Relationships between environmental variables and zooplankton community composition in the current decade (D4)—with particular focus on *A. tonsa* and the congeneric *A. margalefi*, *P. latisetosa*, and *A. clausi*—were studied using an outlying mean index (OMI) analysis through the 'ade4' package in R, extended to the overall zooplankton taxa (dataset DS4). OMI analysis [45] is a multivariate technique to perform niche analyses of species assemblages and explore the relationships between environmental gradients and community structure. The focus is on a specialization criterion called OMI index (i.e., species marginality), which measures the distance between the average habitat conditions used by each species and the average habitat conditions of the sampling area. To characterize the realized niche of each species, the analysis extracts two other terms: the tolerance index (Tol), which measures the habitat breadth of the species, and the residual tolerance (RTol), which represents the variance in the species niche not explained by the measured environmental variables.

3. Results

During the decade before the first record of *A. tonsa* in lagoon (DS1), the mesozooplankton community was composed of more than 40 taxa, with a weak decreasing gradient of overall richness from the inlet (46 taxa) to the intermediate areas (43 taxa) and to the inner zone (41 taxa). Among them, copepods were represented by 24 species, and Acartiidae largely dominated the community with *A. clausi* being the most abundant taxon in every habitat. The species *A. tonsa* was not present in any sample of DS1, even in the innermost areas, where the three dominant species (*A. clausi*, *A. margalefi*, *P. latisetosa*) represented more than 80% of the overall community.

Conversely, over the analysed 1997–2002 period (DS2), *A. tonsa* reached higher abundances at the inner lagoon stations San Giuliano (Figure 2) and Palude della Rosa (Figure 2) and constituted about 90% of the lagoon mesozooplankton community (Figure 3). Seasonal cycles (Figure 2) showed that the species was, generally, nearly absent in the cold season while annual peaks in abundances were reached in summer. The population started to rapidly increase in May, when suitable water temperatures (>15 °C) assisted the growth, and decreased in fall. Annual maxima were reached in different months, depending mainly on the station features. San Giuliano and Palude della Rosa, the stations with larger abundances, had annual peaks in July.

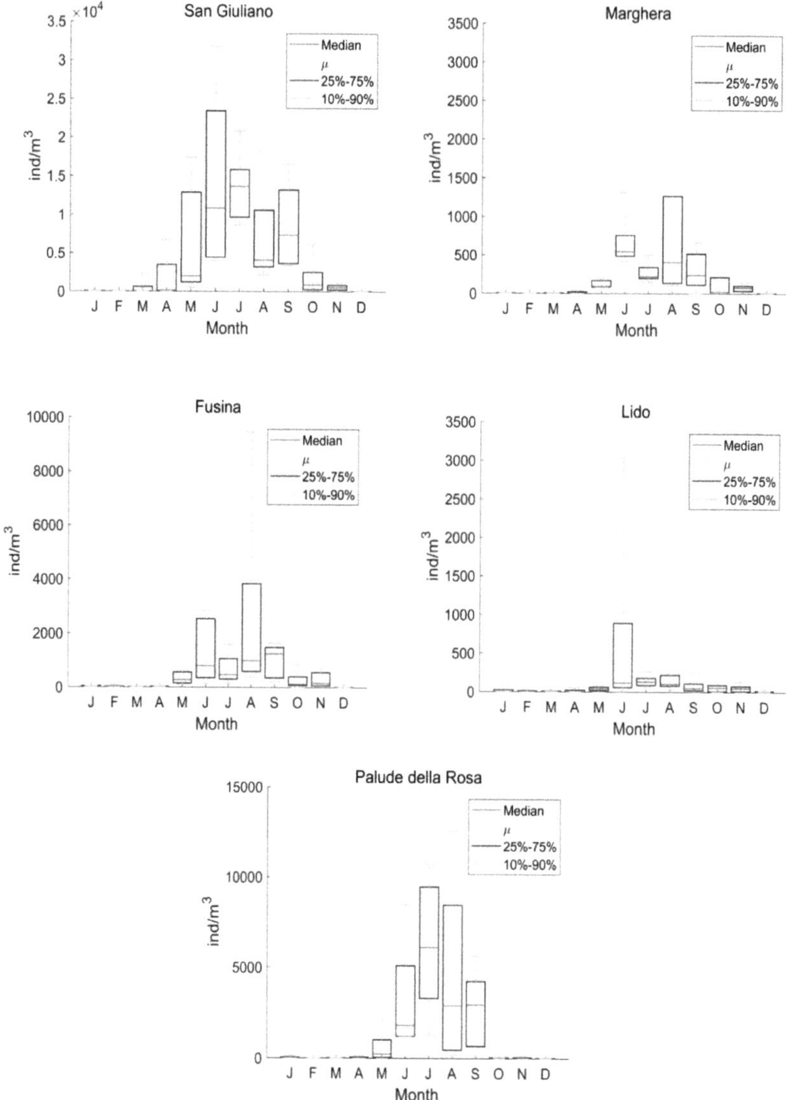

Figure 2. Box plot of monthly *A. tonsa* abundance from the dataset 1997–2002 for the five sampled stations (San Giuliano, Marghera, Fusina, Lido, and Palude della Rosa). The cross (µ) indicates the average, 25–75% indicates the interquartile range, and 10–90% indicates the interdecile range.

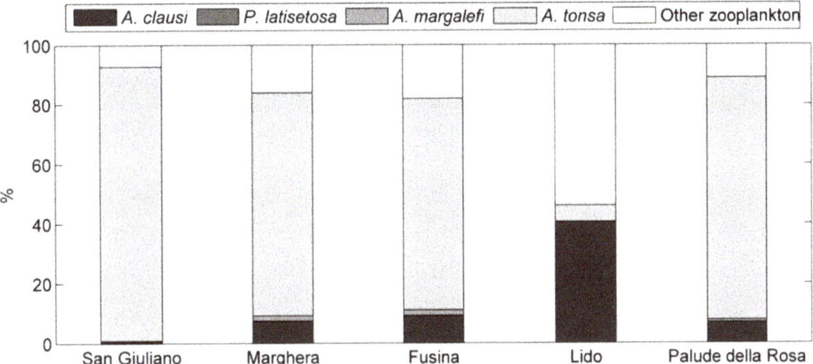

Figure 3. Relative abundance, station by station, of *A. tonsa* with respect to the other *Acartia* species and the remaining zooplankton during the 1997–2002 period.

SK analysis showed a statistically significant increase of the species only at the San Giuliano station (Table 1) along with total zooplankton abundance. The positive trend of total zooplankton appeared to be due to *A. tonsa* exclusively. At Palude della Rosa, the second station in order of abundance, the trend of *A. tonsa* and total zooplankton abundance was also positive, although not statistically significant (Table 1). In the other stations, where *A. tonsa* abundance was generally lower and reached the minimum at Lido station, the trend was found to be negative and not significant (Table 1).

Table 1. Results of the seasonal Kendall test (1997–2002 period).

		Acartia clausi	*Acartia margalefi*	*Acartia tonsa*	Total Zooplankton
San Giuliano	Tau	−0.121	−0.577	0.243	0.300
	p	0.311	0.000	0.040	0.011
Fusina	Tau	−0.125	−0.687	−0.009	−0.008
	p	0.303	0.000	0.943	0.943
Marghera	Tau	−0.346	−0.728	−0.153	−0.142
	p	0.003	0.000	0.199	0.228
Lido	Tau	−0.144	−0.540	−0.094	−0.083
	p	0.225	0.003	0.430	0.480
Palude della Rosa	Tau	−0.059	−0.688	0.078	0.067
	p	0.620	0.000	0.514	0.572

Despite the positive, significant trend of *A. tonsa* at San Giuliano, the estimate of Sen's slope [46] was 30 ind (number of individuals)/m^3/year. This negligible increase over time (Figure 4) may be suggestive of a steady state of the population (i.e., a mature stage of colonization). Alongside the significantly increasing trend of *A. tonsa* abundance at the San Giuliano station, SK highlighted an opposite trend for the other representative species of the *Acartia* genus historically relevant in the lagoon: *A. margalefi* and *A. clausi* (Table 1). In particular, *A. margalefi* significantly (in a statistical sense) decreased at all stations, providing additional evidence of the ongoing decline in abundance of this species.

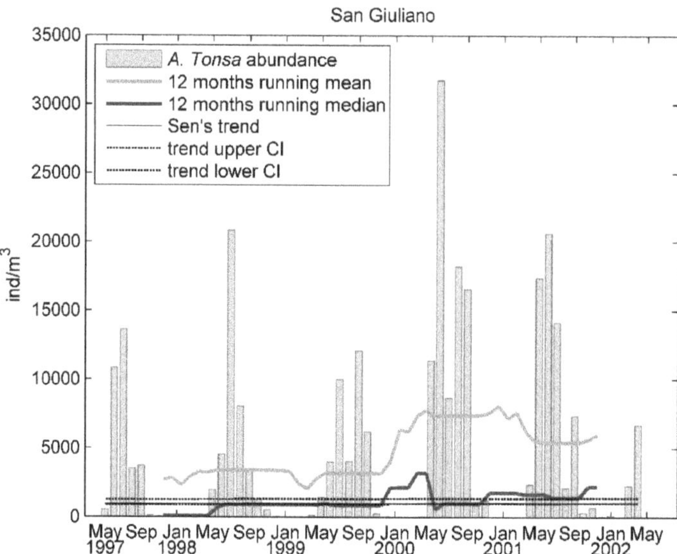

Figure 4. Time series of monthly *A. tonsa* abundance (bars) along with a 12-month running mean and median and Sen's trend line at the San Giuliano station during 1997–2002.

Analyses related to the April 2003 to March 2004 period (DS3) allowed characterization of the actual distribution of *A. tonsa* over an environmental gradient in the Lagoon of Venice. The presence of a decreasing gradient of salinity can be noted, shifting from stations near inlets to stations well inside the lagoon (Table 2). Higher mean salinity values were found at station 14B, which was located near an inlet and, therefore, directly influenced by tidal exchanges, and the marine coastal station 1M (Table 2). Lower salinity values were observed at the innermost stations (2B, 7B, and 9B). The presence of *A. tonsa* at station 1M was negligible and exclusively was due to ebb tidal transport, as this species is usually not resident in marine waters. Indicators of trophic water quality—particulate organic carbon (POC), Chl *a* concentration, and phytoplankton abundance—showed an opposite gradient with respect to salinity, with the larger values at stations 2B, 7B, and 9B (Table 2). The species recorded the largest abundances in those stations (2B and 9B) where POC, Chl *a*, and phytoplankton values were high and where also the mean annual mesozooplankton abundances were highest (6386 ± 10,327 and 4892 ± 11,350 ind m^{-3} respectively; Table 2). Lowest abundance at the site 14B corresponded to the lowest concentrations of POC and Chl *a* as well as low phytoplankton abundance (Table 2).

Percentage contributions of the most abundant taxa to total zooplankton community in each station are reported in Table 3. A few species, such as the copepods *A. tonsa*, *A. clausi*, *Paracalanus parvus* Claus (1863), and *Centropages ponticus* Karavaev (1895), were always present in the zooplankton community over the whole investigated area with a percentage contribution greater than 76%. *A. tonsa* was the most abundant in all lagoon stations with the exception of 14B and 1M stations, which were dominated by *A. clausi*. The percent contributions of the codominants *P. parvus* and *C. ponticus* were generally lower than 5% except for the 14% contribution of *P. parvus* at site 1M. The cladoceran *Penilia avirostris* Dana (1849) was present only at the marine coastal station 1M. The other taxa (i.e., decapods larvae) contributed with low abundances to zooplankton composition.

Table 2. Average and standard deviation values of the hydro-chemical data and biological concentrations (phytoplankton, *A. tonsa*, and main mesozooplankton groups) at the seven stations for the 2003–2004 period.

Stations	1M		2B		7B		9B		11B		14B		17B	
	Avg	SD	Avg	SD	Avg	SD	Avg	SD	Avg	SD	Avg	SD	Avg	SD
Temperature (°C)	16	8	16	9	17	9	18	8	16	10	16	10	16	9
Salinity	35	2	29	4	32	4	32	3	34	3	35	2	33	4
Relative oxygen (%)	96	5	100	9	102	15	90	7	83	19	92	7	95	5
POC (μg L^{-1})	308	73	750	630	747	415	668	386	485	211	345	141	947	886
Chlorophyll *a* (μg L^{-1})	1.78	0.92	4.40	4.32	7.02	6.21	7.81	12.69	2.09	1.46	1.54	0.55	3.84	4.52
Total phytoplankton (cell L^{-1})	2541	3176	10,831	16,019	4382	4592	5716	6874	2082	2406	2128	1993	8318	15,745
Acartia tonsa (ind m^{-3})	23	69	5998	10,333	976	1658	4456	11,209	2823	4986	75	192	1234	2744
Other copepods (ind m^{-3})	3702	5494	316	415	521	1015	257	312	602	588	588	1170	144	167
Cladocerans (ind m^{-3})	186	328	2	5	8	26	4	10	9	14	8	22	0	1
Other zooplankton (ind m^{-3})	142	115	69	110	136	290	176	244	161	272	56	54	42	53
Total zooplankton (ind m^{-3})	4053	5906	6386	10328	1642	2062	4893	11349	3595	5207	727	1233	1419	2781

With respect to the entire zooplankton community, total abundances increased in late spring and summer at all stations except 1M where, instead, spring maxima were observed in May (17×10^3 ind m^{-3}) and June (8×10^3 ind m^{-3}) mainly because of the copepod *A. clausi* (Figure 5). Similar to the datasets 1997–2002 (DS2), the warm season corresponded to the maxima in abundances of *A. tonsa*. In general, the main mesozooplankton groups showed seasonal fluctuations with maxima during warm periods; copepods dominated throughout the year. *A. margalefi* showed very low abundances (maximum mean value of 10 ind m^{-3} in June), while *P. latisetosa* was never found. The rest of the zooplankton community was mainly represented by the meroplankton, especially by decapod, gastropod, and cirriped larvae during the spring and summer periods (Figure 5).

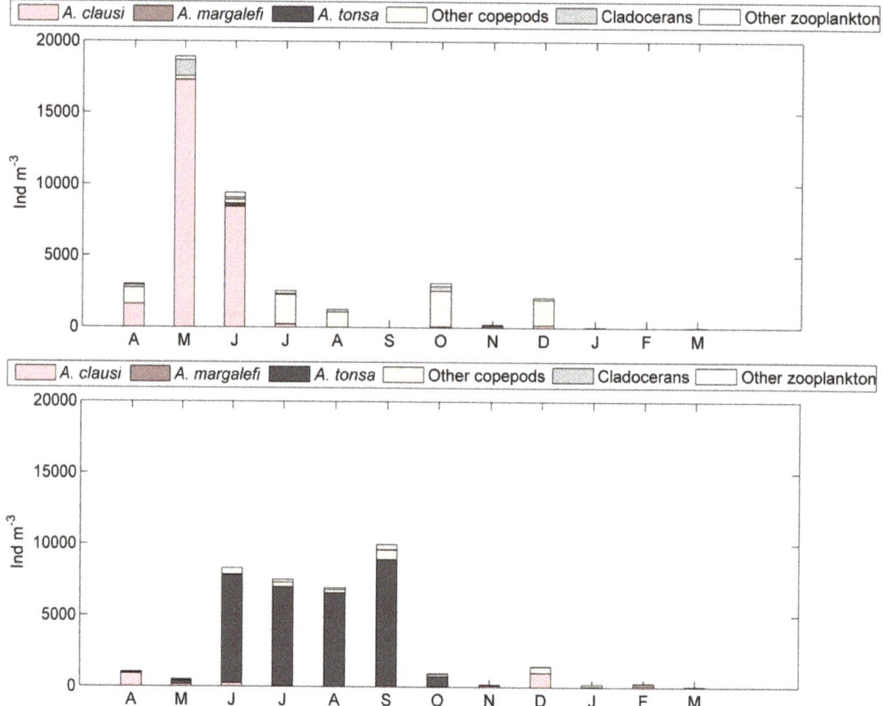

Figure 5. Monthly abundance of *A. tonsa* with respect to other *Acartia* species and remaining zooplankton at coastal station 1M (**top panel**) and at lagoon stations (**bottom panel**) for the 2003–2004 period.

A match between abundances and distribution of *A. tonsa* and environmental parameters emerged from the Kendall rank correlation test for DS3 (Table 4), which showed that *A. tonsa* was the only species of dominant copepods positively and significantly correlated with temperature, phytoplankton, POC and Chl a concentrations, as well as *A. margalefi* species. However, the positive correlation of *A. margalefi* with POC was not significant. With respect to the same parameters, the congener *A. clausi* showed the opposite correlation and was not statistically significant.

Table 3. Mean percentage contributions of the most abundant species/taxa at the seven stations for the 2003–2004 period.

	1M	2B	7B	9B	11B	14B	17B
Acartia clausi	69	1	21	1	8	59	4
Acartia tonsa	1	94	59	91	79	10	87
Centropages ponticus	4	1	2	1	2	2	1
Corycaeus spp.	1						
Harpacticoid spp.			3			2	1
Oithona similis	1					6	
Oncaea spp.						2	
Paracalanus parvus	14	2	4	2	5	5	2
Pseudocalanus elongatus	1				1	1	1
Penilia avirostris	1						
Podon polyphemoides	3						
Chaetognaths	1						
Cirriped larvae			1	1			
Decapod larvae			6	2	3	1	1
Fish eggs	1			1			
Gastropod larvae						2	

Table 4. Results of the Kendall rank correlation test at seven stations for the 2003–2004 period. Statistically significant values are marked in bold.

	Temperature	Salinity	Oxygen	Dissolved Inorganic Nitrogen (DIN)	Chlorophyll *a*	Particulate Organic Carbon (POC)	Phytoplankton
	(°C)		(%)	(μg L^{-1})	(μg L^{-1})	(μg L^{-1})	(cell L^{-1})
Acartia clausi	−0.15	−0.09	0.02	0.09	−0.03	−0.03	−0.10
Acartia margalefi	**0.38**	−0.07	0.09	**−0.31**	**0.31**	0.24	**0.35**
Acartia tonsa	**0.49**	0.12	−0.17	**−0.48**	**0.54**	**0.45**	**0.62**
Centropages ponticus	**0.49**	**0.41**	**−0.25**	**−0.54**	**0.35**	**0.25**	**0.44**
Corycaeus spp.	**−0.54**	−0.05	0.14	**0.39**	**−0.43**	**−0.37**	**−0.53**
Harpacticoid spp.	0.17	−0.07	0.01	−0.10	0.20	**0.23**	0.12
Oithona similis	**−0.43**	−0.07	0.20	**0.24**	**−0.30**	−0.21	**−0.37**
Oncaea spp.	**−0.39**	0.05	0.02	0.23	**−0.28**	**−0.31**	**−0.36**
Paracalanus parvus	0.08	**0.31**	−0.11	−0.19	0.09	0.04	0.07
Pseudocalanus elongatus	**−0.58**	−0.15	0.20	**0.43**	**−0.45**	**−0.33**	**−0.48**
Penilia avirostris	0.19	**0.30**	**−0.25**	−0.24	0.14	0.01	0.16
Podon polyphemoides	0.17	0.06	−0.09	−0.22	0.15	0.01	0.24
Chaetognats	0.13	**0.29**	−0.15	**−0.23**	0.11	−0.03	0.15
Cirriped nauplii	**0.42**	0.10	−0.18	**−0.35**	**0.41**	**0.31**	**0.33**
Decapod larvae	**0.36**	**0.25**	−0.20	**−0.46**	**0.37**	**0.30**	**0.42**
Fish eggs	**0.44**	0.24	−0.17	**−0.41**	**0.39**	0.19	**0.42**
Gastropod larvae	0.19	0.21	−0.11	**−0.24**	0.13	0.05	0.08

The CA was suggestive of a separation of the stations into one main group, which included lagoon stations 2B, 9B, 17B, 11B, and 7B, and the other two isolated stations located (i) in an area filled by seagrass in the central basin close to an inlet (14B), which was heavily influenced by tidal exchanges, and (ii) the coastal station 1M (Figure 6). The main group, associated mainly with inner lagoon sites, included all stations where *A. tonsa* dominated and where all trophic parameters favored its abundance. These stations had common species compositions because their species percentage distributions were similar with respect to all other stations. In fact, the same species percentage distribution considerably differed at the inlet station 14B and coastal station 1M. Indeed, Figure 7 shows that station 14B was characterized by a more heterogeneous community at the taxa level, and that higher percentages of marine taxa prevailed at station 1M such as appendicularians and cladocerans *P. avirostris*, *Evadne* spp., and *Podon polyphemoides* Leuckart (1859). The strictly neritic species *A. clausi* was found in both stations 1M and 14B, sites with high salinity and low inorganic nutrient waters, confirming the preference for coastal waters. Station 7B, the most distant in CA space from the other lagoon stations, was characterized mainly by the presence of *A. margalefi*, fish larvae, and copepod nauplii.

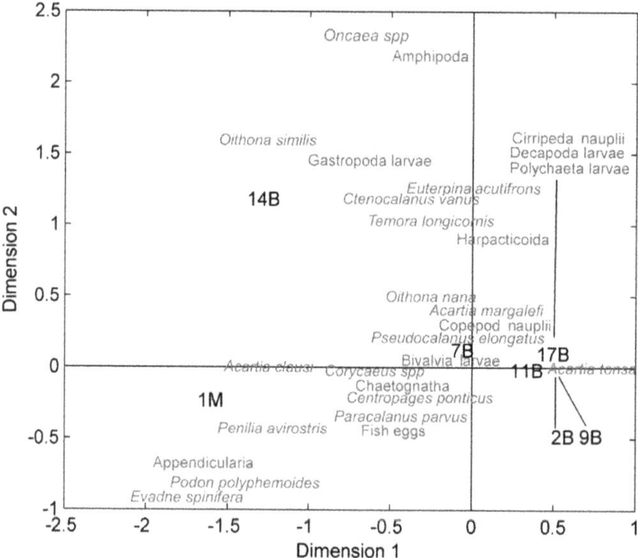

Figure 6. Results of the correspondence analysis (CA) performed on seven stations and total annual abundances (May 2003–April 2004). The first letter of each species' name corresponds to the actual position on the CA space. For enhanced readability, only species with relative abundance >5‰ are shown. Stations B2 and B9 as well as Polychaeta larvae, Decapoda larvae, and Cirripeda nauplii species have been moved from their respective original positions, indicated by the black lines.

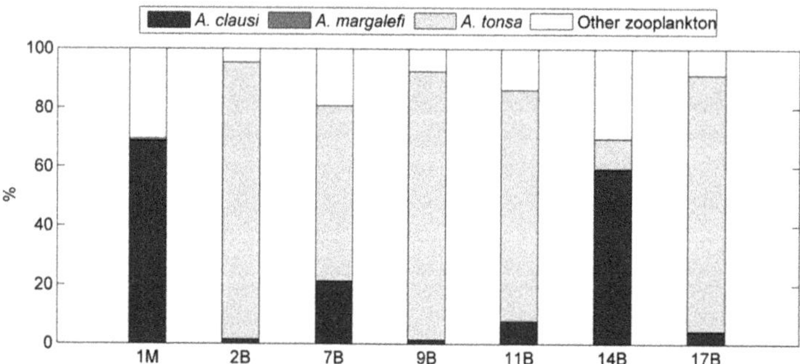

Figure 7. Relative abundance, station by station, of A. tonsa with respect to the other Acartia species and the remaining zooplankton during 2003–2004 period.

Data related to the last period 2014–2017 (DS4), though only seasonal, confirmed the presence and the dominance of three principal *Acartia* species inside the lagoon: *A. margalefi*, *A. clausi* and *A. tonsa*. A different distribution of these species along physical and trophic gradients was also found: *A. margalefi* coexisted with *A. tonsa* mainly in the intermediate lagoon areas (Figure 8). Both virtually disappeared at stations with marine features (Lido and PTF) where *A. clausi* instead dominated. *P. latisetosa* was found with very limited abundance in the inner lagoon stations (Inside, Marghera, and Fusina) only in May 2014 at 4 ind m^{-3} or less.

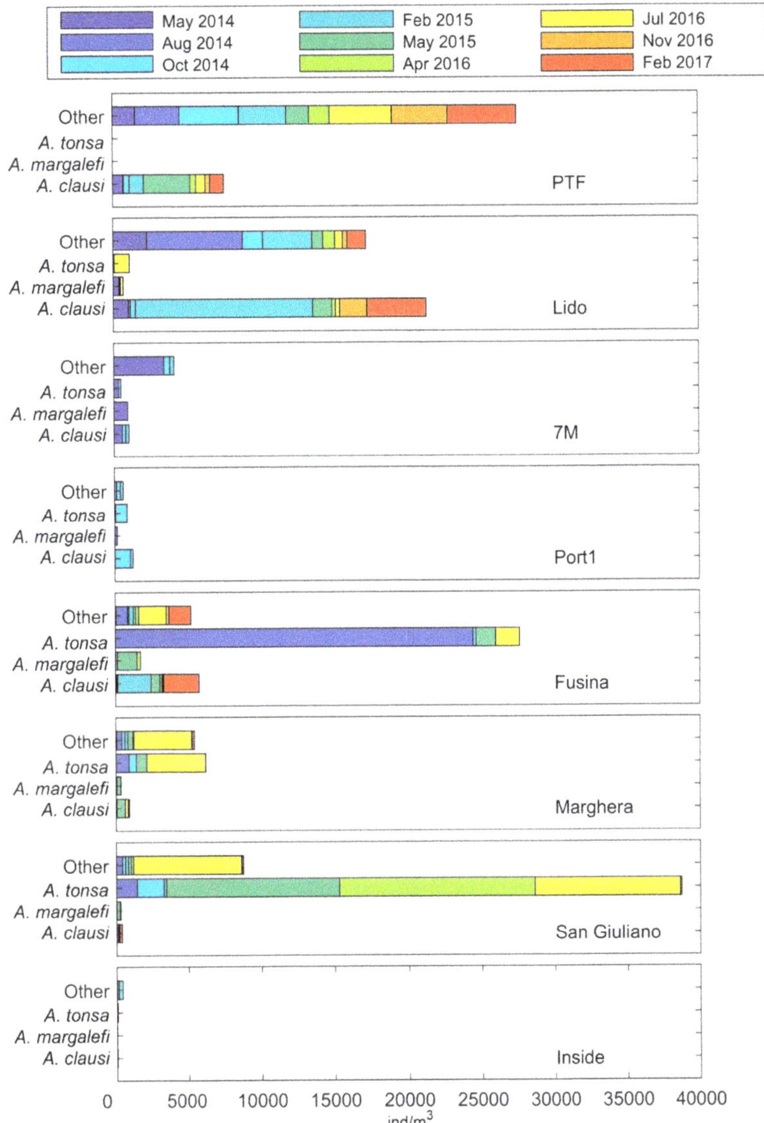

Figure 8. Seasonal relative abundance of *A. tonsa* with respect to other *Acartia* species and remaining zooplankton at stations for the 2014–2017 period. Data related to 7M, Port1, and inside refer only to May 2014, October 2014, and February 2015 sampling dates.

RDA analysis showed that the relative rank distribution of the most representative zooplankton species in the community changed over time and in space by comparing the periods before and after the arrival of *A. tonsa*. First of all, Figure 9 shows that much of the RAD shape was steeply dominated by the arrival of *A. tonsa*, especially in inner areas where the species reached maximum abundance values and exceeded those of the congeneric ones. *A. clausi* maintained the most representative relative abundance of the *Acartia* genus throughout the period prior to the arrival of *A. tonsa* and in all three considered areas (Figure 9). It maintained the same relative rank also in the following periods but only

in the inlet areas. In the years following its appearance, *A. tonsa* replaced *A. clausi* in the intermediate
and inner areas. *P. latisetosa* reached an important distribution rank only in the internal areas and
in the pre-*A. tonsa* period, after which it could always be found on the low-rank curve tail both in
spatial and temporal terms. *A. margalefi*, also with an important relative abundance, before the arrival
of *A. tonsa*, especially in the most typically lagoon areas (intermediate and inner), fell in rank with the
appearance of *A. tonsa* (D3), but then (D4) recovered in terms of presence and abundance mainly in the
intermediate areas (Figure 9).

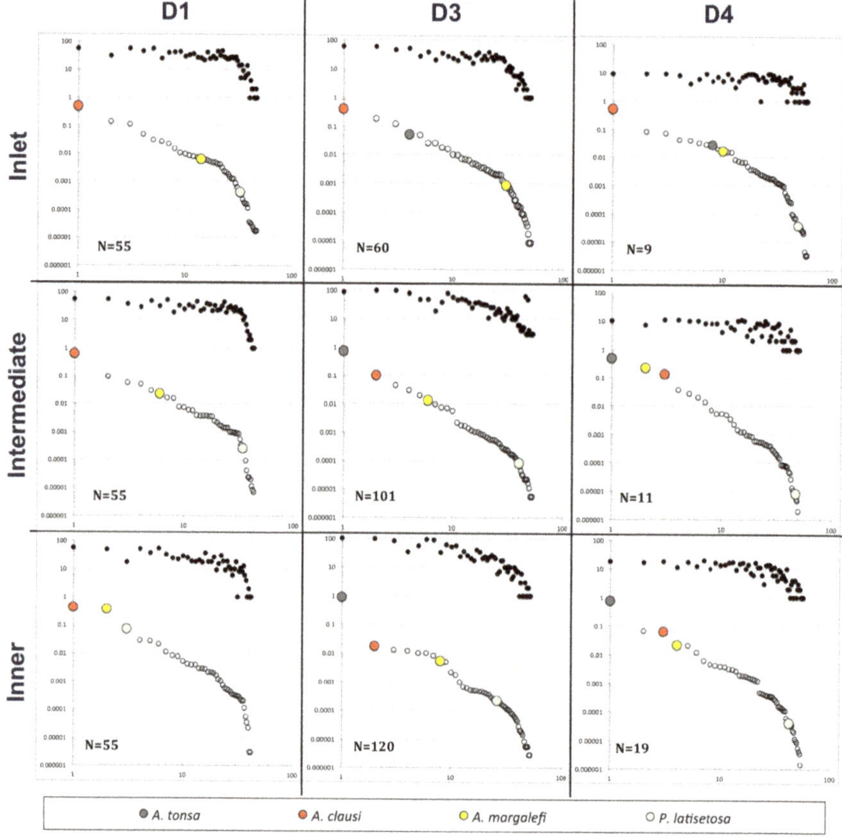

Figure 9. Rank abundance distribution (RAD) and occurrence of the zooplankton community of
the Venice lagoon during the period of study (decade D1: 1975–1985, from data set DS1; decade D3:
1995–2005, from data set DS3; and recent period D4: 2005–2017, from data set DS4). The samples were
lumped according to the three habitat types along a gradient from the mainland to the sea (see text).
In every frame, the upper curve gives the ranked occurrence of each species in the lumped N samples;
the *Acartia* species's relative ranks are highlighted along the lower RAD curve.

With an OMI analysis, we computed and tested niche parameters (Table 5) to describe marginality,
tolerance, and, thus, the variability of responses of *Acartia* species to environmental variables as well
as their possible niche overlap. Figure 10 shows the representation of the statistically significant
species' realized niche position of *Acartia* species on the first factorial plane of the OMI analysis whose
origin represents the average habitat conditions of the sampling area. The two first axes of the OMI
analysis accounted for 81% of the marginality (69% for the first axis). As a consequence, subsequent
graphs used only these two axes. Seventeen out of 66 taxa showed a significant deviation (p < 0.05) of

their niche from the origin (whereas among the Acartiidae, only *A. clausi* was significant, $p < 0.01$) suggesting a significant influence of the environmental conditions for a relevant part of the community (Monte-Carlo krandtest, perm = 999). Furthermore, the between-site analysis confirmed that the environmental gradient of sites (sea, inlet, intermediate, and inner) could be discriminated through the inhabiting zooplankton community (Monte-Carlo rtest, perm = 99, $p < 0.01$). In our case, the tendency of low values of marginality for *Acartia* species indicated that there was no significant difference between the average overall environmental conditions and those where the species were preferentially found. The niches of the three typical lagoon species, *P. latisetosa*, *A. margalefi*, and *A. tonsa*, gravitated around less salty and more trophic environments and also presented an evident overlap. Furthermore, *A. tonsa* and *A. clausi*, the latter having more neritic coastal characteristics, were equidistant from the average environmental conditions. Therefore, this showed a clear affinity, or the opposite, depending on the environment taken into consideration: coastal marine sites for *A. clausi* (sea and inlet in the chart) and lagoon for *A. tonsa* (intermediate and inner). In particular, *A. tonsa* had a low marginality and a high residual tolerance to environmental conditions, whereas *A. clausi* showed an opposite trend but was still relative low OMI values (Table 5). As for the other two typical lagoon species, *A. margalefi* had the lowest value of OMI, while *P. latisetosa* had the lowest RTol in comparison to the rest of the *Acartia* species (Table 5). This meant that *A. clausi* was greatly influenced by environmental conditions, whereas *P. latisetosa* seemed to poorly depend on unconsidered environmental parameters (e.g., POC). Instead, *A. tonsa* showed wider potential ecological tolerance. In general, most common habitat conditions covered by the sampling units corresponded to the ubiquitous or generalist species. In contrast, specialists, which deviated from these general habitat conditions, demonstrated high OMI values, as in the case of *P. latisetosa* with respect to the other two lagoon congeners. Tol values showed a high correlation and dependence of *P. latisetosa* on environmental variables, while *A. tonsa*, with its high RTol, showed a greater adaptability to the variations of the ecosystem in which it gravitated. *A. margalefi* had the lowest OMI and Tol values.

Table 5. Niche parameters of the 20 most abundant zooplankton taxa in the Lagoon of Venice during 2014–2017 (outlying mean index (OMI) analysis). The inertia of each species, OMI, the tolerance index (Tol), and the residual tolerance index (RTol) are indicated. The last column (p) represents the percentage of random permutations (out of 1000) that yielded a higher value than the observed OMI (significant cases are in bold, $p < 0.05$).

Species	Inertia	OMI	Tol	RTol	p
Acartia tonsa	12.76	2.80	1.35	8.61	0.497
Acartia clausi	9.37	3.13	2.08	4.15	**0.012**
Paracalanus parvus	11.78	3.91	2.58	5.28	**0.034**
Decapoda larvae	19.07	11.62	5.09	2.36	0.620
Acartia margalefi	6.37	1.66	0.90	3.80	0.880
Podonidae	7.09	2.36	0.30	4.43	0.818
Penilia avirostris	8.46	5.69	0.49	2.28	**0.030**
Appendicularia	9.84	3.35	2.38	4.09	0.567
Centropages ponticus	8.07	2.95	1.29	3.82	0.214
Ascidiacea larvae	14.44	10.33	1.55	2.56	0.728
Echinodermata larvae	10.35	6.66	0.79	2.91	**0.015**
Oithona similis	10.68	6.38	0.93	3.36	**0.011**
Nauplii copepoda	7.95	1.93	0.90	5.12	0.819
Evadne nordmani	7.08	5.52	0.21	1.35	0.849
Harpacticoida	13.32	1.63	2.48	9.22	0.802
Clausocalanus spp.	12.91	8.93	0.84	3.15	0.268
Oncaeidae	13.21	8.44	1.74	3.03	**0.055**
Bivalvia larvae	12.25	1.72	1.45	9.08	0.690
Polychaeta larvae	9.46	1.77	1.01	6.67	0.214
Paracartia latisetosa [1]	9.80	4.41	3.11	2.29	0.910

[1] *P. latisetosa* does not fall within the 20 higher ranks.

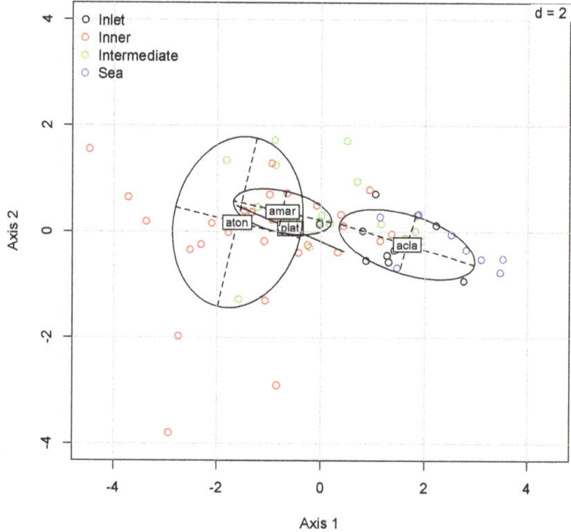

Figure 10. OMI analysis. Realized niche position of *Acartia* species in the Lagoon (2014–2017 dataset). The site scores are projected on the same first two axes of the OMI analysis, and the habitat type is highlighted (Labels: acla = *A. clausi*; aton = *A. tonsa*; plat = *P. latisetosa*; and amar = *A. margalefi*).

4. Discussion

The occurrence of NIS is increasing in marine and estuarine systems. Among the 955 new taxa reported for the Mediterranean Sea, 42 are planktonic copepods [47]. Their invasive behaviour has been recognized as one of the major threats to the conservation of the biodiversity and the functioning of marine ecosystems [48–50]. *A. tonsa* dated its first appearance in the Venice lagoon in 1992. Since then, it never disappeared, and it is now present throughout the year and dominant with the exception of colder months. The species is widespread throughout the lagoon with significant abundance except for the areas closer to the tidal inlets, where the trophic condition is lowered and the hydro-chemical characteristics become less favourable to the species. In the Venice lagoon, *A. tonsa* is currently considered a stabilized species. The most plausible hypotheses about the introduction of this species are that it was brought into the outer port area thanks to its ability to produce resting eggs, then via ballast water from ships [51–53], or released by aquaculture, fisheries, or pet industries [54].

Our study showed how the trophic conditions of the lagoon system were (and still are) the main factors that influenced its adaptive success. Unlike the other main species present within the lagoon zooplankton communities, *A. tonsa* seems strongly dependent on trophic conditions, as it is positively correlated with nutrients, Chl *a*, POC, and phytoplankton concentrations. The favourable environmental conditions that the species found in 1992, fundamentally characterized by high habitat trophic levels [55], are congruent with our current findings and to the ecological characteristics of the species [8,15,56]. This allowed the settlement of *A. tonsa* in the Lagoon of Venice.

Much of the available knowledge about zooplankton communities of the Venice lagoon, mainly copepods, derives from studies along the physical and trophic gradient of the northern lagoon. In this area, *Acartia* genus usually represents a quantitative, important component of the zooplankton communities. It shows spatial and seasonal segregation patterns associated with the hydrological conditions, the seasonal variability, and the trophic status and pollution of the water [31,57,58].

Comparison of the whole zooplankton community throughout the four decades and the different habitats highlighted that in the Venice lagoon, before the arrival of *A. tonsa*, the genus *Acartia* dominated the copepod community with *P. latisetosa* and *A. margalefi* usually found in the innermost and intermediate parts of the lagoon, respectively. In particular, in the more internal areas, *A. margalefi* and *P. latisetosa* ranked second and third, respectively, in terms of presence and abundance, though

P. latisetosa greatly diminished in significance in the intermediate areas. The dominant species *A. clausi* ranked always first, mainly in the coastal area. This picture instead has been profoundly modified with the passage from the first analysed decade (i.e., before the invasion of *A. tonsa*) to the following ones. In fact, the current scenario shows, again, the *Acartia* species among the dominant species of the entire lagoon zooplankton community, and it is always organized along a spatial gradient determined by salinity and trophic conditions but with modified rankings in the different considered areas. *A. clausi* no longer dominates all the considered sub-basins, except for the inlet area where it is, again, the most representative species both in space and in time. *P. latisetosa*, whose presence and dominance were already weak in the intermediate areas, further lost importance in the inner areas and virtually disappeared except in spring 2014. Very little abundance of the species was observed only in the internal channels of the lagoon (Inside, Marghera, and Fusina stations). We argue this could either be due to its segregation in more confined and not yet investigated areas or to the occupation of its original habitat by *A. tonsa*. In the presence of a large salinity and trophic range, the *Acartia* species coexist in time with a strong spatial segregation that shows *A. clausi* limited to the inlet and coastal areas, *A. margalefi* in the intermediate area, and partly overlapping with *A. tonsa* that dominates in the innermost areas.

During the second decade (1995–2005), coinciding with the first phase of stabilization of the species after its first occurrence dated 1992, *A. tonsa* seems to clearly supplant the *A. clausi* and *A. margalefi* congeners in the intermediate and mostly internal areas, whereas occupation of *A. clausi* near the lagoon mouths failed. It is interesting to note that, in more recent times (2005–2017), *A. margalefi* appears to be recovering its original ranking, returning to coexist with *A. tonsa* in the intermediate areas. As reported in other Adriatic lagoons [59,60], the arrival of *A. tonsa* may have caused the disappearance or drastic decrease of congeneric species, and the current coexistence of *A. tonsa* and *A. margalefi* is probably favored by a greater resilience of the latter as well as a return to more favorable environmental conditions always for the latter (decline in nutrient concentration and increase in ecological status) [61].

Therefore, among the main factors that seem to have favoured the adaptive success of *A. tonsa* compared to its congeners, we can ascribe the deterioration of the environmental conditions during the 1970s and the 1980s to when the lagoon of Venice was affected by abnormal inputs of pollutants and trophic loadings, mainly phosphorus and nitrogen compounds, which induced massive macroalgae growth (*Ulva rigida* C. Agardh, 1823) whose decomposition in summer led to frequent dystrophic crises [62–65]. Later, during the 1990s, the Lagoon of Venice experienced significant changes in both primary production and trophic conditions. In particular, *Ulva* coverage rapidly declined [64], mainly because of climatic changes and the increase of sediment resuspension and sedimentation [66]. As a consequence, since the end of 1980s and through the 1990s, the presence patterns of *Acartia* species also changed the mesozooplankton composition of the lagoon.

Within this scenario, and knowing that *A. tonsa* can tolerate low dissolved oxygen concentrations and adapt well to hypoxia [67,68], we hypothesized that the hypoxia and hypertrophy in the inner lagoon area could not have hindered the settlement of *A. tonsa* whose adaptive success almost seemed to have benefited. Our results seem to support the hypothesis that, after a first important collapse of *A. margalefi* coinciding with the deterioration of environmental conditions and with the advent of *A. tonsa*, the first remained segregated in the intermediate areas of the lagoon, leaving *A. tonsa* to the innermost ones. Given the niche overlap that emerged from the OMI analysis, it is quite clear that competition for the same resources may be one of the main factors affecting the coexistence of the different *Acartia* in the same area, and that the observed decline in abundance of *P. latisetosa* and *A. margalefi* could be linked to changes in environmental conditions during the 1980s that favored *A. tonsa* development. *Acartia* genus is known to have both herbivorous and omnivorous feeding habits, depending on the environmental resources [69], although little knowledge exists about the nutritional requirements of *A. margalefi* [70]. The available studies report that *P. latisetosa* probably has an omnivorous or detritivorous diet just like *A. tonsa* [71]. Our analyses suggest for these two congeners a stricter dependence on the considered trophic variables resulting in a stronger competition for food with *A. tonsa.*

Therefore, change in distribution and size of the populations of *A. margalefi* and *P. latisetosa* occurred from the second half of the 1980s to the present. This probably depended on the success of recruitment that, in turn, could have been influenced by the system's trophic conditions but also, consequently, by the presence of effective competitors such as *A. tonsa*, which are also known to be an active predator of early larval forms. Competition with other copepods, especially congeneric, appears to be the main documented impact of *A. tonsa* (e.g., [4,51,52,72,73]) besides being known that *A. tonsa* species may coexist with indigenous Acartiidae (e.g., [6]) or exclude them by competition [5], as suggested in some Mediterranean systems [57,59,60], where it seems that *A. margalefi* and *A. clausi* are not able to settle down in large populations in areas colonized by *A. tonsa* [59,74].

This may indicate that large-scale changes, as well as a biotic interactions with the species and mainly with the "new" species *A. tonsa,* might be responsible for the evolution in the lagoon of Venice of different dominant copepod associations with respect to the past. The triggering phenomenon seems to have been the anthropic action on such a delicate and dynamic ecosystem, whose impact *A. tonsa* took advantage of.

Currently *A. tonsa* dominates the intermediate and internal areas of the lagoon, and it reaches the highest population abundance to the detriment of the indigenous congener species. The process, initially favored by the negative effect of changes in environmental conditions, worsened because a clear niche overlap emerged, which was confirmed by our study, in particular between *A. tonsa* and *A. margalefi* with *A. tonsa* benefiting from the point of view of trophic resources. The extent of its area of distribution in the lagoon confirms its euryhaline species characteristics and its excellent adaptability to instantaneous variations in habitat conditions. At present, the lagoon is oligo-mesotrophic over a great part of its surface [61] and dominated by areas classified as polyhaline (18 to 30 salinity) that correspond to the optimal development of *A. tonsa* (salinity ranging from 15 to 22), where the lowest salinities limit its egg production [75].

The adaptive success of *A. tonsa* with respect to its congener in the Lagoon of Venice corroborate the findings that *A. tonsa* is an opportunistic, tolerant species that can take advantage of eutrophic/impacted ecosystems [4,9,31,76,77] and invade them.

Restriction of *A. tonsa* distribution offshore seems mainly influenced by food availability and less by the salinity. In the Lagoon of Venice, *A. tonsa* can sustain the high energy requirements. Moreover, competition with true marine copepods is reduced because of the habitat selection that takes place along the lagoon–sea gradient. It is also probable that the spatial width and heterogeneity of the Venetian lagoon favored and allowed the gradual recovery of the community of *A. margalefi* in recent years, which we observed in the present work. Succumbing to the highly competitive level of the alien *A. tonsa*, *A. margalefi* seems to have rediscovered a niche that does not completely overlap that of *A. tonsa* right along the saline and trophic gradient that exists in the lagoon.

The reports of new NIS in the Venice lagoon—the copepods *Pseudiaptomus marinus* Sato (1913), *Oithona davisae* Ferrari F.D. and Orsi (1984), and ctenophore *Mnemiopsis leidyi* A. Agassiz (1865) [53,78]—are also very recent. The latter, reported for the first time in the lagoon in 2016, is still present throughout the year, and its strong predator characteristics on the planktonic component [78] could revolutionize again the structure of the *Acartia* genus and in general of the zooplankton communities in a few years. Copepods are important components of the pelagic food web, as they are themselves food resources for numerous bentho-pelagic invertebrates and planktivorous juveniles of fishes. The advent of new species capable of competing for the same resources, as *M. leidyi*, could trigger a process at the base of the trophic levels that would affect the highest ones. So, the story of an invasion could therefore not have ended here.

5. Conclusions

The Lagoon of Venice has been recognized as a hot spot for the introduction of nonindigenous species [25]. Several anthropogenic factors (industrial and urban pollution, mariculture, shipping, and tourism) as well as environmental stressors (e.g., warming) concurred to make this ecosystem ideal for invasion as in other estuaries and coastal areas of the Mediterranean. In this work, four datasets of mesozooplankton were examined, with particular emphasis on Acartiidae and the alien species *A. tonsa*. The first dataset, dated 1975–1980, was used as a term of comparison before *A. tonsa* settlement. The second dataset (during 1997–2002) was used to investigate the seasonal cycle and temporal trends. The third dataset (2003–2004) was used to underpin the spatial variability of abundance and species composition, and the last dataset (the recent period until 2017) was used to confirm the presence of *A. tonsa* several years later. The selected trend test on *A. tonsa* abundance did not point out significant overall increases in the period 1997–2002, but it did point out increases in one single station (the most eutrophic one). This may be suggestive that the population has reached a mature stage of colonization in the lagoon. The annual cycle showed maxima in the warm season (mostly July) and negligible abundances in the cold season. Spatial distribution of *A. tonsa* was found to be significantly, positively associated with temperature, phytoplankton, POC, chlorophyll *a*, and counter gradient of salinity, which confirmed that *A. tonsa* was an opportunistic tolerant species that could take advantage of eutrophic ecosystems. This would seem to have occurred at the expense of the previously dominant species *A. margalefi* and *P. latisetosa*, which clearly declined in abundance over the last years. While *P. latisetosa* almost disappeared, *A. margalefi* was not completely excluded. Stations inside the lagoon showed similar species compositions remarkably different from the station in the outer shelf, which was more representative of coastal conditions and dominated by *A. clausi*. In 2014–2017, *A. tonsa* was found to be still one of the dominant *Acartia* species in the lagoon.

Zooplankton is known to be particularly sensitive to environmental changes, whether resulting from natural or anthropogenic forcing [79–82]. The interaction between anthropogenic activity, climate change, and plankton communities, focusing on systematic changes in plankton community structure, abundance, distribution, and phenology over recent decades, is a key global issue as well as the potential socioeconomic impacts of these changes [83]. In environments of relentless evolution such as the Lagoon of Venice, monitoring the ecological dynamics of species is of the utmost importance. In particular, continuous modifications in zooplankton assemblages in response to anthropogenic and environmental stressors must be considered. Ongoing plankton monitoring in the LTER site of the Lagoon of Venice will act as sentinel research to identify future changes in this complex, heavily impacted ecosystem and its related food web.

Author Contributions: E.C. contributed substantially to the study's conceptualization, data acquisition, analysis, and to the original draft preparation; M.P. contributed to recent data acquisition and analysis and to the review and editing; and A.B. contributed substantially to statistical analyses and to the review of the manuscript. All authors approved the final submitted manuscript.

Funding: This work was supported by the LIFE-WATERS Program, EC-sponsored, for the two-year period 1997–1999. The data set 2002–2003 was collected within the framework of the "Integrazione delle Conoscenze del Sistema Ecologico Lagunare (ICSEL)" project, promoted by the Venice Water Authority (Magistrato alle Acque) through Servizio Ambiente—Consorzio Venezia Nuova, while the dataset 2014–2017 derived from the ADRIATIC IPA BALMAS (Ballast Water Management System for Adriatic Sea Protection) Project and from the LTER.

Acknowledgments: The authors acknowledge Thetis S.p.A. for providing the environmental data during 2003–2004 period and the CNR colleague F. Acri for his help in carrying out biochemical analyses. We wish to thank F. Bernardi Aubry, M. Bastianini, S. Finotto, A. Schröder, S. Pasqual, L. Dametto, and M. Casula for the fieldwork technical support during sampling and the crew of the M/B "Litus" G. Zennaro, D. Penzo, and M. Penzo.

Conflicts of Interest: The authors declare no conflict of interest.

References

1.   IMO (International Maritime Organization). Global Ballast Water Management Project: The Problem. Available online: http://globallast.imo.org/index.asp?page=problem.htm&menu=true (accessed on 7 May 2019).

2. Gaudy, R.; Vinas, M. Premiere signalisation en Méditerrannée du copepode pélagique *Acartia tonsa*. *Rapp. Comm. Int. Mer Mediterr.* **1985**, *219*, 227–229.

3. Seuront, L. First record of the calanoid copepod *Acartia omorii* (Copepoda: Calanoida: Acartiidae) in the southern bight of the North Sea. *J. Plankton Res.* **2005**, *27*, 1301–1306. [CrossRef]

4. David, V.; Sautour, B.; Chardy, P. Successful colonization of the calanoid copepod *Acartia tonsa* in the oligo mesohaline area of the Gironde estuary (SW France) Natural or anthropogenic forcing? *Estuar. Coast. Shelf Sci.* **2007**, *71*, 429–442. [CrossRef]

5. Belmonte, G.; Potenza, D. Biogeography of the family Acartiidae (Calanoida) in the Ponto Mediterranean Province. *Hydrobiologia* **2001**, *453*, 171–176. [CrossRef]

6. Lakkis, S. Coexistence and competition within Acartia (Copepoda, Calanoida) congeners from Lebanese coastal water: Niche overlap measurements. In *Ecology and Morphology of Copepods, Hydrobiologia*; Ferrari, F.D., Bradley, B.P., Eds.; Kluwer Academic Publishers: Brussels, Belgium, 1994; Chapter 292–293; pp. 481–490.

7. Hoffmeyer, M.S. Decadal change in zooplankton seasonal succession in the Bahía Blanca Estuary, Argentina, following introduction of two zooplankton species. *J. Plankton Res.* **1994**, *26*, 1–9. [CrossRef]

8. Durbin, E.G.; Durbin, A.G.; Smayda, T.J.; Verity, P.G. Food limitation of production by adult *Acartia tonsa* in Narragansett bay, Rhode Island. *Limnol. Oceanogr.* **1983**, *28*, 1199–1213. [CrossRef]

9. Brylinski, J.M. Report on the presence of *Acartia tonsa* Dana (Copepoda) in the harbour of Dunkirk (France) and its geographical distribution in Europe. *J. Plankon Res.* **1981**, *3*, 255–260. [CrossRef]

10. Gruszka, P. The River Odra Estuary as a Gateway for Alien Species Immigration to the Baltic Sea Basin. *Acta Hydrochim. Hydrobiol.* **1999**, *27*, 374–382. [CrossRef]

11. Farabegoli, A.; Ferrari, I.; Manzoni, C.; Pugnetti, A. Prima segnalazione nel Mare Adriatico del copepode calanoide *Acartia tonsa*. *Nov. Thalass.* **1989**, *10*, 207–208.

12. Quarta, S.; Belmonte, G.; Caroppo, C.; Pacifico, P.; Petraroli, P. Zooplankton seasonal trend in the Lesina and Varano Lagoons (Apulian coast of Italy). *Oebalia* **1992**, *17*, 403–404.

13. Comaschi, A.; Cavalloni, B. Variabilità stagionale dello zooplancton nella Palude della Rosa (laguna di Venezia bacino settentrionale). *SITE* **1995**, *16*, 75–78.

14. Durbin, A.G.; Durbin, E.G. Standing stock and estimated production rates of phytoplankton and zooplankton in Narragansett Bay, Rhode Island. *Estuaries* **1981**, *4*, 24–41. [CrossRef]

15. Roman, M.R. Utilization of detritus by the copepod *Acartia tonsa*. *Limnol. Oceanogr.* **1984**, *5*, 949–959. [CrossRef]

16. Paffenöfer, G.A.; Stearn, D.E. Why is *Acartia tonsa* (Copepoda: Calanoida) restricted to nearshore environment? *Mar. Ecol. Prog. Ser.* **1988**, *42*, 33–38. [CrossRef]

17. Zhang, J.; Ianora, A.; Wu, C.; Pellegrini, D.; Esposito, F.; Buttino, I. How to increase productivity of the copepod *Acartia tonsa* (Dana): Effects of population density and food concentration. *Aquac. Res.* **2015**, *46*, 2982–2990. [CrossRef]

18. Tagliapietra, D.; Sigovini, M.; Volpi Ghirardini, A. A review of terms and definitions to categorise estuaries, lagoons and associated environments. *Mar. Freshw. Res.* **2009**, *60*, 497–509. [CrossRef]

19. Tagliapietra, D.; Zanon, V.; Frangipane, G.; Umgiesser, G.; Sigovini, M. Physiographic zoning of the Venetian Lagoon. In *Scientific Research and Safeguarding of Venice, 2010*; CORILA: Venice, Italy, 2007; pp. 161–164.

20. Occhipinti Ambrogi, A.; Marchini, A.; Cantone, G.; Castelli, A.; Chimenz, C.; Cormaci, M.; Froglia, C.; Furnari, G.; Gambi, M.C.; Giaccone, G.; et al. Alien species along the Italian coasts: An overview. *Biol. Invasion* **2011**, *13*, 215–237. [CrossRef]

21. Keppel, E.; Sigovini, M.; Tagliapietra, D. A new geographical record of *Polycera hedgpethi* Er. Marcus, 1964 (Nudibranchia: Polyceridae) and evidence of its established presence in the Mediterranean Sea, with a review of its geographical distribution. *Mar. Biol. Res.* **2012**, *8*, 969–981. [CrossRef]

22. Sfriso, A.; Campolin, M.; Sfriso, A.; Buosi, A.; Facca, C. Change of aquatic flora and vegetation in ecological gradients from the lagoon inlets to some fishing ponds of the Venice lagoon. *Biol. Mar. Mediterr.* **2012**, *19*, 53–56.

23. Sfriso, A.; Facca, C. Annual growth and environmental relationships of the invasive species *Sargassum muticum* and *Undaria pinnatifida* in the lagoon of Venice. *Estuar. Coast. Shelf Sci.* **2013**, *129*, 162–172. [CrossRef]

24. Tagliapietra, D.; Keppel, E.; Sigovini, M. First record of the colonial ascidian *Didemnum vexillum* Kott, 2002 in the Mediterranean: Lagoon of Venice (Italy). *BioInvasions Rec.* **2012**, *1*, 247–254. [CrossRef]

25. Marchini, A.; Galil, B.S.; Occhipinti Ambrogi, A. Recommendations on standardizing lists of marine alien species: Lessons from the Mediterranean Sea. *Mar. Pollut. Bull.* **2015**, *101*, 267–273. [CrossRef] [PubMed]

26. Solidoro, C.; Pastres, R.; Cossarini, G.; Ciavatta, S. Seasonal and spatial variability of water quality parameters in the lagoon of Venice. *J. Mar. Syst.* **2004**, *51*, 7–18. [CrossRef]

27. Gačić, M.; Kovačević, V.; Mancero Mosquera, I.; Mazzoldi, A.; Cosoli, S. Water fluxes between the Venice Lagoon and the Adriatic Sea. In *Flooding and Environmental Challenges for Venice and Its Lagoon: State of Knowledge*; Fletcher, C.A., Spencer, D.T., Eds.; Cambridge University: Cambridge, UK, 2005; pp. 431–444.

28. Cucco, A.; Umgiesser, G. Modelling the Venice Lagoon residence time. *Ecol. Model.* **2006**, *193*, 34–51. [CrossRef]

29. Ravera, O. The Lagoon of Venice: The result of both natural factors and human influence. *J. Limnol.* **2000**, *59*, 19–30. [CrossRef]

30. Solidoro, C.; Bandelj, V.; Aubry Bernardi, F.; Camatti, E.; Ciavatta, S.; Cossarini, G.; Facca, C.; Franzoi, P.; Libralato, S.; Melaku Canu, D.; et al. Response of Venice Lagoon Ecosystem to Natural and Anthropogenic Pressures over the Last 50 Years. In *Coastal Lagoons Critical Habitats of Environmental Change*; Kennish, M., Paerl, H.W., Eds.; CRC Press: Boca Raton, FL, USA, 2010; pp. 483–511.

31. Bianchi, F.; Acri, F.; Bernardi Aubry, F.; Berton, A.; Boldrin, A.; Camatti, E.; Cassin, D.; Comaschi, A. Can plankton communities be considered as bio indicators of water quality in the Lagoon of Venice? *Mar. Pollut. Bull.* **2003**, *46*, 964–971. [CrossRef]

32. Bianchi, F.; Socal, G.; Alberighi, L.; Cioce, F. Cicli nictemerali dell'ossigeno disciolto nel bacino centrale della laguna di Venezia. *Biol. Mar. Mediterr.* **1996**, *3*, 628–630.

33. Socal, G.; Bianchi, F.; Alberighi, L. Effects of thermal pollution and nutrient discharges on a spring phytoplankton bloom in the industrial area of the lagoon of Venice. *Vie Milieu* **1999**, *49*, 19–31.

34. Harris, R.P.; Wiebe, P.H.; Lenz, J.; Skjoldal, H.R.; Huntley, M. *ICES Zooplankton Methodology Manual*; Academic Press: Cambridge, MA, USA, 2000; p. 684.

35. Camatti, E.; Ferrari, I. Mesozooplankton. In *Metodologie di Studio Del Plancton Marino*; Socal, G., Buttino, I., Cabrini, M., Mangoni, O., Penna, A., Totti, C., Eds.; Manuali e Linee Guida: Ispra, Italy, 2010; Volume 56, pp. 489–506.

36. Throndsen, J. Preservation and storage. In *Phytoplankton Manual*; Sournia, A., Ed.; UNESCO: Paris, France, 1978.

37. Utermöhl, H. Zur Vervollkommnung der quantitativen Phytoplankton Methodik. *Int. Ver. Theor. Angew. Limnol.* **1958**, *9*, 1–38. [CrossRef]

38. Zingone, A.; Honsell, G.; Marino, D.; Montresor, M.; Socal, G. Fitoplankton. *Nov. Thalass.* **1990**, *11*, 183–198.

39. EPA (Environmental Protection Agency). *Methods for Determination of Chemical Substances in Marine and Estuarine Environmental Matrices*, 2nd ed.; EPA: Cincinnati, OH, USA, 1997.

40. IRSA (Istituto di Ricerca sulle Acque). Metodi di Analisi Per Acque di Mare. *Quad. Ist. Ric. Acque* **1997**, *59*, 570–571.

41. Hirsch, R.M.; Slack, J.R.; Smith, R.A. Techniques of trend analysis for monthly water quality data. *Water Resour. Res.* **1982**, *18*, 107–121. [CrossRef]

42. Kendall, M.G. *Rank Correlation Methods*, 4th ed.; Charles Griffin: London, UK, 1975.

43. Legendre, P.; Legendre, L.F. *Numerical Ecology*, 3rd ed.; Elsevier: Amsterdam, The Netherlands, 2012; Volume 24, p. 1006.

44. Davis, J.C. *Statistics and Data Analysis in Geology*, 3rd ed.; Wiley, J.S., Ed.; Wiley: Hoboken, NJ, USA, 1986; p. 646.

45. Dolédec, S.; Chessel, D.; Gimaret-Carpentier, C. Niche separation in community analysis: A new method. *Ecology* **2000**, *81*, 2914–2927. [CrossRef]

46. Sen, P.K. Estimates of the regression coefficient based on Kendall's Tau. *J. Am. Stat. Assoc.* **1968**, *63*, 1379–1389. [CrossRef]

47. Zenetos, A.; Gofas, S.; Verlaque, M.; Cinar, M.E.; Garcia Raso, J.E.; Bianchi, C.N.; Morris, C.; Azzurro, E.; Bilecenoglu, M.; Froglia, C.; et al. Alien species in the Mediterranean Sea by 2010. A contribution to the application of European Union's Marine Strategy Framework Directive (MSFD). Part I. Spatial distribution. *Mediterr. Mar. Sci.* **2010**, *11*, 381–493. [CrossRef]

48. Carlton, J.T. Biological invasions and cryptogenic species. *Ecology* **1996**, *77*, 1653–1655. [CrossRef]

49. Ruiz, G.M.; Fofonoff, P.; Hines, A. Non- indigenous species as stressors in estuarine and marine communities: Assessing invasion impacts and interactions. *Limnol. Oceanogr.* **1999**, *44*, 950–972. [CrossRef]

50. Mack, R.N.; Simberloff, D.; Lonsdale, W.M.; Evans, H.; Clout, M.; Bazzaz, F.A. Biotic invasions: Causes, epidemiology, global consequences and control. *Ecol. Appl.* **2000**, *10*, 689–710. [CrossRef]

51. Gubanova, A. Occurrence of *Acartia tonsa* Dana in the Black Sea. Was it introduced from the Mediterranean? *Mediterr. Mar. Sci.* **2000**, *1*, 105–109. [CrossRef]

52. Gubanova, A.; Altukhova, D.; Stefanova, K.; Arashkevichc, E.; Kamburskad, L.; Prusovaa, I.; Svetlichnya, L.; Timoftee, F.; Uysalfand, Z. Species composition of Black Sea marine planktonic copepods. *J. Mar. Syst.* **2014**, *135*, 44–52. [CrossRef]

53. Vidjak, O.; Bojanić, N.; de Olazabal, A.; Benzi, M.; Brautović, I.; Camatti, E.; Hure, M.; Lipej, L.; Lučić, D.; Pansera, M.; et al. Zooplankton in Adriatic port environments: Indigenous communities and non-indigenous species. *Mar. Pollut. Bull.* **2008**. [CrossRef]

54. Ruiz, G.M.; Carlton, J.T.; Grosholz, E.D.; Hines, A.H. Global invasions of marine and estuarine habitats by non-indigenous species: Mechanisms, extent and consequences. *Am. Zool.* **1997**, *37*, 621–632. [CrossRef]

55. Comaschi, A.; Acri, F.; Alberghi, L.; Bastianini, M.; Bianchi, F.; Cavalloni, B.; Socal, G. Presenza di *Acartia tonsa* (Copepoda: Calanoida) nella laguna di Venezia. *Biol. Mar. Mediterr.* **1994**, *1*, 273–274.

56. Bandelj, V.; Socal, G.; Park, Y.S.; Lek, S.; Coppola, J.; Camatti, E.; Capuzzo, E.; Milani, L.; Solidoro, C. Analysis of multitrophic plankton assemblages in the lagoon of Venice. *Mar. Ecol. Prog. Ser.* **2008**, *368*, 23–40. [CrossRef]

57. Comaschi, A.; Acri, F.; Bianchi, F.; Bressan, M.; Camatti, E. Temporal changes of species belonging to Acartia genus (Copepoda: Calanoida) in the northern basin of the Venice lagoon. *Boll. Mus. Civ. Stor. Nat. Venezia* **2000**, *50*, 189–193.

58. Camatti, E.; Comaschi, A.; Coppola, J.; Milani, L.; Minocci, M.; Socal, G. Analisi dei popolamenti zooplanctonici nella laguna di Venezia dal 1975 al 2004. *Biol. Mar. Mediterr.* **2006**, *13*, 46–53.

59. Sei, S.; Ferrari, I. First report of the occurrence of *Acartia tonsa* (Copepoda, Calanoida) in the Lesina lagoon (South Adriatic Sea–Mediterranean Sea). *Mar. Biod. Rec.* **2008**, *1*, 37. [CrossRef]

60. Sei, S.; Rossetti, G.; Villa, F.; Ferrari, I. Zooplankton variability related to environmental changes in a eutrophic coastal lagoon in the Po Delta. *Hydrobiologia* **1996**, *329*, 45–55. [CrossRef]

61. Sfriso, A.; Buosi, A. Trophic status changes in the Venice lagoon during the last 40 years. In Proceedings of the 49th Congresso della Società Italiana di Biologia Marina Cesenatico (FC), Rome, Italy, 4–8 June 2018.

62. Sfriso, A. Flora and vertical distribution of macroalgae in the lagoon of Venice: A comparison with previous studies. *Plat Biosyst.* **1987**, *121*, 69–85. [CrossRef]

63. Sfriso, A.; Birkemeyer, T.; Ghetti, P.F. Benthic macrofauna changes in areas of Venice lagoon populated by seagrasses or seaweeds. *Mar. Environ. Res.* **2001**, *52*, 323–349. [CrossRef]

64. Sfriso, A.; Facca, C.; Ghetti, P.F. Temporal and spatial changes of macroalgae and phytoplankton in a Mediterranean coastal area: The Venice lagoon as a case study. *Mar. Environ. Res.* **2003**, *56*, 617–636. [CrossRef]

65. Sfriso, A.; Facca, C.; Ceoldo, S.; Silvestri, S.; Ghetti, P.F. Role of macroalgal biomass and clam fishing on spatial and temporal changes in N and P sedimentary pools in the central part of the Venice lagoon. *Oceanol. Acta* **2003**, *26*, 3–13. [CrossRef]

66. Sfriso, A.; Marcomini, A. Decline of Ulva growth in the Lagoon of Venice. *Bioresour. Technol.* **1996**, *58*, 299–307. [CrossRef]

67. Decker, M.B.; Breitburg, D.L.; Marcus, N.H. Geographical differences in behavioral responses to hypoxia: Local adaptation to an anthropogenic stressor? *Ecol. Appl.* **2003**, *13*, 1104–1109. [CrossRef]

68. Kimmel, D.G.; Boicourt, W.C.; Pierson, J.J.; Roman, M.R.; Zhang, X. A comparison of the mesozooplankton response to hypoxia in Chesapeake Bay and the northern Gulf of Mexico using the biomass size spectrum. *J. Exp. Mar. Biol. Ecol.* **2009**, *381*, 65–73. [CrossRef]

69. Mauchline, J. The biology of calanoid copepods. *Adv. Mar. Biol.* **1998**, *33*, 1–710.

70. Castro-Longoria, E. Egg Production and Hatching Success of Four Acartia Species under Different Temperature and Salinity Regimes. *J. Crustac. Biol.* **2003**, *23*, 289–299. [CrossRef]

71. El Shabrawy, G.M.; Belmonte, G. Abundance and affirmation of *Paracartia latisetosa* (copepoda, calanoida) in the inland lake Qarun (Egypt). *Thalass. Sal.* **2004**, *27*, 151–160.

72. Kritchagin, N. Otchet o faunisticheskikh issledovaniyakh proizvedennikh v etom godu po porutcheniyu kievskogo obsh estva estestvoispitateley na beregakh Tchernogo morya. *Zap. Kiev. Obs. Estestvoispitatalei* **1873**, *3*, 346–430.

73. Aravena, G.; Villate, F.; Uriarte, I.; Iriarte, A.; Ibáñez, B. Response of Acartia populations to environmental variability and effects of invasive congenerics in the estuary of Bilbao, Bay of Biscay. *Estuar. Coast. Shelf Sci.* **2009**, *83*, 621–628. [CrossRef]

74. Gaudy, R.; Cervetto, G.; Pagano, M. Comparison of the metabolism of *Acartia clausi* and *Acartia tonsa*: Influence of temperature and salinity. *J. Exp. Mar. Biol. Ecol.* **2000**, *247*, 51–65. [CrossRef]

75. Cervetto, G.; Gaudy, R.; Pagano, M. Influence of salinity on the distribution of *Acartia tonsa* (Copepoda, Calanoida). *J. Exp. Mar. Biol. Ecol.* **1999**, *239*, 33–45. [CrossRef]

76. Biancalana, F.; Dutto, M.; Berasategui, A.; Kopprio, G.; Hoffmeyer, M. Mesozooplankton assemblages and their relationship with environmental variables: A study case in a disturbed bay (Beagle Channel, Argentina). *Environ. Monit. Assess.* **2014**, *186*, 8629–8647. [CrossRef] [PubMed]

77. Dunbar, F.N.; Webber, M.K. The phytoplankton distribution in the eutrophic Kingston Harbour, Jamaica. *Bull. Mar. Sci.* **2003**, *73*, 343–359.

78. Malej, A.; Tirelli, V.; Lučić, D.; Paliaga, P.; Vodopivec, M.; Goruppi, A.; Ancona, S.; Benzi, M.; Bettoso, N.; Camatti, E.; et al. *Mnemiopsis leidyi* in the northern Adriatic: Here to stay? *J. Sea Res.* **2017**, *124*, 10–16. [CrossRef]

79. Irvine, K.; Moss, B.; Bales, M.; Snook, D. The changing ecosystem of a shallow, brackish lake, Hickling Broad, Norfolk, UK II. Trophic relationships with special reference to the role of Neomysis integer. *Freshw. Biol.* **1993**, *29*, 119–139. [CrossRef]

80. Li, M.; Gargett, A.; Denman, K. What determines seasonal and interannual variability of phytoplankton and zooplankton in strongly estuarine systems? Application to the semi enclosed estuary of strait of Georgia and Juan de Fuca Strait. *Estuar. Coast. Shelf Sci.* **2000**, *50*, 467–488. [CrossRef]

81. Poggiale, J.C.; Dauvin, J.C. Long term dynamics of three benthic Ampelisca (Crustacea Amphipoda) populations from the Bay of Morlaix (western English Channel) related to their disappearance after the "Amoco Cadiz" oil spill. *Mar. Ecol. Prog. Ser.* **2001**, *214*, 201–209. [CrossRef]

82. Gucu, A.C. Can overfishing be responsible for the successful establishment of *Mnemiopsis leidyi* in the Black sea? *Estuar. Coast. Shelf Sci.* **2002**, *54*, 439–451. [CrossRef]

83. Hays, G.C.; Richardson, A.J.; Robinson, C. Climate change and marine plankton. *Trends Ecol. Evol.* **2005**, *20*, 337–344. [CrossRef]

Article

Effects of Asian Dust and Phosphorus Input on Abundance and Trophic Structure of Protists in the Southern Yellow Sea

Xi Chen [1], Guang-Xing Liu [1,2], Xiao Huang [1,3], Hong-Ju Chen [1,2], Chao Zhang [1,2] and Yang-Guo Zhao [1,2,*]

[1] College of Environmental Science and Engineering, Ocean University of China, Qingdao 266100, China; asdchenxi180@sina.com (X.C.); gxliu@ouc.edu.cn (G.-X.L.); huangxiao901231@126.com (X.H.); hongjuc@ouc.edu.cn (H.-J.C.);zhangchao@ouc.edu.cn (C.Z.)

[2] Key Lab of Marine Environmental Science and Ecology, Ministry of Education, Ocean University of China, Qingdao 266100, China

[3] School of Civil and Environmental Engineering, Harbin Institute of Technology (Shenzhen), Shenzhen 518055, China

* Correspondence: ygzhao@ouc.edu.cn

Received: 16 April 2019; Accepted: 3 June 2019; Published: 7 June 2019

Abstract: To reveal the effects of Asian dust and phosphorus (P) input on the structure and function of micro-food web in the Yellow Sea, an experiment was conducted onboard the southern Yellow Sea where P was deficient. The response of the abundance and trophic structure of planktonic protists to different concentrations of dust and P were studied. The results showed that the sand-dust deposition presented variable effects on different sizes of protists during incubation periods. At the initial stage of incubation with dust, the amount of all sizes of autotrophic protists, especially 10–20 μm, were improved; on the contrary, the heterotrophic and mixotrophic protists were inhibited. At the late period, the increase of autotrophic protists was restricted, while the 2–5 μm heterotrophic and mixotrophic protists obviously increased. Similarly, adding P demonstrated the obviously positive effect on the 10–20 μm autotrophic protists at the initial period, and then the growth was restricted at the late period. These results were consistent with that of sand-dust deposition. Hence, it could be presumed that the positive effect of sand-dust deposition on autotrophic protists in the southern Yellow Sea was achieved by the release of P from dust. P in the early stage of sand-dust deposition promotes the growth of large-size autotrophic protists, which may accelerate red tides in eutrophic ocean. The stimulation of small-size heterotrophic protists at the late period of sand-dust deposition contributed to the material cycle and food transmission in the ocean. Therefore, the effects of sand-dust deposition on the abundance and trophic structure of different sizes of planktonic protists could change the structure of micro-food web in the southern Yellow Sea and further affected the ecological function of planktonic protists.

Keywords: Yellow Sea; sand-dust deposition; protists; trophic structure

1. Introduction

The Yellow Sea, as one of the four marginal seas in China, is an area where land, ocean and atmosphere interact more intensely with concentrated human activities and marine economic development. It owns abundant natural resources and developed coastal economy. Previous studies found that the Yellow Sea was intermittently restricted by phosphorus (P) [1,2], which leads to a constant changing process of protists trophic structure in this ocean area, and then altered the ecological functions of protists, such as nutrients utilization and transformation [3,4]. Sand-dust deposition is an important pathway for transporting land-based nutrients and pollutants to the ocean and providing nutrients for the marine planktonic protists. Researchers found that sand-dust deposition held a significant positive correlation with chlorophyll *a* and primary productivity [5,6]. Asian dust is an important part of global dust, and the Yellow Sea, located in the downwind zone of the Asian dust source area, is the greatest probability of being affected by Asian dust in China's offshore waters. Some studies showed that Asian dust deposited into the Yellow Sea [6–8] through a long-distance transportation in sand-dust weather, and obvious "fertilization" phenomenon was observed to affect the primary productivity. Thus, Asian sand-dust deposition is an important factor that affects the primary productivity of the Yellow Sea. The nutrients carried by sand-dust [9–12] could alleviate the phosphorus deficiency in the Yellow Sea, thereby affected the abundance and trophic structure of marine biota. However, the effect of sand-dust on the growth of nanoplanktonic protists in the P-limited Yellow Sea has not been reported.

Marine nanoplankton with size ranging 2–20 μm, is a key component of marine micro-food web and they play an irreplaceable role in maintaining primary productivity and material cycle [13–15]. It is remarkable that different size groups of nanoplankton (e.g., 2–4, 4–5, 5–7, 7–10 and >10 μm) revealed different roles in the food web due to the variations of species and proportion [16]. As such, three size groups, i.e., 2–5, 5–10, and 10–20 μm of nanoplankton were proposed in this study. According to the mechanisms of energy and nutrient acquisition, nanoplanktonic protists are divided into autotrophic, heterotrophic (i.e., protozoa) and mixotrophic protists [2,17–19]. Autotrophic nanoplanktonic protists are the key contributor to marine primary productivity [17], Heterotrophic nanoplanktonic protists affect the community structure and function of nanoplankton by preying on bacteria, cyanobacteria or smaller protists [18,19], and being preyed by medium-sized zooplankton [20,21], and then flow to a higher trophic level from the bottom of the food web [22,23]. Mixotrophic nanoplanktonic protists own both the autotrophic and heterotrophic mode, and improve the utilization of nutrients by changing nutritional habits [2,3,17]. Hence, the material conversion and energy flow of the marine plankton ecosystem depends on different trophic marine nanoplankton protists. Therefore, it is of great significance to investigate the effects of dust and phosphorus input on the abundance and trophic structure of planktonic protists.

However, there are several blind spots that need to be further revealed in the relationship between the sand-dust and planktonic protists. Firstly, the effect of sand-dust input on nanoplanktonic protists is unknown. Though the sand-dust can provide nutrients during the transport process, it also adsorbs substantial heavy metals such as copper, cadmium, plumbum, and other land-based pollutants, which present a strong toxic effect on plankton [24,25]. Secondly, whether the sand-dust, as a supplement of nutrients, can effectively compensate for P limitation in the Yellow Sea is still unclear. Thirdly, it is difficult to accurately distinguish the ecological functions of different species due to the complexity of marine nanoplanktonic protists.

Hence, in this study, to make a better understanding of the effects of sand-dust and phosphorus on the abundance and trophic structure of different sizes of planktonic protists, an experiment was conducted onboard in the southern Yellow Sea where P was deficient., the corresponding changes of nanoplanktonic protists were following investigated according to the size groups (i.e., 2–5, 5–10, and 10–20 μm) and trophic types (i.e., autotrophic, heterotrophic and mixotrophic protists) [3,18,22,26–30]. The results could provide a scientific supplement for revealing the effects of sand-dust deposition on the trophic structure and ecological function of marine planktonic protists. Meanwhile, this study supply a supplement for further analyzing the effects of sand-dust deposition on marine ecosystems.

2. Materials and Methods

2.1. Sand-Dust Preparation

In this study, sand-dust samples were collected from the surface soil of Hunshandake Sandy Land (42°22′28″ N, 112°58′34″ E) in May, 2011. The samples were stored at −20 °C until use. The large stone in sand-dust samples was eliminated by a sieve with 20 μm of pore in diameter and then aged artificially as the previous method [31]. To analyze the concentrations of nutrients, DOC and trace metals in the sand-dust, 30 min of ultra-sonication treatment for the sand-dust was conducted under the temperature of about 0 °C. The contents of nutrients were analyzed by an ion chromatograph (ICS-1100, Dionex Corporation, Bannockburn, IL, USA). Heavy metals were analyzed by an inductively coupled plasma mass spectrometry (ICP-MS-7500c, Agilent Technologies, Palo Alto, CA, USA). Dissolved organic carbon (DOC) was determined by the high temperature combustion oxidation method (TOC-V, Shimadzu Corporation, Kyoto, Japan). The contents of nutrients, DOC and dissolved trace elements in sand-dust are shown in Table 1.

Table 1. Concentrations of nutrients, Dissolved organic carbon (DOC), dissolved trace metals in the sand-dust.

Ingredients	Nutrients and DOC (μmol mg^{-1})					Dissolved Trace Metals (μg g^{-1})							
	NO$_3^-$	NO$_2^-$	NH$_4^+$	PO$_4^{3-}$	DOC	Fe	Cu	Zn	Pb	As	Ni	Cd	Cr
Concentration	0.53	0.03	0.01	0.004	7.20	473.12	0.23	4.27	0.24	0.09	7.08	0.04	0.13

2.2. Onboard Culturing Experiment

In November 2014, an onboard artificial sand-dust deposition experiment was carried out on the 'Dongfanghong NO.2' scientific research ship at H03 station (36°06′00″ N, 121°39′00″ E) in the southern Yellow Sea. The location of sampling station H03 is shown in Figure 1. Surface seawater was collected by using a shipborne CTD water collector (USA). The collected seawater was immediately filtered by a silk membrane with pore of 20 μm, to remove micro, small, medium and large plankton. The filtered seawater was filled into 1.5 L sterile polyethylene terephthalate (PET) culture bottles and afterwards the bottles were placed in a water tank. The tank was tightly fixed on the deck without shading. The surface seawater from the station H03 was fed into the outer layer tank to maintain the inner bottle temperature and ensure it was the same as the surface water.

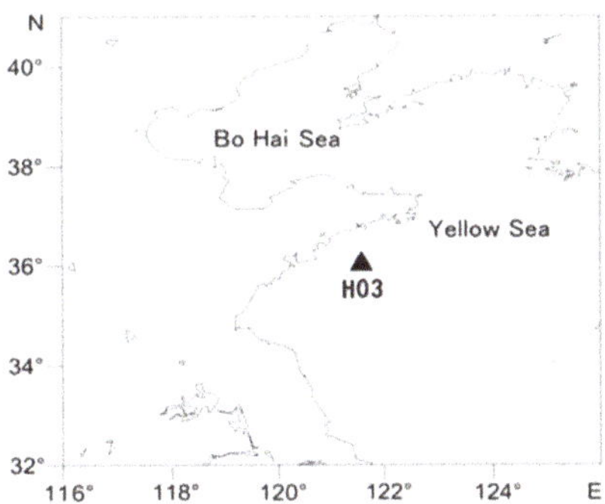

Figure 1. Incubation Station.

The culture experiment was comprised of five groups, i.e., control group, low dust group (LD), high dust group (HD), low phosphorus group (LP), and high phosphorus group (HP). The amounts of added dust or P (as NaH$_2$PO$_4$) referred to the observation data on the Yellow Sea as the previous report [32]. The amount of added dust and P in each experiment group is shown in Table 2. Dust or

NaH$_2$PO$_4$ were added only one time immediately after the establishment of the culture system. All culture experiments were conducted triplicate. Samples were taken every day to determine the protists, DOC and inorganic nutrients during the five days' culture period.

Table 2. Experiment design and concentrations of dust and phosphorus.

Groups	Treatment	Concentration
Control	none	none
Low dust group (LD)	dust	2 mg/L
High dust group (HD)	dust	20 mg/L
Low phosphorus group (LP)	NaH$_2$PO$_4$	0.2 μmol/L
High phosphorus group (HP)	NaH$_2$PO$_4$	1 μmol/L

2.3. Sample Collection and Measurement

2.3.1. Nanoplanktonic Protists Samples

Nanoplanktonic protists samples were collected at 8 a.m. each day during the incubation period. Before sampling, PET culture bottles were slowly inverted for three times to uniformly mix nanoplankton in the incubation system. There was 10 mL seawater obtained from the culture bottle and then transferred into the sterile freezing tube. Immediately, 10% (*w/v*) paraformaldehyde was added into the tube to fix protists (for paraformaldehyde, the final concentration was 0.5% (*w/v*)). After slightly mixing, the tubes were quickly frozen in a liquid nitrogen tank and then transferred to the refrigerator at −80 °C for storage. Three parallel samples were taken at one time.

The abundances of different sizes of nanoplanktonic protists were counted by a flow cytometry (BD C6 plus) at different wavelength of fluorescence, as explained by Christaki et al. (2011) [16] and Zubkov et al. (2007) [33]. The sample was stained by SYBR Green I, then determined by flow cytometry with excitation wavelength 488 nm. The trophic mode of the protists was determined by green fluorescence (FL1, 530 ± 20 nm) and red fluorescence (FL3, >630 nm), the protists which showed high FL1 value and low FL3 value were heterotrophic, the protists with high FL1 value and high FL3 value were autotrophic, the rest of protists were mixotrophic. The size of protists were measured by calibration beads (2, 5 and 10 μm).

2.3.2. Dissolved Organic Carbon (DOC)

The seawater samples were filtered with a high-temperature treated GF/F filter (Whatman, Maidstone, UK) and stored in a high-temperature treated glass bottle at 4 °C in the dark. The samples were determined by using a total organic carbon meter (TOC-VCPN) [34].

2.3.3. Inorganic Nutrients

The seawater samples were filtered by acid treated 0.45 μm acetic acid fibre membrane and frozen at −20 °C after adding the fixative. The concentrations of different nutrients were determined by QuAAtro nutrient automatic analyzer. NH$_4^+$ was oxidized by sodium hypochlorite with indigo-phenol blue (660 nm), NO$_3^-$ reduced by copper-cadmium column with naphthalene ethylenediamine hydrochloride (550 nm), NO$_2^-$ with naphthalene ethylenediamine hydrochloride (550 nm) and PO$_4^{3-}$ with phosphomolybdenum blue (880 nm) [32,34].

2.4. Data Processing and Analysis

Multivariate ANOVA in SPSS Inc. (Chicago, IL, USA) was used to analyze the significant difference of different groups in the culture system.

3. Results and Analysis

At H03 Station, the concentration of DOC is 79.68 µmol/L, the concentration of DIN is 2.71 µmol/L, and the concentration of PO_4^{3-} is 0.13 µmol/L. The N:P ration is 21 (>16) showed that H03 station was P-limited.

3.1. Effects of Dust and Phosphorus Addition on the Abundance of Nanoplanktonic Protists

3.1.1. Changes of 10–20 µm Protists

Changes of 10–20µm protists under sand-dust and phosphorus addition are shown in Figure 2.

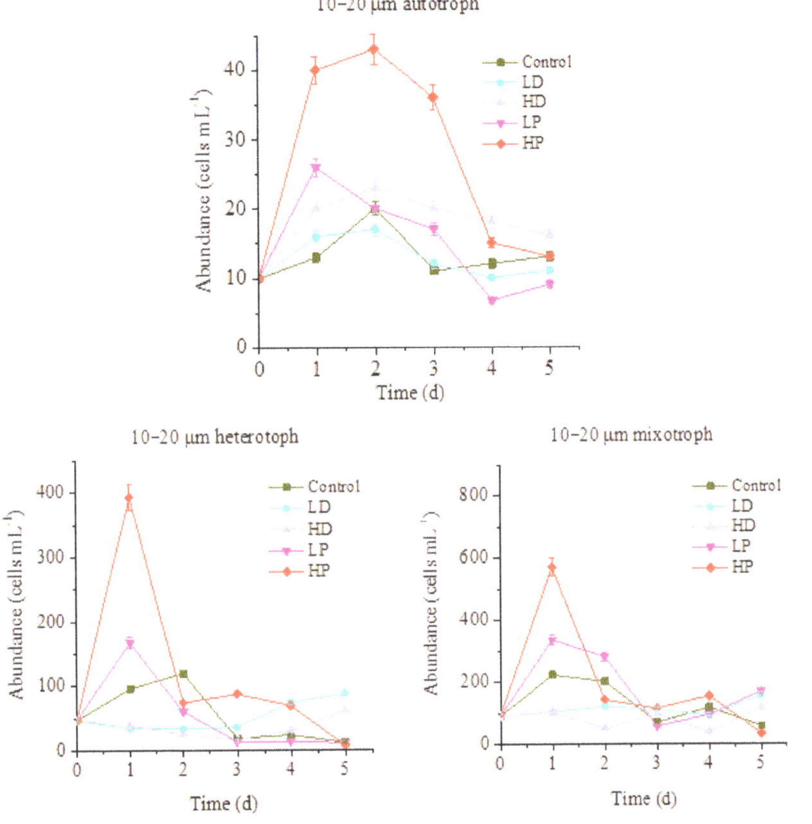

Figure 2. Variations in abundance of 10–20 µm protists.

The changes of 10–20 µm of protists in different concentrations of sand-dust and phosphorus supplementation groups were different from the control. The maximum values of autotrophic and heterotrophic protists in control group were 20 and 119 cells mL^{-1}, respectively, appeared on the 2nd day of incubation. The maximum values of the mixotrophic protists were 223 cells mL^{-1} on the first day of incubation. The peak of HP treated autotrophic protists was 43 cells mL^{-1}, appeared on the 2nd day of incubation, which was significantly higher than that of control ($p < 0.05$). The peaks of HP treated heterotrophic and mixotrophic protists appeared on the 1st day of incubation, 393 and 571 cells mL^{-1}, respectively, which were 3.30 and 2.56 times higher than that of control ($p < 0.05$). The high value lasted for one day and then declined rapidly. The maximum values of LP treatment for autotrophic, heterotrophic and mixotrophic protists were 26, 167, and 336 cells mL^{-1} on the first day of incubation, respectively, which were 1.30, 1.40 and 1.51 times of the control ($p < 0.05$). At the end of incubation, the abundance of mixotrophic protists was 2.98 times higher than that of control ($p < 0.05$).

In summary, the abundances of 10–20 µm protists in P supplying groups were higher than the control, and the maximum value of HP treated autotrophic protists was the largest. It showed that the addition of P will promote the growth of all planktonic protists, especially when the ocean area was P-limited. In the early stage of incubation, the higher concentration of P led to a stronger growth promoting effect on autotrophic protists; they increased and lasted for a longer time. At the end of incubation, the abundance of all kinds of protists in P-supplemented groups was lower than that in the early stage, and some were even lower than that at the control group. This might be caused by the more rapid consumption of nutrients at the early stage of incubation, and the nutritional deficiency that appeared at the late stage. Therefore, one-time addition of P could only promote the growth of protists in a short time. If nutrients are not supplied continuously, their growth will be limited, due to lack of nutrients in the later period. Compared with that, the abundance of LP treated mixotrophic protists was higher than that at control group at the late stage, which indicated that the P utilization efficiency of mixotrophic protists in this condition might be higher.

The differences were observed between the three trophic types of protists at the sand-dust addition groups. The abundance of autotrophic protists at HD and LD groups increased first then gradually decreased. The maximum values were 23 and 17 cells mL^{-1} on the 2nd day after culture. The maximum value of HD treatment group was 1.15 times higher than the control (<0.05). Although the peak value of HD treatment group was lower than that of P treated group, the value on the last day of incubation was significantly higher than that of P treated group (<0.05). The results showed that the addition of high concentration sand-dust could promote the growth of 10–20 µm autotrophic protists, which might be caused by the continuously supplement P from the sand-dust. The abundances of heterotrophic and mixotrophic protists decreased first and then increased at the sand-dust adding groups. At the initial stage of incubation, they were significantly lower than that of the control and P adding groups ($p < 0.05$). At the late stage of incubation, the abundance of heterotrophic protists in the two sand-dust groups was higher than that at control and P adding groups. The abundance of mixotrophic protists was higher than that at control and HP groups, but lower than that at LP group. The abundances of heterotrophic and mixotrophic protists in HD treatment group were always higher than that of LD treatment group. Hence, the early sand-dust deposition presented an inhibiting effect on the growth of 10–20 µm heterotrophic and mixotrophic protists, which was ascribed to the toxic effect of the heavy metals and other harmful substances dissolved from the sand-dust [5]. The inhibition effect of sand-dust was much greater than the promotion of P dissolution. At the late stage of incubation, the sand-dust stimulated the growth of heterotrophic and mixotrophic protists, especially heterotrophic protists. This indicated that the heterotrophic protists gradually tolerated the harmful substances from sand-dust. Furthermore, the nutrients dissolved from the sand-dust including trace elements such as Fe [8,10,11] (Table 1), supplemented the nutrients requirement at the late stage of incubation. The effect of dust was more significant for 10–20 µm heterotrophic and mixotrophic protists at the late stage of culture.

3.1.2. Response of 5–10 µm of Protists to Sand-Dust and Phosphorus Addition

Changes in the abundance of 5–10 µm of nanoprotists in different trophic modes under the stress of sand and phosphorus addition are shown in Figure 3.

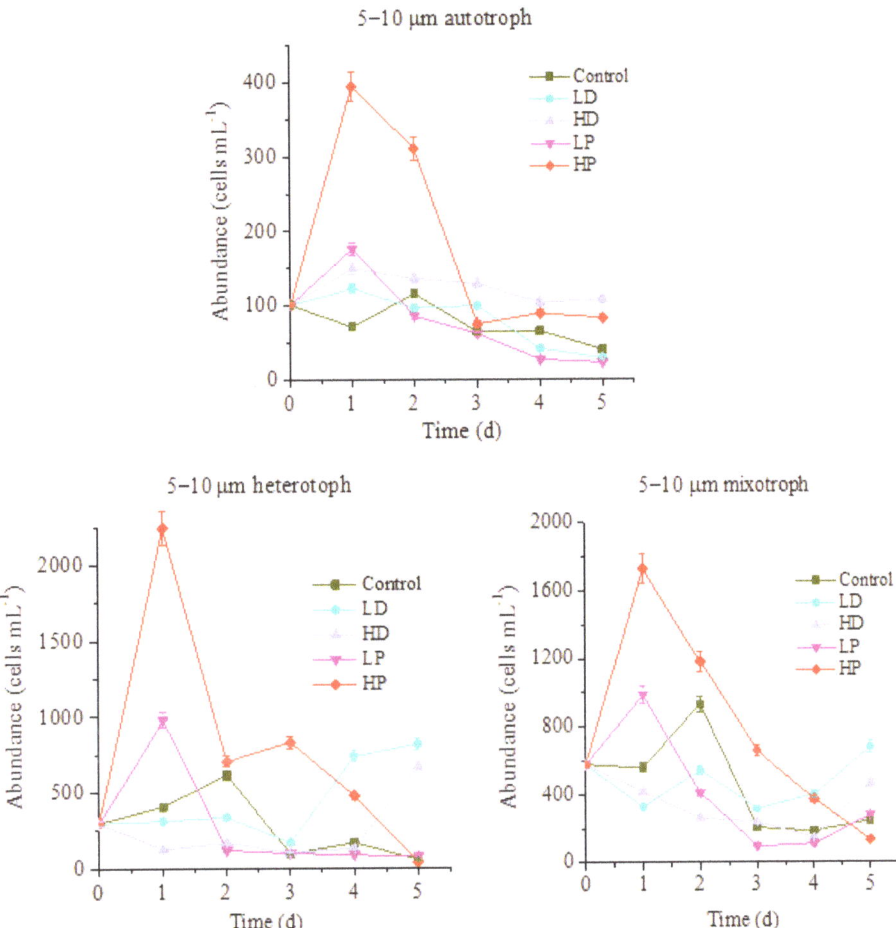

Figure 3. Variation of protists abundance in 5–10 μm size.

The variation trend of the abundance for 5–10 μm protists during incubation is basically the same as that of 10–20 μm protists, while the variation range is quite different. For the control group, the peak values of autotrophic, heterotrophic and mixotrophic protists appeared on the 2nd day of incubation with the cell concentrations of 116, 615 and 930 cells mL^{-1}, respectively. Compared with that, at HP treatment group, the peak values of autotrophic, heterotrophic and mixotrophic protists were 395, 2245 and 1732 cells mL^{-1}, which were 3.41, 3.85 and 1.86 times higher than those of control ($p < 0.05$). The LP treatment groups followed the peaks of HP treatment. The maximum concentrations of autotrophic, heterotrophic and mixotrophic protists were 176, 980, and 989 cells mL^{-1}, respectively, higher than those at control groups ($p < 0.05$), but they were significantly lower than those at HP treatment group ($p < 0.05$). The abundance decrease of protists in P-supplemented groups at the late stage of incubation might be related to the nutrient deficiency due to their rapid growth at the early stage. As such, it can be referred that the growth of 5–10 μm planktonic protists in this ocean is also limited by P. Addition of P can promote the reproduction of all of the planktonic protists. The higher the concentration of P, the stronger the promotion effect on reproduction of planktonic protists. Compared with responding results of 10–20 μm protists to P addition, it showed that the response of 5–10 μm protists to P addition is faster, but the duration is shorter.

Similarly, there were also differences between three trophic types of 5–10 μm protists in the sand-dust supplementation groups during the cultivation period. The abundance of autotrophic protists in HD and LD treatment groups increased first and then decreased. Same as P-adding group, the largest value of HD and LD groups appeared on the 1st day of incubation, with the 150 and 123 cells mL^{-1}, respectively. The peak value of autotrophic protists abundance in HD group was 1.29 times higher than that at control group all the time. Nevertheless it was lower than that of P-adding group in the early stage of incubation ($p < 0.05$), and higher than that of P-adding group in the final stage. It showed that the addition of high concentration of sand-dust stimulated the growth of 5–10 μm autotrophic protists for a longer time, which might be ascribed to the release of P from sand-dust [6]. The heterotrophic and mixotrophic protists abundance of dust addition groups decreased at the early stage of incubation, which was significantly lower than that at control group and P-adding groups ($p < 0.05$). The higher concentration of dust resulted in a sharper decline, which was confirmed by the fact that heterotrophic and mixotrophic protists abundance of HD treatment groups were always lower than of LD group during all the culture period. Interestingly, at the late stage of incubation, the abundance of two trophic modes protists was higher than that at P-adding groups and control group. Hence, the growth of heterotrophic and mixotrophic protists was significantly inhibited by the sand-dust deposition in the early stage. The higher concentration of dust presented the stronger inhibition effect, which was related to the toxic effect of heavy metals and other harmful substances dissolved from the dust [5]. At the late stage, the growth of protists was promoted, especially for heterotrophs. This might be related to the increase of tolerance for protists to harmful substances and the supplementation of nutrients dissolved from the dust.

3.1.3. Response of 2–5 μm of Protists to Sand-Dust and Phosphorus Addition

Figure 4 shows that the variation trend of 2–5 μm protists abundance is similar to that of above mentioned two sizes of protists, but the variation ranges are quite different. For the control group, the maximum values of autotrophic, heterotrophic and mixotrophic protists appeared on the 2nd day of culture, which were 4071, 9491 and 11,965 cells mL^{-1}, respectively. However, the peak values of P supplementation groups mainly emerged on the 1st day of culture. The highest values of HP treated autotrophic, heterotrophic and mixotrophic protists were 9303, 18,867 and 23,501 cells mL^{-1}, which were significantly higher than those of control ($p < 0.05$). The maximum values of LP treated three trophic modes protists were 6245, 11618 and 15,240 cells mL^{-1}, respectively, higher than those of control ($p < 0.05$), but lower than those at HP group ($p < 0.05$). It can be seen that the growth of 2–5 μm planktonic protists in this ocean area was also restricted by P. The addition of P could promote the growth of planktonic protists of different trophic modes. Comparably, the response of 2–5 μm planktonic protists to P addition was faster than that of 10–20 μm protists, and the duration lasts longer on HP treated heterotrophic and mixotrophic protists.

The changes of 2–5 μm protists abundance of different trophic modes in the sand-dust addition groups were different. The highest abundance of HD and LD treated autotrophic protists appeared on the 1st day of culture, same as that of P-adding groups, with the maximum values of 4860 and 4255 cells mL^{-1}, respectively. The peak value of HD group was significantly higher than that at control group, but lower than that at P-adding group ($p < 0.05$). At the late stage of incubation, HD group was higher than those at control and P-adding groups, indicating that the addition of high concentration sand-dust benefits to the growth of 2–5 μm autotrophic protists. This result might be related to the dissolution of P from sand-dust [6]. For the heterotrophic and mixotrophic protists, the concentration decreased at the early stage of incubation, which was significantly lower than that at control group and P-adding groups ($p < 0.05$). At the end of cultivation, the abundance of heterotrophic and mixotrophic protists increased, and got higher than those at control and the P-adding groups, especially heterotrophic protists in the LD group was 2.98 times higher than at control group. The result showed that the growth of 2–5 μm heterotrophic and mixotrophic protists, especially heterotrophic protists, was also inhibited by the sand-dust at the initial of incubation, then been promoted in the late stage.

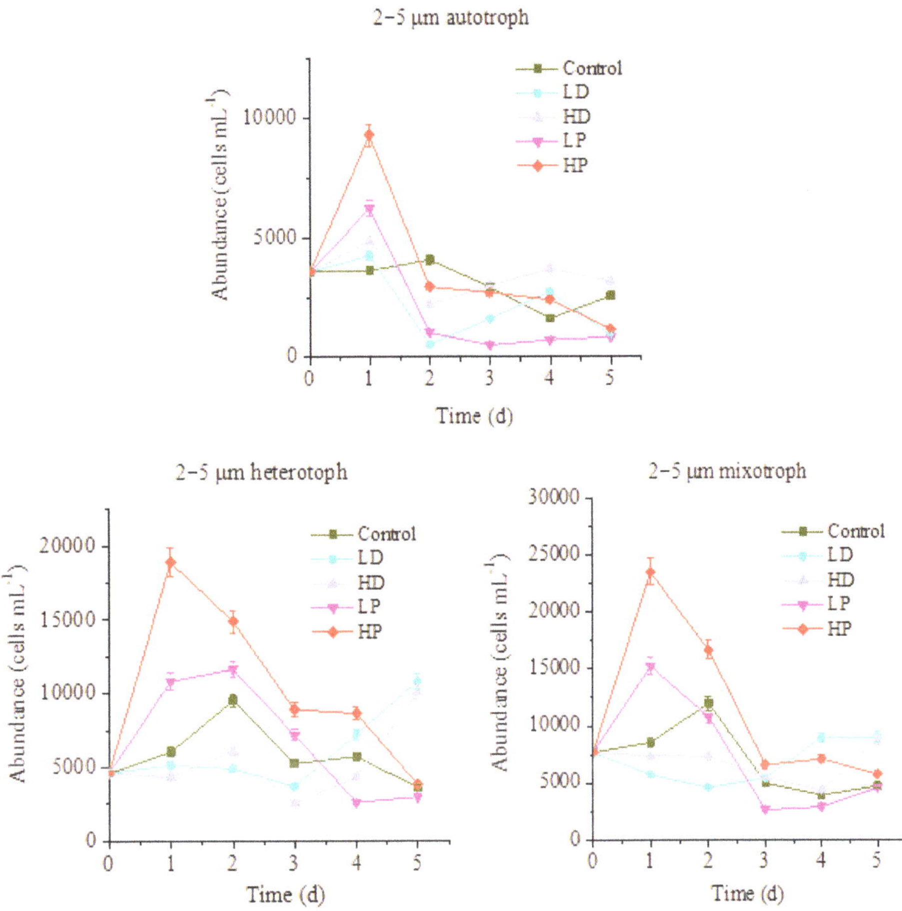

Figure 4. Variation of protists abundance in 2–5 μm size.

3.2. Effects of Sand-Dust and Phosphorus Addition on the Composition of Protists

3.2.1. Effects of Dust and Phosphorus Dosage on Trophic Structure of 10–20 μm Protists

The changes of trophic structure of 10–20 μm protists under the addition of dust and phosphorus are shown in Figure 5. The mixotrophic protists predominated in each group, accounting for 63.33% on average, followed by heterotrophic protists, accounting for 30.68% on average. On the 1st day of culture, the peak value of heterotrophic protists was appeared at P-supplemented groups and the growth rate was higher than that at other groups. On the 1st day of HP addition culture, the proportion of heterotrophic protists was higher than that at control group with 10.23%, and the proportion of mixotrophic protists was lower than that of the control group with 10.30% ($p < 0.05$), the proportion of autotrophic protists did not change obviously. At the end of the culture, the proportion of autotrophic protists at HP group increased by 8.02% ($p < 0.05$) because its high growth rate maintained for a long time, and the proportion of mixotrophic protists decreased by 7.41% ($p < 0.05$). The proportions of heterotrophic protists had no significant difference among different experimental groups ($p > 0.05$). The results showed that the autotrophic protists over competed heterotrophic protists and mixotrophic protists under the P dosage condition and thus the proportion of autotrophic protists increased significantly.

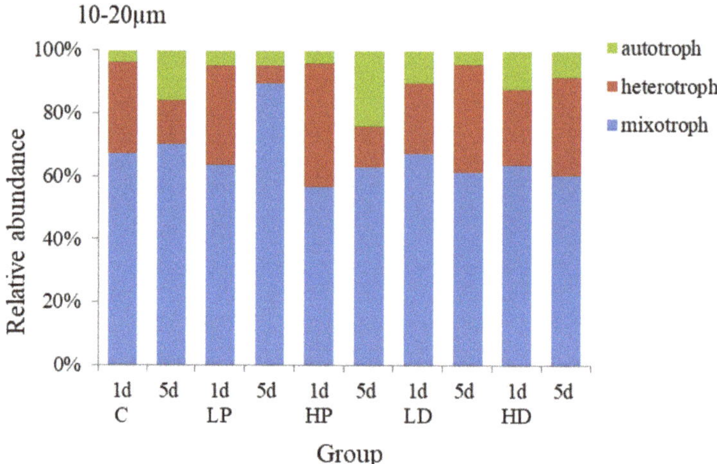

Figure 5. Variation in the trophic structure of protists in 10–20 μm size.

The trophic structure of 10–20 μm protists in sand-dust addition groups was different from that of P addition groups. Compared with the control group, the proportion of autotrophic protists increased at the initial stage of cultivation, while the growth of heterotrophic and mixotrophic protists was inhibited, and the higher the concentration of sand-dust lead to the stronger effect. On the 1st day, the proportion of autotrophic protists at HD group increased by 8.58% ($p < 0.05$), while the proportion of heterotrophic protists and mixotrophic decreased compared with that of control group ($p < 0.05$). However, at the end of culture, the growth rate of autotrophic protists decreased due to the significant increase of the heterotrophic growth rate at sand-dust group, which made the proportion of heterotrophic protists in sand-dust group higher than that at control group by 17.86% ($p < 0.05$), and significantly higher than that of P addition, and the proportion of autotrophic and mixotrophic protists decreased by 7.80% and 10.06%, respectively, compared with the control group. The results showed that the sand-dust could stimulate the growth of autotrophic protists at early stage, but inhibit the heterotrophic protists significantly. At the late stage, sand-dust presented a stronger promoting effect on the heterotrophic protists.

3.2.2. Effects of Dust and Phosphorus Dosage on Trophic Structure of 5–10 μm Protists

The response of trophic structure of 5–10 μm protists to dust and phosphorus dosage is shown in Figure 6. It showed that mixotrophic protists and heterotrophic protists were the dominant species, accounting for 45.74% and 44.88% of the total abundance. The relative abundance of autotrophic protists was relatively small, accounting for 9.38% on average. Compared with 10–20 μm groups, the proportion of 5–10 μm mixotrophic protists decreased by 17.59% ($p < 0.05$), while the proportion of heterotrophic protists increased by 14.20% ($p < 0.05$). On the first day of incubation, the growth of heterotrophic protists was higher, the proportion of mixotrophic protists reduced. At HP treatment group, heterotrophic protists increased by 12.16% ($p < 0.05$), mixotrophic protists decreased by 14.34% ($p < 0.05$), and autotrophic protists decreased slightly ($p > 0.05$), compared with control group. At the end of culture, at HD addition group, the proportion of heterotrophic protists increased by 20.47% ($p < 0.05$), the proportion of mixotrophic protists decreased by 20.58% ($p < 0.05$), and the proportion of heterotrophic protists presented little change ($p > 0.05$), compared with the control groups, respectively. Hence, the 5–10 μm heterotrophic protists were more sensitive to P addition in the early stage of culture and more competitive than the mixed and autotrophic protists. However, at the end of culture, the autotrophic protists became more competitive and their proportion increased significantly, indicating that P dosage had a stronger and longer effect on the growth of autotrophic protists.

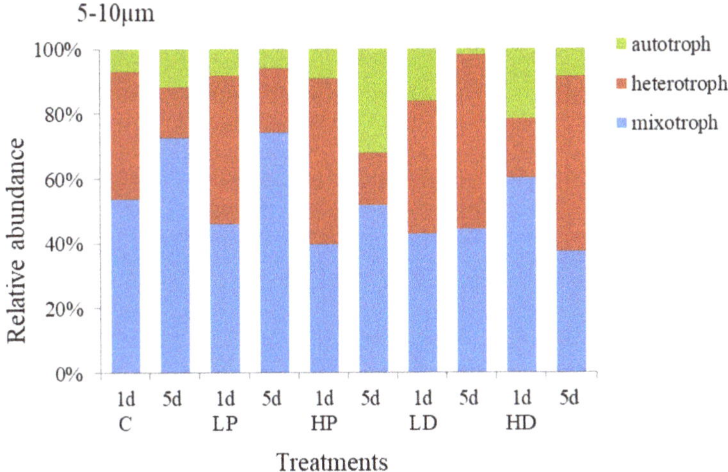

Figure 6. Variation in the trophic structure of protists in 5–10 μm size.

The trophic structure of 5–10 μm protists under sand-dust addition was different from that of P addition. The change of autotrophic protists in sand-dust adding group was the same as that of the P adding group. On the first day of culture, the autotrophic protists increased significantly and the higher concentration sand-dust led to the higher scale increase. At the initial stage of culture, the proportion of autotrophic protists at LD and HD groups increased to 16.10% and 21.65% respectively, which was 9.25% and 14.79% higher than that at control group ($p < 0.05$), but the growth of heterotrophic and mixotrophic protists were significantly inhibited ($p < 0.05$). At the end of culture, the proportion of heterotrophic protists at LD and HD groups increased to 53.50% and 53.76% respectively, 37.71% and 37.97% higher than that at control group ($p < 0.05$). These increases were significantly higher than that at P-supplemented groups ($p < 0.05$). Meanwhile, the proportion of autotrophic and mixotrophic protists decreased, especially for the HD and LD treatments, they were 34.91% and 27.91% lower than those at control group ($p < 0.05$). The results showed that the early stage of sand-dust dosage had obvious promoting effect on the growth of autotrophic protists, but inhibited the heterotrophic protists. At the late stage, it presented stronger promoting effect on the heterotrophic protists.

3.2.3. Effects of Dust and Phosphorus Dosage on Trophic Structure of 2–5 μm Protists

Changes in the composition and trophic structure of 2–5 μm protists under sand-dust and phosphorus addition stress are shown in Figure 7. The results showed that the mixotrophic protists dominated in 2–5 μm protists, accounting for 45.11% of the total abundance, the heterotrophic followed by 37.00%, and the autotrophic was about 17.89%. Compared with the 10–20 μm and 5–10 μm protists, the autotrophic increased by 11.89% and 8.50%, respectively, meaning which reflected that the trophic structure of 2–5 μm protists were more balanced than other sizes. On the 1st day, the protists trophic structure in P adding groups did not change obviously compared with control group, but at the end of culture, the proportion of HP and LP treated mixotrophic protists increased by 10.22% and 11.43% respectively. While the autotrophic protists decreased by 12.55% and 13.41%, respectively. The results showed that P addition presented little effect on the trophic structure of protists at the early stage, but decreased the proportion of the autotrophic and increased the proportion of the mixotrophic at the late stage.

The trophic structure of 2–5 μm protists under sand-dust adding changed largely. At the early stage, the effect of sand-dust addition on the autotrophic presented the same trend as that of P addition compared with the control. Both P and sand-dust adding improved the proportion of the autotrophic protists. The relative abundance of autotrophic protists at LD and HD treated groups were 28.15% and 29.26%, respectively, which increased by 8.23% and 9.15% comparing with the control. However, the growth of heterotrophic and mixotrophic protists was inhibited. At the end of incubation, the proportion of heterotrophic protists increased to 52.04% and 45.94% at LD and HD treatment groups, respectively, 18.80% and 12.70% higher than that at control group. The proportion of the autotrophic protists at LD and HD treatment groups decreased to 4.40% and 14.49%, respectively, which was 19.05% and 8.97% lower than that at control group. The results showed that the sand-dust deposition had a positive effect on the growth of 2–5 μm autotrophic protists at the early stage, but increased the proportion of the heterotrophic protists at the late stage.

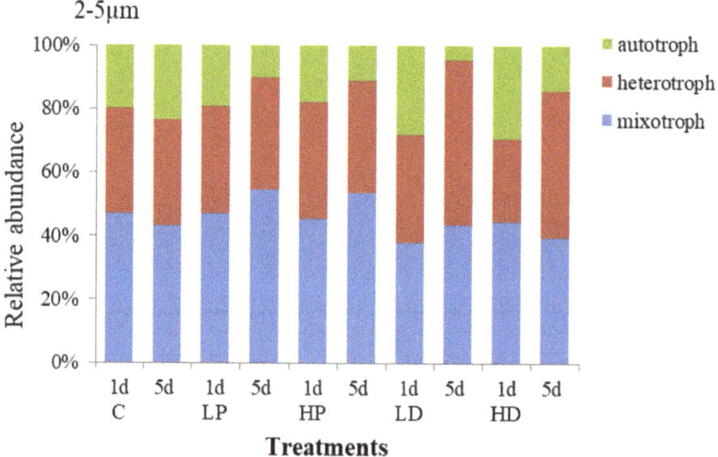

Figure 7. Variation in the trophic structure of protists in 2–5 μm size.

4. Discussion

According to the changes of relative abundance and trophic structure of protists by the rank sum test of Kruskal-wallis and Nemenyi-Wilcoxon-Wilcox, it showed that P promoted the growth of protists with the different particle sizes and the trophic modes at the early stage of culture. While the sand-dust presented different effects on the growth of different trophic modes. It stimulated the autotrophic but inhibited the heterotrophic and mixotrophic protists. The order of relative abundance of autotrophic protists with different particle sizes was 10–20 μm > 5–10 μm > 2–5 μm. The order of decreasing the proportion of heterotrophic protists with different particle sizes was 5–10 μm > 10–20 μm > 2–5 μm; and the order of inhibiting the proportion for mixotrophic protists with different particle sizes was 10–20 μm > 5 μm > 2–5 μm. The early stage of sand-dust deposition obviously inhibited the heterotrophic and mixotrophic protists by the dissolution of toxic and harmful substances in sand-dust. Mixotrophic and heterotrophic protists are the primary predator of autotrophic protists; Pearce et al. (2011) found 42%~82% primary production was consumed by mixotrophic and heterotrophic protists [35]. The decrease of mixotrophic and heterotrophic protists could accelerate the growth of autotrophic protists. In this study, sand-dust addition promoted the growth of autotrophic protists at the early stage, and the decrease of mixotrophic and heterotrophic protists accelerated this process. The further study confirmed that sand-dust storm could significantly induce the occurrence of red tide in the southern Yellow Sea [18,30]. Besides, the harmful inhibiting effect of sand-dust on heterotrophic protists community would weaken the marine matter cycle and food transfer efficiency.

At the late stage of culture, P presented a prohibitive effect on the growth of protists, which was related to the rapid growth of protists at the earlier stage, which resulted in nutritional deficiency and growth restriction at the late stage. In the late stage of sand-dust deposition, the autotrophic protists were limited due to the excessive growth of autotrophic protists in the earlier stage. At the late stage, sand-dust demonstrated stimulating effect for heterotrophic and mixotrophic protists, and the order was heterotrophic > mixotrophic protists. As for the heterotrophic protists with different particle sizes, the order was 5–10 μm > 2–5 μm > 10–20 μm, and for mixotrophic protists, the order was 10–20 μm > 5–10 μm > 2–5 μm. In the late stage of sand-dust deposition, the toxicity of harmful substances to heterotrophic protists gradually weakened. These protists slowly adapted to the sand-dust environment and dissolution of organic substances and other nutrients in sand-dust provided abundant nutrients for the heterotrophic protists [17]. As such, their abundance maintained stable and their proportion gradually increased. The effects of low sand-dust group and high sand-dust group on the proportion of heterotrophic and mixotrophic protists were basically coincident. However, the effect of high sand-dust group was stronger and the nutritional supplementation lasted longer in the later period.

For the trophic structure, at the early stage of incubation, P addition could promote all kinds of phytoplankton protists, especially the 10–20 μm autotrophic protists, which were consistent with that of autotrophic protists at the early stage of sand-dust deposition. At the late stage of incubation, both P addition and sand-dust deposition restricted all sizes of autotrophic protists. Therefore, it is speculated that the effect of sand-dust deposition on the autotrophic protists in this ocean were ascribed to the dissolution of P from sand-dust. The early promotion of sand-dust was to supplement the P in the sea area, while at the late stage, the inhibition for the autotrophic protists was mainly related to the early rapid growth of autotrophic protists by consuming up the P and other nutrients. The limiting capacity of the sand-dust group at the late stage is less than that at P-adding groups. This reflected that the dust could continuously supply the P and other nutrients.

The abundant P at the early stage of sand-dust deposition promoted the rapid growth of large-size autotrophic protists. This phenomenon will further accelerate the occurrence of red tides in eutrophic sea areas. Sand-dust deposition in the late stage stimulated the small-size heterotrophic protists and accelerated the material cycle efficiency and food transfer capacity in the sea. Therefore, the influence of sand-dust on the structure of different particle size and trophic protists will change the structure of the micro-food web in the ocean. It will also change ability of material transformation in the water body and ultimately, and affect the ecological function of protists in the transforming matter and producing food in the sea.

5. Conclusions

(1) Sand-dust deposition affected the trophic structure of different particle sizes of planktonic protists in the southern Yellow Sea. This could lead to change the structure of micro-food webs in the sea, and affect the ecological functions of micro-food webs in material transformation and food production.

(2) The growth of planktonic protists of all trophic modes in this ocean was restricted by P. The early addition of P could promote the growth of planktonic protists of all trophic modes in the southern Yellow Sea. The effect on the 10–20 μm autotrophic protists was most obviously, while the late addition of P mainly restricted the growth of different sizes of protists.

(3) The effect of initial sand-dust deposition on autotrophic protists was the same as that of P, it inhibited heterotrophic and mixotrophic protists. The positive effect of sand-dust deposition on heterotrophic and mixotrophic protists was strong at the late stage, and it improved the abundance of small-sized heterotrophic protists.

(4) The positive effect of sand-dust deposition on autotrophic protists in the Yellow Sea might be related to the dissolution of P from the sand-dust. The promotion of small-size heterotrophic protists in the late stage of sand-dust deposition could accelerate the material circulation efficiency and food transformation in the sea.

Author Contributions: Conceptualization, X.C. and Y.-G.Z.; methodology, X.C., G.-X.L. and H.-J.C.; investigation, X.C., H.-J.C., and C.Z.; formal analysis, X.C. and C.Z.; project administration; G.-X.L. and H.-J.C.; resources, G.-X.L. and H.-J.C.; software, X.C.; supervision, G.-X.L.; writing—original draft preparation, X.C., and Y.-G.Z.; writing—review and editing, X.C., Y.-G.Z. and X.H.; funding acquisition, G.-X.L.; validation, G.-X.L.; visualization, X.C and X.H.

Funding: Financial support for this project was provided by the National Natural Science Foundation of China (NSFC; 41210008) and the Major State Basic Research Development Program of China (973 Program; 2014CB953701). This study was conducted at the Ocean University of China.

Conflicts of Interest: The authors declare no conflict of interest.

References

1. Wang, B.D.; Wang, G.Y. Horizontal distributions and transportation of nutrients in the southern Huanghai Sea. *Acta Oceanol. Sin.* **1999**, *21*, 124–129.

2. Xie, L.P.; Sun, X.; Wang, B.D.; Xin, M. Temporal and spatial distributions of trophic structure and potential nutrient limitation in the Bohai Sea and the Yellow Sea. *Mar. Sci.* **2012**, *36*, 45–53.

3. Ghyoot, C.; Lancelot, C.; Flynn, K.J.; Mitra, A.; Gypens, N. Introducing mixotrophy into a biogeochemical model describing an eutrophied coastal ecosystem. *Prog. Oceanogr.* **2017**, *157*, 1–11. [CrossRef]

4. Stoecker, D.K. Conceptual models of mixotrophy in planktonic protists and some ecological and evolutionary implications. *Eur. J. Protistol.* **1998**, *34*, 281–290. [CrossRef]

5. Griffin, D.W.; Kellogg, C.A. Dust storms and their impact on ocean and human health: Dust in earth's atmosphere. *EcoHealth* **2004**, *1*, 284–295. [CrossRef]

6. Richon, C.; Dutay, J.; Dulac, F.; Dulac, F.; Wang, R.; Balkanski, Y.; Nabat, P.; Aumont, O.; Desboeufs, K.; Laurent, B.; et al. Modeling the impacts of atmospheric deposition of nitrogen and desert dust-derived phosphorus on nutrients and biological budgets of the Mediterranean Sea. *Prog. Oceanogr.* **2018**, *163*, 21–39. [CrossRef]

7. Zhang, K.; Gao, H.W. The characteristics of Asian-dust storms during 2000–2002: From the source to the sea. *Atmos. Environ.* **2007**, *41*, 9136–9145.

8. Tan, S.C.; Shi, G.Y.; Wang, H. Long-range transport of spring dust storms in Inner Mongolia and impact on the China seas. *Atmos. Environ.* **2012**, *46*, 299–308. [CrossRef]

9. Tan, S.C.; Shi, G.Y.; Shi, J.H.; Gao, H.W.; Yao, X.H. Correlation of Asian dust with chlorophyll and primary productivity in the coastal seas of China during the period from 1998 to 2008. *J. Geophys. Res.* **2011**, *116*, G02029. [CrossRef]

10. Tan, S.C.; Shi, G.Y. The relationship between satellite-derived primary production and vertical mixing and atmospheric inputs in the Yellow Sea cold water mass. *Cont. Shelf. Res.* **2012**, *48*, 138–145. [CrossRef]

11. Tan, S.C.; Wang, H. The transport and deposition of dust and its impact on phytoplankton growth in the Yellow *Sea. Atmos. Environ.* **2014**, *99*, 491–499. [CrossRef]

12. Gao, Y.; Arimoto, R.; Duce, R.A.; Lee, D.S.; Zhou, M.Y. Input of atmospheric trace elements and mineral matter to the Yellow Sea during the spring of a low-dust year. *J. Geophys. Res.* **1992**, *97*, 3767–3777. [CrossRef]

13. Azam, F.; Fenchel, T.; Field, J.G.; Gray, J.S.; Meyer-Reil, L.A.; Thingstad, F. The ecological role of water-column microbes in the sea. *Mar. Ecol. Prog. Ser.* **1983**, *10*, 257–263. [CrossRef]

14. Hagström, A.; Azam, F.; Anderson, A.; Wikner, J.; Rassoulzadegan, F. Microbial loop in an oligotrophic pelagic marine ecosystem: Possible roles of cyanobacteria and nanoflagellates in the organic fluxes. *Mar. Ecol. Prog. Ser.* **1988**, *49*, 171–178. [CrossRef]

15. Sherr, E.B.; Sherr, B.F. Bacterivory and herbivory: Key roles of phagotrophic protists in pelagic food webs. *Microb. Ecol.* **1994**, *28*, 223–235. [CrossRef] [PubMed]

16. Christaki, U.; Courties, C.; Massana, R.; Catala, P.; Lebaron, P.; Gasol, J.M.; Zubkov, M. Optimized routine flow cytometric enumeration of heterotrophic flagellates using SYBR Green I. *Limnol. Oceanogr. Meth.* **2011**, *9*, 329–339. [CrossRef]

17. Lin, Y.C.; Tsai, A.Y.; Chiang, K.P. Trophic coupling between Synechococcus and pigmented nanoflagellates in the coastal waters of Taiwan, Western Subtropical Pacific. *J. Oceanogr.* **2009**, *65*, 781–789.

18. Mitra, A.; Flynn, K.J.; Tillmann, U.; Raven, J.A.; Caron, D.; Stoecker, D.K.; Not, F.; Hansen, P.J.; Hallegraeff, G.; Sander, R.W.; et al. Defining planktonic protist functional groups on mechanisms for energy and nutrient acquisition: Incorporation of diverse mixotrophic strategies. *Protists* **2016**, *167*, 106–120. [CrossRef] [PubMed]

19. Sherr, E.B.; Sherr, B.F. Significance of predation by protists in aquatic microbial food webs. *Antonie Van Leeuwenhoek* **2002**, *81*, 293–308. [CrossRef] [PubMed]

20. Stoecker, D.K.; Capuzzo, J.M. Predation on protozoa: Its importance to zooplankton. *J. Plankton Res.* **1990**, *12*, 891–908. [CrossRef]

21. Jeong, H.J.; Song, J.E.; Kang, N.S.; Kim, S.; Yoo, Y.D.; Park, J.Y. Feeding by heterotrophic dinoflagellates on the common marine heterotrophic nanoflagellate Cafeteria sp. *Mar. Ecol. Prog. Ser.* **2007**, *333*, 151–160. [CrossRef]

22. Sato, M.; Yoshikawa, T.; Takeda, S.; Furuya, K. Application of the size-fractionation method to simultaneous estimation of clearance rates by heterotrophic flagellates and ciliates of pico- and nanophytoplankton. *J. Exp. Mar. Bio. Ecol.* **2007**, *349*, 334–343. [CrossRef]

23. Sinistro, R. Top-down and bottom-up regulation of planktonic communities in a warm temperate wetland. *J. Plankton Res.* **2010**, *32*, 209–220. [CrossRef]

24. Cheng, T.T.; Lü, D.R.; Chen, H.B.; Wang, G.C. Size distribution and element composition of dust aerosol in Chinese Otindag Sandland. *Chin. Sci. Bull.* **2005**, *50*, 788–792. [CrossRef]

25. Duan, J.C.; Tan, J.H. Atmospheric heavy metals and Arsenic in China: Situation, sources and control policies. *Atmos. Environ.* **2013**, *4*, 93–101. [CrossRef]

26. Sherr, B.F.; Sherr, E.B. Proportional distribution of total numbers, biovolume, and bacterivory among size classes of 2-20μm nonpigmented marine flagellates. *Mar. Microb. Food Webs* **1991**, *5*, 227–237.

27. Christaki, U.; Courties, C.; Joux, F.; Jeffrey, W.H.; Neveux, J.; Naudin, J. Community structure and trophic role of ciliates and heterotrophic nanoflagellates in Rhone River diluted mesoscale structures (NW Mediterranean Sea). *Aquat. Microb. Ecol.* **2009**, *57*, 263–277. [CrossRef]

28. Mitra, A.; Flynn, K.J.; Burkholder, J.; Berge, T.; Calber, A.; Raven, J.A.; Granéli, E.; Gilvert, P.M.; Hansen, P.J.; Stoecker, D.K.; et al. The role of mixotrophic protists in the biological carbon pump. *Biogeosciences* **2014**, *11*, 995–1005. [CrossRef]

29. Caron, D.A.; Alexander, C.H.; Allen, A.E.; Archibald, J.M.; Arrmbrust, E.V.; Bachy, C.; Bell, C.J.; Bharti, A.; Dyhrman, S.T.; Guida, S.M.; et al. Probing the evolution, ecology and physiology of marine protists using transcriptomics. *Nat. Rev. Microbiol.* **2017**, *15*, 6–20. [CrossRef]

30. Stibor, H.; Stockenreiter, M.; Nejstgaard, J.C.; Ptacnik, R.; Sommer, U. Trophic switches in pelagic systems. *Curr. Opin. Syst. Biol.* **2019**, *13*, 108–114. [CrossRef]

31. Guieu, C.; Dulac, F.; Desboeufs, K.; Wagener, T.; Pulido-Villena, E.; Grisoni, J.; Louis, F.; Ridame, C.; Blain, S.; Brunet, C.; et al. Large clean mesocosms and simulated dust deposition: A new methodology to investigate responses of marine oligotrophic ecosystems to atmospheric inputs. *Biogeosciences* **2010**, *7*, 2765–2784. [CrossRef]

32. Shi, J.H.; Gao, H.W.; Zhang, J.; Tan, S.C.; Ren, J.L.; Liu, C.G.; Liu, Y.; Yao, X.H. Examination of causative link between a spring bloom and dry/wet deposition of Asian dust in the Yellow Sea. *Chin. J. Geophys. Res.* **2012**, *117*, D17304. [CrossRef]

33. Zubkov, M.V.; Burkill, P.H.; Topping, J.N. Flow cytometric enumeration of DNA-stained oceanic planktonic protists. *J. Plankton Res.* **2007**, *29*, 79–86. [CrossRef]

34. Knap, A.; Michaels, A.; Close, A. Protocols for the Joint Global Ocean Flux Study (JGOFS) Core Measurement. In *Scientific Committee on Oceanic Research Manual and Guides*; UNESCO: Paris, France, 1994; Volume 29, pp. 210–243.

35. Pearce, I.; Davidson, A.T.; Thomson, P.G.; Wright, S.; Enden, R. Marine microbial ecology in the sub-Antarctic Zone: Rates of bacterial and phytoplankton growth and grazing by heterotrophic protists. *Deep Sea Res. Part II* **2011**, *58*, 2248–2259. [CrossRef]

Article

Stable Isotope Analysis and Persistent Organic Pollutants in Crustacean Zooplankton: The Role of Size and Seasonality

Roberta Piscia [1,*], **Michela Mazzoni** [1,2], **Roberta Bettinetti** [1,2], **Rossana Caroni** [1], **Davide Cicala** [1] **and Marina Manca** [1]

[1] Water Research Institute—National Research Council, IRSA-CNR, Largo Tonolli 50, 28922 Verbania, Italy
[2] DiSUIT, Department of Human Sciences and Innovation for the Territory, Via S. Abbondio 12, 22100 Como (CO), Italy
* Correspondence: roberta.piscia@irsa.cnr.it; Tel.: +39-0323-518300

Received: 12 June 2019; Accepted: 14 July 2019; Published: 18 July 2019

Abstract: Zooplankton is crucial for the transfer of matter, energy, and pollutants through aquatic food webs. Primary and secondary consumers contribute to the abundance and standing stock biomass, which both vary seasonally. By means of taxa- and size-specific carbon and nitrogen stable isotope analysis, the path of pollutants through zooplankton is traced and seasonal changes are addressed, in an effort to understand pollutant dynamics in the pelagic food web. We analyzed zooplankton plurennial changes in concentration of polychlorinated biphenyls (PCBs) and dichlorodiphenyltrichloroethane and its relatives (DDTs) and in taxa-specific $\delta^{15}N$ signatures in two size fractions, $\geq$450 µm and $\geq$850 µm, representative of the major part of zooplankton standing stock biomass and of the fraction to which fish predation is mainly directed, respectively. Our work is aimed at verifying: (1) A link between nitrogen isotopic signatures and pollutant concentrations; (2) the predominance of size versus seasonality for concentration of pollutants; and (3) the contribution of secondary versus primary consumers to carbon and nitrogen isotopic signatures. We found a prevalence of seasonality versus size in pollutant concentrations and isotopic signatures. The taxa-specific $\delta^{15}N$ results correlated to pollutant concentrations, by means of taxa contribution to standing stock biomass and $\delta^{15}N$ isotopic signatures. This is a step forward to understanding the taxa-specific role in pollutant transfer to planktivores and of zooplankton enrichment in PCBs and DDTs.

Keywords: stable isotope analysis; persistent organic pollutants; crustacean zooplankton; freshwater; size fractions; seasonality

1. Introduction

In lakes, crustacean zooplankton are low-order consumers and represent an important link between the base of the pelagic food web and the organisms at higher trophic levels, which may have economic or conservation value. Although considered a homogeneous compartment, zooplankton is composed of organisms that differ substantially from each other not only in their taxonomy, but also in body size, metabolic rate, and ecological roles [1]. While heterogeneity of the zooplankton community has been fairly well-documented in basic ecological studies, it is often overlooked in ecotoxicological ones, particularly in models, which are usually focused mainly on sources and top-levels of pollution patterns, the latter being directly related to human health. In particular, it is a crucial step in the flow of pollutants through aquatic ecosystems [2–5]. Both these pathways are involved, because zooplankton is directly related to primary producers and is able to process organic matter to feed on phytoplankton and to actively contribute to the microbial loop [6,7]. Zooplankton also represents a food source for all fish types at the larval stage and for zooplanktivorous fish at the adult stage [8,9]. Predation by fish and vertebrates is, in general, visual, therefore depending on prey size and visibility [10]. Both primary and secondary consumers may contribute to standing stock biomass in each size fraction considered. The contribution of primary and secondary consumers to zooplankton standing stock biomass generally varies with the season in deep, temperate, subalpine lakes, such as Lake Maggiore. In deep, temperate lakes, earlier spring warming and conditions of thermal water stratification influenced by climate change promote an early spring peak in phytoplankton biomass, followed by a peak in zooplankton biomass, mainly due to primary consumers, after about one month. The subsequent slow decline of zooplankton primary consumers is concomitant with an increasing contribution of zooplankton secondary consumers to the total zooplankton biomass in late summer. This pattern of change in zooplankton community composition also includes a change in organism size; in fact, a prevalence of larger sizes is observed when the contribution of predatory zooplankters increases.

Lake Maggiore has a long history of severe pollution brought about by persistent organic pollutants, such as dichlorodiphenyltrichloroethane and its metabolites. The International Commission for the Protection of the Italian–Swiss Common Waters (CIPAIS) supports monitoring activities in this lake, with a particular focus on priority substances, such as DDx and other organic chemicals of industrial origin, such as polychlorinated biphenyls (PCBs) and polybrominated diphenyl ethers (PBDEs), known for their harmful effects on human health and on environment [11,12]. Changes in physical and chemical conditions may cause their partial release into the dissolved phase, with increased bioavailability for organisms [13]. Because of their physico-chemical properties, persistent organic pollutants (POPs) can be detected everywhere, including worldwide remote regions far away from emission sources, such as polar or Alpine ecosystems [14–17]. In this study, we analyzed the seasonality of zooplankton in relation to persistent organic pollutant concentrations (i.e., DDTs and PCBs) in two different size fractions, one representative of most zooplanktonic biomass ($\geq$450 µm) and the other representative of fish food source ($\geq$850 µm). The former is representative of the major part of the crustacean zooplankton, while the latter represents the size fraction to which fish predation is mainly directed [10,18]. Despite its importance, zooplankton body size is usually neglected in studies on trophic pollutant transfer. Both seasonality and size are expected to drive zooplankton pollutant concentration over the season and a proper biomass estimation is essential to determine whether there is a prevalence of either of the two drivers. Furthermore, in order to study the modifications in the food web, we used carbon and nitrogen stable isotope analysis (^{13}C, ^{15}N; SIA). By this method, food sources (e.g., littoral versus pelagic, deep versus shallow) can be clearly identified in a system, as can the time-specific contribution of organisms to food webs. The method was originally applied to marine environments and to fish in particular, not least for fingerprinting their origin. Only recently it has been increasingly applied in freshwater environments, not only regarding fish or other organisms of direct human consumption, but also regarding those which sustain their production and growth, thus allowing for a more reliable reconstruction of the food web [19]. SIA is applied to investigate ecosystems in the making, but also to predict, by means of organism nitrogen isotopic signature, time- and space-specific trophic position, from which, among others, biomagnification depends [20–22]. The trophic role and food sources of organisms can be evaluated by carbon and nitrogen stable isotope analysis [23–25]. Isotopic signatures provide information on the resources exploited by a consumer within an environment and its trophic positions in the food web [24,26]. In particular, the isotopic carbon signature is almost constant between food source and consumer, whereas a nitrogen isotopic enrichment is observed with an increase of the trophic status [27]. In the present work, we related the trophic role of zooplankton with persistent organic pollutant (POPs) concentrations in Lake Maggiore (northern Italy). Previous long term studies in this deep, large Italian lake suggested that zooplankton seasonal dynamics and the different primary and secondary consumer contributions to zooplankton biomass are reflected by different contributions of the two important native predatory cladocerans (*Bythotrephes* and *Leptodora*), of diaptomids and cyclopoid copepods, and of the cladoceran *Daphnia*.

Our purpose is to investigate, in particular, how DDTs and PCBs vary with the season and in the two different size fractions, the former representing the bulk of crustacean zooplankton biomass and the latter representing the size fraction on which fish predation is mainly oriented. We hypothesize that variations are related to changes in $\delta^{15}N$ signature of the two fractions and that the latter results from the taxa-specific contribution to zooplankton biomass. We expect to find a correlation between taxa-specific signatures and taxa contribution to measured zooplankton standing stock biomass.

2. Study Site

Lake Maggiore is the second deepest ($Depth_{max}$ = 370 m) and largest (area = 212 km^2, volume 37.5 km^3) subalpine lake located in the northwestern part of Italy (Figure 1). By its nature oligotrophic, the lake underwent eutrophication during the 1960s and 1970s. The eutrophication reversal and the return to oligotrophic conditions were due to a reduction in total phosphorus discharge into the lake after wastewater treatment facilities were opened and thanks to the ban of phosphorus-containing detergents.

Figure 1. Map of Lake Maggiore. Sampling station location is shown with the red dot (from Google Earth).

Since 1996, contamination by dichlorodiphenyltrichloroethane and its relatives (DDTs), produced by a farm plant located in the lake catchment basin was detected [28]. Polychlorinated biphenyls (PCBs) in a steady-state condition were also reported [29–32].

Lake Maggiore is one of the best-studied lakes in Italy, with the first ecological research, including studies on zooplankton population dynamics, dating back to the early 1900s [33,34]. Long-term studies on zooplankton seasonal dynamics have been conducted over the last ten years by C and N stable isotope analyses, aimed at identifying trophic interactions and taxa-specific roles in the transfer of matter and pollutants within the pelagic food web.

3. Materials and Methods

Zooplankton sampling was performed seasonally at Ghiffa station (45°57′ N; 8°38′ E; Figure 1) in correspondence with the lake's deepest area (370 m). We collected two samples every day using two zooplankton nets that were 58 cm in diameter (450 μm and 850 μm in mesh size), which were hauled from 0 to 50 m deep several times in order to obtain a sufficient amount of organisms needed to perform the analyses (min = 2, max = 15). The total volume filtered was approximately 26 m^3 per zooplankton collection. Samples of the size fraction ≥450 μm were collected from January 2012 to January 2016, while samplings of the size fraction ≥850 μm started later (May 2013). All samples were collected in duplicate. Live organisms of one of the two samples were concentrated in a laboratory setting in approximately 1 L of lake water, frozen at −20 °C, and subsequently used for carbon (^{13}C) and nitrogen (^{15}N) stable isotope analysis (SIA). One third of the second zooplankton sample was immediately fixed and preserved in 95% ethanol for microscopic analysis at 6.3×. We calculated the biomass of the different taxa by way of length–weight regression equations [35,36]. On the basis of the taxa, zooplankton organisms were grouped into primary consumers (herbivores, i.e., *Daphnia, Eubosmina, Diaphanosoma,* and copepod diaptomids) and secondary consumers (predators, i.e., *Bythotrephes, Leptodora,* and copepod cyclopoids). The remaining portion of the second sample (two-thirds) was filtered on a 1.2 μm pore glass–fiber-filter (GF/C, 4.7 cm of diameter) and then frozen at −20 °C. As a consequence of using large net mesh sizes, rotifers and early developmental stages of crustacean zooplankton (i.e., the smallest body size organisms) and large phytoplankton colonies were not included in our samples. However, they are of only marginal importance for the purpose of our study, in which large zooplankton was regarded as an important link in the transfer of pollutants from primary producers to fish owing to selective predation on visible prey, particularly large-bodied zooplankton [37]. SIA was performed on pooled samples of the two size fractions (≥450 μm and ≥850 μm) and on the already-listed crustacean taxa present in sufficient amounts in the sample. Under the dissecting microscope, we pooled individuals of each taxon in order to reach the minimum dry weight (DW) of 2 mg per sample (approximately 70–700 individuals depending on the biomass). Samples were oven-dried for 24 h at 60 °C, before being homogenized and transferred into tin capsules of 5 × 9 mm in size. The isotopic compositions of organic carbon and nitrogen were determined by Ján Veizer Stable Isotope Laboratory (Ottawa University, Ontario, Canada) following methods already described in Visconti and Manca [38]. Isotope ratios were expressed as parts per thousand (‰) difference from a standard reference of PeeDee Belemnite for carbon and atmospheric N_2 for nitrogen, according to the equation:

$$^{13}\text{C, } ^{15}\text{N} = \left[\left(\frac{R_{\text{sample}}}{R_{\text{standard}}} \right) - 1 \right] \times 1000 \tag{1}$$

where R is the isotopic ratio ^{13}C/^{12}C and ^{15}N/^{14}N.

For determination of organic compounds (OCs), the materials and methods described in Mazzoni et al. [1] were followed. Quantitative DDT homologue analyses were performed by the external standard method using a solution containing pp'DDT, pp'DDE, and pp'DDD as the reference standard, prepared from single pure compounds (Pestanal, Sigma–Aldrich, Steinheim, Germany) in iso-octane (Carlo Erba, pesticide analysis grade, Val de Reuil, France). Arochlor 1260 (Alltech, IL, USA), while the addition of PCB 28, 52, and 118, was used for PCB quantification. The analyzed congeners consisted of PCB 28, 52, 101, 118, 138, 149, 153, 170, and 180. The detection limit for each OC was 1 ng·g^{-1} lipid weight (l.w.).

We used the software SigmaPlot 11.0 (version 11, Systat Software Inc., San Jose, CA, USA) to perform statistical analyses. After verifying that the datasets were normally distributed, we applied two-way ANOVA in order to identify which factor, size or seasonality, drove changes throughout the year.

4. Results

A strong seasonality characterized stable isotope data and zooplankton biomass composition. In Figures 2 and 3, we reported data of taxa/groups biomass percentage composition, δ^{13}C‰ and δ^{15}N‰ values and total crustacean biomass of the two size fractions analyzed (≥450 μm and ≥850 μm). A remarkably higher isotopic carbon (or less depleted) signature in summer and a higher nitrogen signature in autumn and winter was observed in the samples of both size fractions. The most depleted carbon signatures in the fraction ≥850 μm occurred with high relative abundance of primary consumers to total zooplankton biomass, concomitant to the spring peak or abundance of the cladoceran filter feeder *Daphnia*. In the ≥450 µm fraction, where most depleted carbon signatures were recorded, cyclopoids and diaptomids were also important contributors to the total biomass, together with Cladocera filter feeders.

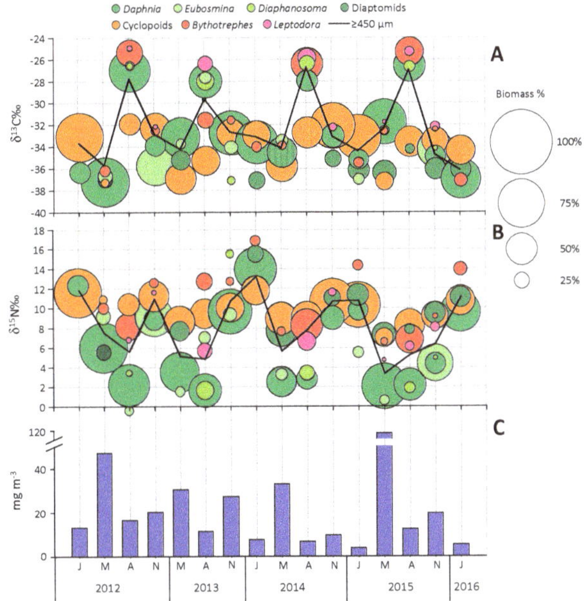

Figure 2. (**A**) Biomass percentage composition of planktonic crustacean taxa (bubbles) of Lake Maggiore and their δ^{13}C‰ values (bubble centers) and δ^{13}C‰ values of pooled samples (black line); (**B**) biomass percentage composition of planktonic crustacean taxa (bubbles) of Lake Maggiore and their δ^{15}N‰ values (bubble centers) and δ^{15}N‰ values of pooled samples (black line); (**C**) total biomass of Lake Maggiore planktonic crustacean taxa. All data refer to the fraction ≥450 µm.

Zooplankton standing stock biomass largely changed with the seasons. In samples belonging to the ≥450 µm size fraction, a spring peak was followed by a second phase of biomass increase in autumn.

In the ≥850 µm size fraction, peak biomass values were generally recorded in August. The unusual increase in May 2015 was related to the detection of the peak population growth of *Daphnia*. As already highlighted, *Daphnia* can be an important contributor to standing stock biomass, due to its peak biomass in spring being related to the development of young, smaller specimens.

We observed a clear seasonality of δ^{13}C isotopic signatures for the ≥450 µm size fraction during the studied years. The least depleted values in δ^{13}C‰ were regularly measured in August, while more depleted values characterized the spring samples (May), when the contribution to the peak biomass of primary consumers, particularly *Daphnia*, was at its peak. Limited seasonal variation regarding the contribution to the total biomass was characteristic of the cyclopoid copepods, whose contribution to carbon and nitrogen isotopic signatures showed similar values during the year. In the >450 µm size fraction, cyclopoids and calanoids were related to depleted values of carbon isotope signatures and to the highest values of nitrogen signatures. The biomass contributions of different zooplankton taxa, both primary and secondary consumers, drove patterns of both isotopic signatures; therefore, the measured values, e.g., in January 2012, integrated isotopic signatures of the two taxa present, i.e., cyclopoid copepods and, to a lesser extent, *Daphnia*.

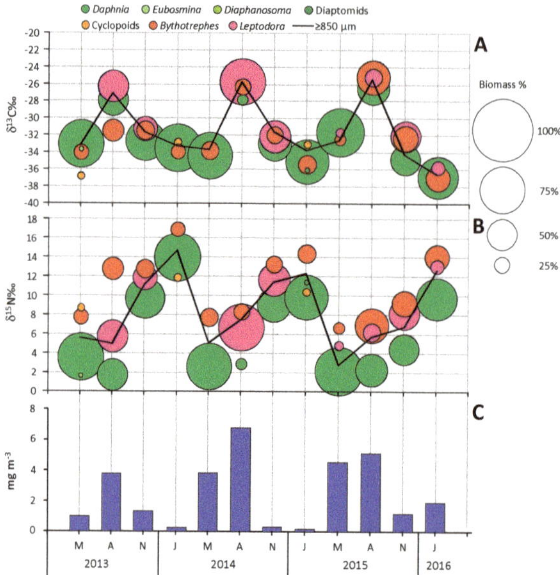

Figure 3. (**A**) Biomass percentage composition of planktonic crustacean taxa (bubbles) of Lake Maggiore and their δ^{13}C‰ values (bubble centers) and δ^{13}C‰ values of pooled samples (black line); (**B**) biomass percentage composition of planktonic crustacean taxa (bubbles) of Lake Maggiore and their δ^{15}N‰ values (bubble centers) and δ^{15}N‰ values of pooled samples (black line); (**C**) total biomass of Lake Maggiore planktonic crustacean taxa. All data refer to the fraction ≥850 µm.

A marked seasonality was also characteristic of the δ^{13}C and δ^{15}N signatures in the ≥850 µm size fraction. The least carbon-depleted values in August were related to the high contribution of secondary consumers, while more carbon-depleted values in spring and winter were related to a higher contribution of primary consumers. Values tended to be most enriched in winter, with secondary consumers more enriched than primary ones. The signatures of δ^{15}N‰ measured in the ≥850 µm size fraction were related to the relative biomass contribution of primary and secondary consumers. Among predatory cladocerans, *Bythotrephes* was generally much more δ^{15}N enriched than *Leptodora*. For example, in August 2013, the nitrogen enrichment of *Bythotrephes* was higher in comparison to *Leptodora* and *Daphnia*. Differences in δ^{13}C signatures suggest that *Leptodora* was living in deeper waters than *Bythotrephes* and *Daphnia*. Changes in isotopic signature and contribution to total biomass were mainly related to cladoceran primary and secondary consumers, thus enabling the investigation of infra-Cladocera food relationships.

As hypothesized, we found that carbon and nitrogen isotopic signatures of the two pooled size fractions could be reconstructed from taxa-specific signatures weighted on contribution to total biomass, by applying the following equation:

$$\delta^{15}\text{N}, \delta^{13}\text{C‰}_{\text{pooled sample}} = (i \times biomass\%)_{taxon1} + \ldots + (i \times biomass\%)_{taxonN}, \tag{2}$$

where $i = \delta^{15}$N‰ and δ^{13}C‰ are taxa-specific signatures.

Relationships between measured and estimated isotopic signatures were significant (δ^{13}C‰: $R^2 = 0.904$, N = 28, F = 244.682, $P < 0.001$; δ^{15}N‰: $R^2 = 0.942$, N = 28, F = 438.413, $P < 0.001$; Figure 4). Such relationships are the result of the accuracy of estimation of both biomass and taxa-specific isotopic signatures. In fact, taxa-specific biomass estimation is based on length–weight regression equations applied to a consistent number of specimens (at least 25 specimens per taxa) and taxa-specific isotopic signatures analyses are based on a very high number of specimens (cfr. Section 3), which is necessary for Isotope Ratio Mass Spectrometry (IRMS) analysis (at least 1 mg of dry weight). The accuracy of standing stock biomass determination from length–weight regression of taxa-specific analysis was also confirmed by the comparison with the direct biomass dry weight estimation, which was shown to be statistically significant ($R^2 = 0.814$, N = 8, F = 26.251 $P = 0.002$).

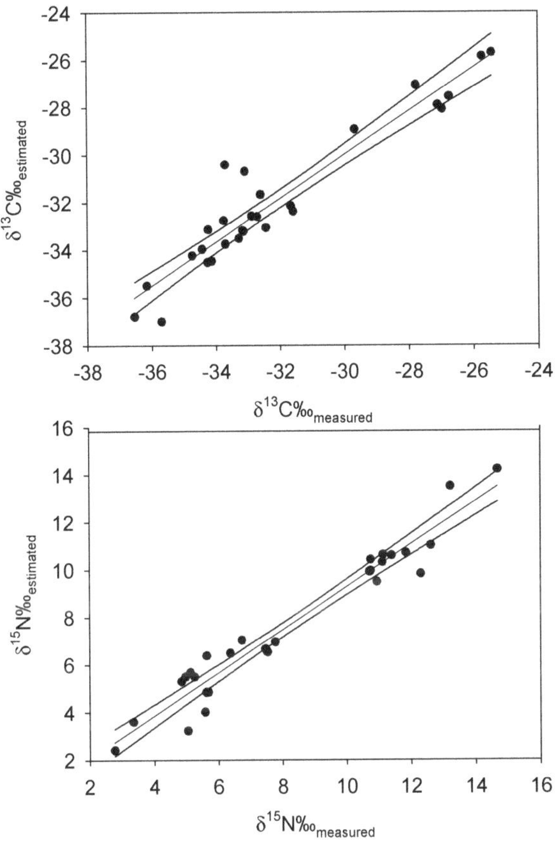

Figure 4. Relationship between isotopic fingerprint (carbon and nitrogen) of pooled zooplankton samples belonging to ≥450 µm and ≥850 µm size fractions determined by instrumental analysis (measured) and reconstructed values (estimated) following Equation (2).

The difference between the smaller and the larger size fraction was mainly determined by cladoceran taxa composition. In the case of DDTs, the main effects cannot be properly interpreted because a weak, but significant, interaction between size and seasonality was determined when two-way ANOVA was applied (N = 28, F = 3.149, $P < 0.048$), so that the size of a factor's effect depends upon the level of the other factor. Changes in PCBs and $\delta^{15}N$ values were shown to be strongly driven by seasonality (PCB: N = 28, F = 14.312, $P < 0.001$; $\delta^{15}N$: N = 28, F = 29.169, $P < 0.001$) and for both datasets there was no statistically significant interaction between factor size and seasonality ($P = 0.181$ and $P = 0.596$, respectively). Results of the two-way ANOVA are reported in Table 1.

A regression analysis was performed with $\delta^{15}N$ data and pollutant concentrations (Figure 5) with ≥450 µm and ≥850 µm size fractions. Changes in both DDT and PCB concentrations on a logarithmic scale were related to zooplankton $\delta^{15}N$‰ isotopic signatures (DDTs: $R^2 = 0.541$, N = 26, F = 28,340, $P < 0.001$; PCBs: $R^2 = 0.502$, N = 27, F = 25.159, $P < 0.001$). The data in the upper right side of the graph refer to autumn and winter samples, while those in the lower side of the graph are related to spring and summer samples.

Table 1. Results of the two-way ANOVA tests performed on DDTsPCBs concentrations and $\delta^{15}N\%_0$ fingerprint of zooplankton samples of the two size fractions ($\geq$450 μm and $\geq$850 μm).

	DF	SS	MS	F	P
DDTs					
Size	1	3.672	3.672	58.456	<0.001
Season	3	4.948	1.649	26.256	<0.001
Size X Season	3	0.593	0.198	3.149	0.048
PCBs					
Size	1	0.002	0.002	0.011	0.918
Season	3	8.550	2.850	14.312	<0.001
Size X Season	3	1.052	0.351	1.760	0.187
$\delta^{15}N\%_0$					
Size	1	3.239	3.239	1.195	0.287
Season	3	7.226	79.075	26.169	<0.001
Size X Season	3	5.229	1.743	0.643	0.596

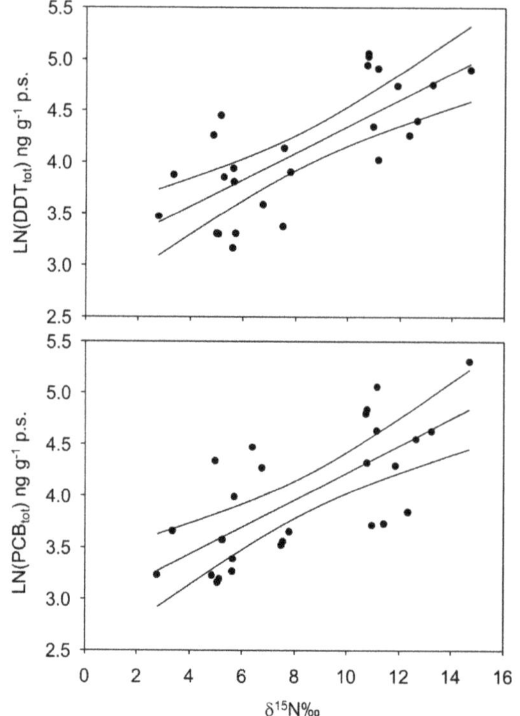

Figure 5. Relationship between nitrogen isotopic fingerprint of pooled zooplankton samples from Lake Maggiore, belonging to $\geq$450 μm and $\geq$850 μm size fractions, and persistent organic pollutant (POPs) concentrations, expressed as logarithmic scale.

POP concentrations in the $\geq$850 μm size fraction were generally lower than in the $\geq$450 μm size fraction (Figure A1).

5. Discussion

The pelagic zooplankton is composed of a large variety of organisms differing in taxonomic composition and body size. Together, protists, monogonont rotifers, Cladocera, and Copepoda contribute to the zooplankton community, the latter with both individuals at their adult stage and a number of developmental stages that are different in size and trophic role. Regarding net zooplankton, as a result of the mesh size, only rotifers and crustaceans were reliably sampled. The former can be numerically dominant during the spring phase, along with naupliar and early copepodite stages of copepods. Their contribution to zooplankton biomass is little, however, with respect to Cladocera and adult or sub-adult copepods [38–40]. Altogether, the latter are ≥450 μm in size.

Zooplanktivorous fish, however, tend to select larger (≥850 μm) and more visible prey, such as *Bythotrephes*, large, ovigerous *Daphnia*, and *Leptodora*. Therefore, these are a component of the zooplankton population, which is active in transferring POPs to fish [10,41–44].

By choosing to focus on the two size fractions of crustacean zooplankton, we aimed to investigate components which are directly involved in the transfer of pollutants in the trophic chain. In addition, investigating the role of micro-zooplankton is almost impossible, given that POP concentrations and $\delta^{13}C$ and $\delta^{15}N$ analyses require high weight amounts of zooplankton material (approximately 1 μg dry weight and 1 mg dry weight, respectively). Further studies are required to elucidate the role of micro-zooplankton as carrier of pollutants.

As expected, zooplankton biomass composition largely varied with the season in both size fractions analyzed, because in deep subalpine lakes, the abundance and composition of zooplankton populations vary along the season [45–47]. A spring peak abundance of phytoplankton population triggers the growth of primary zooplankton consumers, such as the large filter-feeder *Daphnia*, which, in Lake Maggiore, usually reaches its peak population density in May [47]. The increase of *Daphnia* and other primary consumer cladocerans (*Eubosmina*, *Diaphanosoma*) in spring is followed by an increase in predatory cladocerans (*Bythotrephes* and *Leptodora*), able to selectively exploit primary consumers and contribute to their decline [48,49]. In turn, the decrease in zooplankton primary consumers promotes a second phytoplankton peak in summer, which leads to a second phase of increase in primary consumers in autumn.

In this study, seasonality was also observed in isotopic carbon and nitrogen zooplankton signatures. In pelagic zooplankton, $\delta^{13}C$‰ tended toward less depleted values in summer, likely mirroring phytoplankton isotopic signatures [38,39,50]. During thermal stratification, the high growth rates of phytoplankton cells can lead to the consumption of dissolved atmospheric CO_2, causing a shift in carbon exploitation sources toward the uptake of the HCO_3^- ion [51–54]. The unselective filter feeder *Daphnia* feeds on seston particles, which fit the distance between their intersetular filtering combs (i.e., 0.2–50 μm) [55]. Since Lake Maggiore seston is mainly composed of phytoplankton cells and carbon fractionation is negligible, *Daphnia* isotopic signature specimens matched variations of the phytoplankton carbon fingerprint [45,50]. The pelagic zooplankton nitrogen isotopic signature also varied with the season, with the maximum enrichment in winter. Isotopic nitrogen tended to increase along the trophic food chain and stepwise enrichment varied in different environments and with the structure of the trophic web [45,56,57]. Isotopic nitrogen enrichment is diet-dependent [58]. In eutrophic lakes, where high values of $\delta^{15}N$‰ are usually found, zooplankton organisms do not feed only on phytoplankton, but also exploit alternative organic particulate matter, such as bacteria, protozoa, exuviae, and fecal pellets of zooplankton specimens, which are more $\delta^{15}N$-enriched than phytoplankton algae. We can hypothesize that, in winter, when the abundance of phytoplankton cells is low, alternative, $\delta^{15}N$-enriched food sources (e.g., bacteria, protozoa, exuviae) became more important, leading to the observed enrichment in isotopic nitrogen content. In addition, under food shortage conditions, the zooplankton might be able to exploit $\delta^{15}N$‰-enriched lipids which accumulate in high food concentration conditions [59,60].

Distinguishing between primary and secondary consumers is crucial for the reconstruction of patterns of matter and energy through the pelagic food web. As primary and secondary consumers largely overlap each other in body size, such distinction cannot be based on size fraction analyses of zooplankton. We found that the isotopic signature of pooled zooplankton samples of the two size analyzed fractions was tightly related to taxa-specific contribution to total biomass, thus allowing for the identification of taxa-specific contribution to measured $\delta^{15}N\%$ isotopic signatures. The latter is, in turn, correlated to POP concentrations, therefore allowing for a better definition of primary and secondary zooplankton consumer taxa in the transfer of toxicants through the pelagic food web. The significant relationship between pooled zooplankton samples and taxa-specific contributions to the total biomass in Lake Maggiore results from the fact that other zooplankton particles are excluded when samples are filtered through 450 µm nets. In Lake Maggiore, detritus is usually <126 µm and the contribution of phytoplankton colonies >76 µm is largely negligible. Therefore, any other zooplankton contribution to the total biomass in the samples of the two size fractions analyzed was certainly negligible. This is not the case in eutrophic or mesotrophic lakes, where zooplankton is smaller and detritus and phytoplankton colonies can be found in zooplankton samples in non-negligible amounts [1,61].

We hypothesized that both size and seasonality can influence POP concentrations in zooplankton populations and that the two parameters are related to the $\delta^{15}N\%$ isotopic signature. Our results highlighted a predominance of seasonality versus size. The ≥450 µm fraction was composed of primary (*Daphnia*, *Eubosmina*, *Diaphanosoma* and diaptomids) and secondary consumers (cyclopoids, *Bythotrephes* and *Leptodora*) throughout the year, while in the ≥850 µm fraction, the biomass of secondary consumers is dominant in August. As secondary consumers are expected to be higher in POP concentrations, we would have expected to record the highest values of POP concentrations in summer; on the contrary, we observed high POP concentrations in winter, when zooplankton nitrogen enrichment was at its maximum. In fact, the $\delta^{15}N\%$ and POP results directly correlated, with both tending to increase from spring to winter. Ecotoxicological studies on marine trophic chains demonstrated a similar relationship along the food chain [62,63]. In Lake Maggiore, the seasonal pattern of POP concentrations was very similar terms of values between years of the study period [28], suggesting a steady-state condition. In these conditions, the major $\delta^{15}N$-enrichment of carnivorous species was probably hidden by seasonal variations in $\delta^{15}N\%$, which was dependent on quality of organic particles, i.e., the bacteria, protozoa, or organic particles deriving from dead organisms should have higher POP concentrations than phytoplankton cells.

The predominance of seasonality versus size was confirmed also by DDT concentrations in the smallest size fraction. Theoretically, the big predators, *Bythotrephes* and *Leptodora*, and the largest *Daphnia* specimens of the ≥850 µm size fraction should be characterized by higher values of POP concentrations, the former due to this specimen belonging to an upper trophic level and the latter because adults of a larger size can accumulate more pollutants in their tissues. However, our results likely reflect the fact that copepods rich in lipids (substance where POPs are mainly stored) were not present in the larger size fraction. However, because zooplanktivorous fish mainly select large cladocerans [43], the ≥850 µm fraction should be more representative of fish food sources.

6. Conclusions

Zooplankton play a crucial activity in matter and energy transfer in the food web. In temperate lakes, biomass, taxa, and size composition largely vary with the season. The contribution of primary and secondary zooplankton consumers also varies seasonally, along with the $\delta^{15}N\%$ isotopic signature. By investigating seasonal changes in biomass and taxa-specific contributions and in carbon and nitrogen isotopic signatures along with DDT and PCB concentrations, we aimed at quantifying the relative importance of size versus seasonality. We found that the $\delta^{15}N\%$ of two size fractions, representative of bulk crustacean zooplankton biomass and of the fraction on which fish mainly prey, respectively, significantly correlated to POP concentration, as both varied with the season. The bulk zooplankton nitrogen isotopic signature resulted from a biomass-weighted taxa-specific isotopic signature, thus enabling us to distinguish between the contributions of secondary and primary consumers to measured nitrogen enrichment, from which the concentration of pollutants depends. DDT concentration was higher in the >450 μm fraction, while no difference was found in PCB concentration in the two size fractions. The difference between the two was mainly related to copepod adults, which were entirely lacking in the larger, >850 μm size fraction, the one on which fish selectively prey. Overall, seasonality was largely predominant over size for the dynamics of DDTs and PCBs, therefore suggesting that, in temperate lakes, more than one time spot studies are required; indeed, multi-year studies are ideal for verifying, among others, the condition of a steady state which must be fulfilled before applying models regarding the transfer of pollutants through an ecosystem.

The oligotrophic, subalpine Lake Maggiore was ideal for investigating the role of zooplankton in pollutant transfer, as size spectra distinction allows for the neglect of background noise caused by micro-zooplankton, protists, and phytoplankton algae. This is not the case for more productive lakes, in which such size-based distinction cannot be applied in the same way.

Author Contributions: Formal analysis, M.M.M. and R.P.; SIA and zooplankton investigation, M.M.M. and R.P.; POPs investigations, M.M. and R.B.; writing—original draft preparation, M.M.M., R.P., R.C., and D.C.; Review, R.B. and M.M.

Funding: This research was funded by CIPAIS (Commissione Internazionale per la Protezione delle Acque Italo-Svizzere).

Conflicts of Interest: The authors declare no conflict of interest.

Appendix A

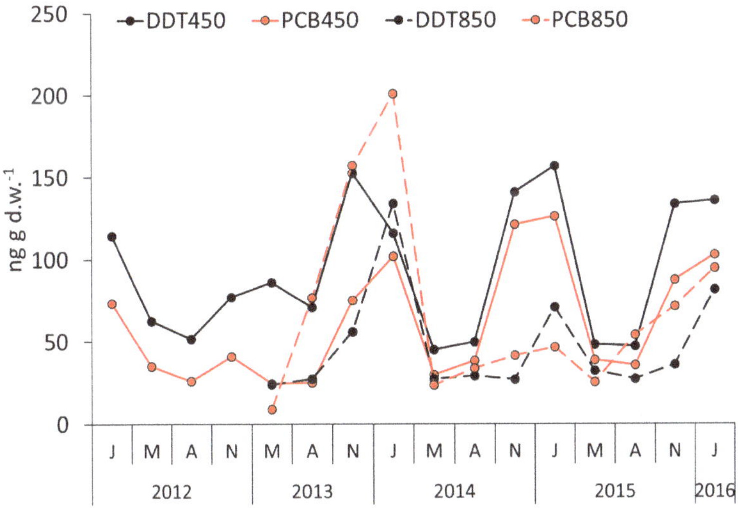

Figure A1. Concentration of POPs in two crustacean zooplankton size fractions (≥450 μm and ≥850 μm) of Lake Maggiore.

References

1. Mazzoni, M.; Boggio, E.; Manca, M.; Piscia, R.; Quadroni, S.; Bellasi, A.; Bettinetti, R. Trophic transfer of persistent organic pollutants through a pelagic food web: The case of Lake Como (Northern Italy). *Sci. Total Environ.* **2018**, *640*, 98–106. [CrossRef] [PubMed]
2. Bettinetti, R.; Manca, M. Understanding the role of zooplankton in transfer of pollutants through trophic food webs. In *Zooplankton: Species Diversity, Distribution and Seasonal Dynamics*; Kehayias, G., Ed.; Nova Science Publishers, Inc.: Hauppauge, NY, USA, 2013; pp. 1–18.

3. Bettinetti, R.; Galassi, S.; Guzzella, L.; Quadroni, S.; Volta, P. The role of zooplankton in DDT biomagnification in a pelagic food web of Lake Maggiore (Northern Italy). *Environ. Sci. Pollut. Res.* **2010**, *17*, 1508–1518. [CrossRef] [PubMed]

4. Loseto, L.L.; Stern, G.A.; Ferguson, S.H. Size and biomagnification: How habitat selection explains beluga mercury levels. *Environ. Sci. Technol.* **2008**, *42*, 3982–3988. [CrossRef] [PubMed]

5. Forest, A.; Sampei, M.; Makabe, R.; Sasaki, H.; Barber, D.G.; Gratton, Y.; Wassmann, P.; Fortier, L. The annual cycle of particulate organic carbon export in Franklin Bay (Canadian Artica): Environmental control and food web implications. *J. Geophys. Res.* **2008**, *113*, C03S05. [CrossRef]

6. Peduzzi, P.; Herndl, G.J. Zooplankton activity fueling the microbial loop: Differential growth response of bacteria from oligotrophic and eutrophic waters. *Limnol. Oceanogr.* **1992**, *37*, 1087–1092. [CrossRef]

7. Zingel, P.; Agasild, H.; Karus, K.; Kangro, K.; Tammert, H.; Tõnno, I.; Tõnu, F.; Nõges, T. The influence of zooplankton enrichment on the microbial loop in a shallow, eutrophic lake. *Eur. J. Protistol.* **2016**, *52*, 22–35. [CrossRef] [PubMed]

8. Confer, J.L.; Howick, G.L.; Corzette, M.H.; Kramer, S.L.; Fitzgibbon, S.; Landesberg, R. Visual predation by planktivores. *Oikos* **1978**, *31*, 27–37. [CrossRef]

9. Bandara, K.; Varpe, Ø.; Ji, R.; Eiane, K. Artificial evolution of behavioral and life history strategies of high-latitude copepods in response to bottom-up and top-down selection pressures. *Progr. Oceanogr.* **2019**, *173*, 134–164. [CrossRef]

10. de Bernardi, R.; Giussani, G.; Manca, M. Cladocera: Predators and prey. *Hydrobiologia* **1987**, *145*, 225–243. [CrossRef]

11. Eljarrat, E.; Barcelo, D. Priority lists for persistent organic pollutants and emerging contaminants based on their relative toxic potency in environmental samples. *TrAC Trends Anal. Chem.* **2003**, *22*, 655–665. [CrossRef]

12. Ruiz-Fernández, A.C.; Ontiveros-Cuadras, J.F.; Sericano, J.L.; Sanchez-Cabeza, J.A.; Kwong, L.L.W.; Dunbar, R.B.; Mucciarone, D.A.; Pérez-Bernal, L.H.; Páez-Osuna, F. Long-range atmospheric transport of persistent organic pollutants to remote lacustrine environments. *Sci. Total Environ.* **2014**, *493*, 505–520. [CrossRef]

13. Pisanello, F.; Marziali, L.; Rosignoli, F.; Poma, G.; Roscioli, C.; Pozzoni, F.; Guzzella, L. In situ bioavailability of DDT and Hg in sediments of the Toce River (Lake Maggiore basin, Northern Italy): Accumulation in benthic invertebrates and passive samplers. *Environ. Sci. Pollut. Res.* **2016**, *23*, 10542–10555. [CrossRef]

14. Tremolada, P.; Villa, S.; Bazzarin, P.; Bizzotto, E.; Comolli, R.; Vighi, M. POPs in mountain soils from the Alps and Andes: Suggestions for a 'precipitation effect' on altitudinal gradients. *Water Air Soil Pollut.* **2008**, *188*, 93–109. [CrossRef]

15. Hung, H.; Kallenborn, R.; Breivik, K.; Su, Y.; Brorström-Lundén, E.; Olafsdottir, K.; Thorlacius, J.M.; Leppänen, S.; Bossi, R.; Skov, H.; et al. Atmospheric monitoring of organic pollutants in the Arctic under the Arctic Monitoring and Assessment Programme (AMAP): 1993–2006. *Sci. Total Environ.* **2010**, *408*, 2854–2873. [CrossRef]

16. van Drooge, B.; Garriga, G.; Koinig, K.; Psenner, R.; Pechan, P.; Grimalt, J. Sensitivity of a remote alpine system to the Stockholm and LRTAP regulations in POP emissions. *Atmosphere* **2014**, *5*, 198–210. [CrossRef]

17. Guzzella, L.; Salerno, F.; Freppaz, M.; Roscioli, C.; Pisanello, F.; Poma, G. POP and PAH contamination in the southern slopes of Mt. Everest (Himalaya, Nepal): Long-range atmospheric transport, glacier shrinkage, or local impact of tourism? *Sci. Total Environ.* **2016**, *544*, 382–390. [CrossRef]

18. Zhang, C.; Zhong, R.; Wang, Z.; Montaña, C.G.; Song, Y.; Pan, K.; Wang, S.; Wu, Y. Intra-annual variation of zooplankton community structure and dynamics in response to the changing strength of bio-manipulation with two planktivorous fishes. *Ecol. Indic.* **2019**, *101*, 670–678. [CrossRef]

19. Perga, M.E.; Gerdeaux, D. Seasonal variability in the δ13C and δ^{15}N values of the zooplankton taxa in two alpine lakes. *Acta Oecol.* **2006**, *30*, 69–77. [CrossRef]

20. Hobson, K.A.; Fisk, A.; Karnovsky, N.; Holst, M.; Gagnon, J.M.; Fortier, M. A stable isotope (δ^{13}C, δ^{15}N) model for the North Water food web: Implications for evaluating trophodynamics and the flow of energy and contaminants. *Deep Sea Res. Part II Top. Stud. Oceanogr.* **2002**, *49*, 5131–5150. [CrossRef]

21. Bettinetti, R.; Quadroni, S.; Manca, M.; Piscia, R.; Volta, P.; Guzzella, L.; Roscioli, C.; Galassi, S. Seasonal fluctuations of DDTs and PCBs in zooplankton and fish of Lake Maggiore (Northern Italy). *Chemosphere* **2012**, *88*, 344–351. [CrossRef]

22. Romero-Romero, S.; Herrero, L.; Fernández, M.; Gómara, B.; Acuña, J.L. Biomagnification of persistent organic pollutants in a deep-sea, temperate food web. *Sci. Total Environ.* **2017**, *605*, 589–597. [CrossRef]

23. Peterson, B.J.; Fry, B. Stable isotopes in ecosystem studies. *Annu. Rev. Ecol. Syst.* **1987**, *18*, 293–320. [CrossRef]

24. Post, D.M. Using stable isotopes to estimate trophic position: Models, methods, and assumptions. *Ecology* **2002**, *83*, 703–718. [CrossRef]

25. Minagawa, M.; Wada, E. Stepwise enrichment of ^{15}N along food chains: Further evidence and the relation between ^{15}N and animal age. *Geochim. Cosmochim. Acta* **1984**, *48*, 1135–1140. [CrossRef]

26. Inger, R.; Bearhop, S. Applications of stable isotope analyses to avian ecology. *IBIS* **2008**, *150*, 447–461. [CrossRef]

27. Piscia, R.; Boggio, E.; Bettinetti, R.; Mazzoni, M.; Manca, M. Carbon and Nitrogen Isotopic Signatures of Zooplankton Taxa in Five Small Subalpine Lakes along a Trophic Gradient. *Water* **2018**, *10*, 94. [CrossRef]

28. Guzzella, L.M.; Novati, S.; Casatta, N.; Roscioli, C.; Valsecchi, L.; Binelli, A.; Parolini, M.; Solcà, N.; Bettinetti, R.; Manca, M.; et al. Spatial and temporal trends of target organic and inorganic micropollutants in Lake Maggiore and Lake Lugano (Italian-Swiss water bodies): Contamination in sediments and biota. *Hydrobiologia* **2018**, *824*, 271–290. [CrossRef]

29. Nizzetto, L.; Gioia, R.; Li, J.; Borgå, K.; Pomati, F.; Bettinetti, R.; Dachs, J.; Jones, K.C. Biological pump control of the fate and distribution of hydrophobic organic pollutants in water and plankton. *Environ. Sci. Technol.* **2012**, *46*, 3204–3211. [CrossRef]

30. Riva, C.; Binelli, A.; Parolini, M.; Provini, A. The case of pollution of Lake Maggiore: A 12-year study with the bioindicator mussel *Dreissena polymorpha*. *Water Air Soil Pollut.* **2010**, *210*, 75–86. [CrossRef]

31. Guilizzoni, P. Indagini su DDT e sostanze pericolose nell'ecosistema del Lago Maggiore. In *Rapporto Annuale 2012, Rapporto Finale 2008–2012*; Commissione Internazionale per la protezione delle acque italo-svizzere, Ed.; C.N.R.-I.S.E. Sede di Verbania: Verbania, Italy, 2013; p. 151.

32. Volta, P.; Tremolada, P.; Neri, M.C.; Giussani, G.; Galassi, S. Age-dependent bioaccumulation of organochlorine compounds in fish and their selective biotransformation in top predators from Lake Maggiore (Italy). *Water Air Soil Pollut.* **2009**, *197*, 193–209. [CrossRef]

33. de Bernardi, R.; Giussani, G.; Manca, M.; Ruggiu, D. Long-term dynamics of plankton communities in Lago Maggiore (N. Italy). *Verh. Int. Ver. Theor. Angew. Limnol.* **1988**, *23*, 729–733. [CrossRef]

34. Ruggiu, D.; Morabito, G.; Panzani, P.; Pugnetti, A. Trends and relations among basic phytoplankton characteristics in the course of the long-term oligotrophication of Lake Maggiore (Italy). In *Phytoplankton and Trophic Gradients*; Springer: Dordrecht, The Netherlands, 1998; pp. 243–257.

35. McCauley, E. The estimation of the abundance and biomass of zooplankton in samples. In *A Manual on Methods for the Assessment of Secondary Productivity in Fresh Waters*; Downing, J.A., Rigler, F.H., Eds.; Blackwell Scientific Pubblication: Oxford, UK, 1984; Volume 17, pp. 228–265.

36. Manca, M.; Comoli, P. Biomass estimates of freshwater zooplankton from length-carbon regression equations. *J. Limnol.* **2000**, *59*, 15–18. [CrossRef]

37. Zaret, T.M. *Predation and Freshwater Communities*; Yale Univ. Press: New Haven, CT, USA, 1980; p. 187.

38. Visconti, A.; Manca, M. Seasonal changes in the δ^{13}C and δ^{15}N signatures of the Lago Maggiore pelagic food web. *J. Limnol.* **2011**, *70*, 263–271. [CrossRef]

39. Visconti, A.; Volta, P.; Fadda, A.; Di Guardo, A.; Manca, M. Seasonality, littoral versus pelagic carbon sources, and stepwise ^{15}N-enrichment of pelagic food web in a deep subalpine lake: The role of planktivorous fish. *Can. J. Fish. Aquat. Sci.* **2013**, *71*, 436–446. [CrossRef]

40. Fadda, A.; Manca, M.; Camin, F.; Ziller, L.; Buscarino, P.; Mariani, M.; Padedda, B.M.; Sechi, N.; Virdis, T.; Lugliè, A. Study on the suspended particulate matter of aMediterranean artificial lake (Sos Canales Lake) using Stable Isotope Analysis of carbon and nitrogen. *Ann. Limnol. Int. J. Limnol.* **2016**, *52*, 401–412. [CrossRef]

41. Giussani, G.; de Bernardi, R. Food selectivity in *Coregonus* sp. of Lago Maggiore: An energetical approach. *Mem. Ist. Ital. Idrobiol.* **1977**, *34*, 121–130.

42. Coulas, R.A.; Macisaac, H.J.; Dunlop, W. Selective predation on an introduced zooplankter (*Bythotrephes cederstroemi*) by lake herring (*Coregonus artedii*) in Harp Lake, Ontario. *Freshwater Biol.* **1998**, *40*, 343–355. [CrossRef]

43. Manca, M.; Vijverberg, J.; Polishchuk, L.V.; Voronov, D.A. Daphnia body size and population dynamics under predation by invertebrate and fish predators in Lago Maggiore: An approach based on contribution analysis. *J. Limnol.* **2008**, *67*, 15–21. [CrossRef]

44. Bonecker, C.C.; Azevedo, F.D.; Simões, N.R. Zooplankton body-size structure and biomass in tropical floodplain lakes: Relationship with planktivorous fishes. *Acta Limnol. Bras.* **2011**, *23*, 217–228. [CrossRef]

45. Mackas, D.L. Seasonal cycle of zooplankton off southwestern British Columbia: 1979–89. *Can. J. Fish. Aquat. Sci.* **1992**, *49*, 903–921. [CrossRef]

46. Kenitz, K.M.; Visser, A.W.; Mariani, P.; Andersen, K.H. Seasonal succession in zooplankton feeding traits reveals trophic trait coupling. *Limnol. Oceanogr.* **2017**, *62*, 1184–1197. [CrossRef]

47. Manca, M.; Rogora, M.; Salmaso, N. Inter-annual climate variability and zooplankton: Applying teleconnection indices to two deep subalpine lakes in Italy. *J. Limnol.* **2015**, *74*, 123–132. [CrossRef]

48. Manca, M.; DeMott, W.R. Response of the invertebrate predator *Bythotrephes* to a climate-linked increase in the duration of a refuge from fish predation. *Limnol. Oceanogr.* **2009**, *54*, 506–512. [CrossRef]

49. Manca, M. Invasions and re-emergences: An analysis of the success of *Bythotrephes* in Lago Maggiore (Italy). *J. Limnol.* **2011**, *70*, 76–82. [CrossRef]

50. Fadda, A.; Rawcliffe, R.; Padedda, B.M.; Luglié, A.; Sechi, N.; Camin, F.; Ziller, L.; Manca, M. Spatio-temporal dynamics of C and N isotopic signature of zooplankton: A seasonal study on a man-made lake in the Mediterranean region. *Ann. Limnol. Int. J. Limnol.* **2014**, *50*, 279–287. [CrossRef]

51. Zohary, T.; Erez, J.; Gophen, M.; Berman-Frank, I.; Stiller, M. Seasonality of stable carbon isotopes within the pelagic food web of Lake Kinneret. *Limnol. Oceanogr.* **1994**, *39*, 1030–1104. [CrossRef]

52. Leggett, M.F.; Servos, M.R.; Hesslein, R.; Johannsson, O.; Millard, E.S.; Dixon, D.G. Biogeochemical influences on the carbon isotope signatures of Lake Ontario biota. *Can. J. Fish. Aquat. Sci.* **2009**, *56*, 2211–2218. [CrossRef]

53. Leggett, M.F.; Johannsson, O.; Hesslein, R.; Dixon, D.G.; Taylor, W.D.; Servos, M.R. Influence of inorganic nitrogen cycling on the δ^{15}N of Lake Ontario biota. *Can. J. Fish. Aquat. Sci.* **2000**, *57*, 1489–1496. [CrossRef]

54. Caroni, R.; Free, G.; Visconti, A.; Manca, M. Phytoplankton functional traits and seston stable isotopes signature: A functional-based approach in a deep, subalpine lake, Lake Maggiore (N. Italy). *J. Limnol.* **2012**, *71*, 84–94. [CrossRef]

55. Geller, W.; Müller, H. The filtration apparatus of Cladocera: Filter mesh-sizes and their implications on food selectivity. *Oecologia* **1981**, *49*, 316–321. [CrossRef]

56. France, R.L.; Peters, R.H. Ecosystem differences in the trophic enrichment of ^{13}C in aquatic food webs. *Can. J. Fish. Aquat. Sci.* **1997**, *54*, 1255–1258. [CrossRef]

57. Tewfik, A.; Bell, S.S.; McCann, K.S.; Morrow, K. Predator diet and trophic position modified with altered habitat morphology. *PLoS ONE* **2016**, *11*, e0147759. [CrossRef]

58. Adams, T.S.; Sterner, R.W. The effect of dietary nitrogen content on trophic level ^{15}N enrichment. *Limnol. Oceanogr.* **2000**, *45*, 601–607. [CrossRef]

59. Gulati, R.; Demott, W. The role of food quality for zooplankton: Remarks on the state-of-the-art, perspectives and priorities. *Freshw. Biol.* **1997**, *38*, 753–768. [CrossRef]

60. Elser, J.J.; Kyle, M.; Learned, J.; McCrackin, M.L.; Peace, A.; Steger, L. Life on the stoichiometric knife-edge: Effects of high and low food C: P ratio on growth, feeding, and respiration in three *Daphnia* species. *Inland Waters* **2016**, *6*, 136–146. [CrossRef]

61. Leoni, B. Zooplankton predators and prey: Body size and stable isotope to investigate the pelagic food web in a deep lake (Lake Iseo, Northern Italy). *J. Limnol.* **2017**, *76*, 85–93. [CrossRef]

62. Borgå, K.; Fisk, A.T.; Hoekstra, P.F.; Muir, D.C.G. Biological and chemical factors of importance in the bioaccumulation and trophic transfer of persistent organochlorine contaminants in arctic marine food webs. *Environ. Toxicol. Chem.* **2004**, *23*, 2367–2385. [CrossRef]

63. Walters, D.M.; Jardine, T.D.; Cade, B.S.; Kidd, K.A.; Muir, D.C.G.; Leipzig-Scott, P. Trophic magnification of organic chemicals: A global synthesis. *Environ. Sci. Technol.* **2016**, *50*, 4650–4658. [CrossRef]

Article

Response of Cladocera Fauna to Heavy Metal Pollution, Based on Sediments from Subsidence Ponds Downstream of a Mine Discharge (S. Poland)

Agnieszka Pociecha [1,*], Agata Z. Wojtal [1], Ewa Szarek-Gwiazda [1], Anna Cieplok [2], Dariusz Ciszewski [3] and Andrzej Kownacki [1]

1 Department of Freshwater Biology, Institute of Nature Conservation, Polish Academy of Sciences, Adama Mickiewicza Av. 33, 31-120 Kraków, Poland; wojtal@iop.krakow.pl (A.Z.W.); szarek@iop.krakow.pl (E.S.-G.); kownacki@iop.krakow.pl (A.K.)
2 Department of Hydrobiology, University of Silesia, Bankowa St. 9, 40-007 Katowice, Poland; anna.cieplok@us.edu.pl
3 Faculty of Geology, Geophysics and Environmental Protection, University of Science and Technology AGH, Adama Mickiewicza Av. 30, 30-059 Kraków, Poland; ciszewski@geol.agh.edu.pl
* Correspondence: pociecha@iop.krakow.pl

Received: 5 March 2019; Accepted: 17 April 2019; Published: 18 April 2019

Abstract: Mining is recognized to deeply influence invertebrate assemblages in aquatic systems, but different invertebrates respond in different ways to mining cessation. Here, we document the response of the cladoceran assemblage of the Chechło river, S. Poland (southern Poland) to the cessation of Pb-Zn ore mining. The aquatic system includes the river and associated subsidence ponds in the valley. Some ponds were contaminated during the period of mining, which ceased in 2009, while one of the ponds only appeared after mining had stopped. We used Cladocera to reveal how the cessation of mine water discharge reflected on the structure and density of organisms. A total of 20 Cladocera taxa were identified in the sediment of subsidence ponds. Their density ranged from 0 to 109 ind./1 cm^3. The concentrations of Zn, Cd, Cu and Pb were much higher in sediments of the ponds formed during peak mining than in the ponds formed after the closure of the mine. Statistical analysis (CCA) showed that *Alonella nana*, *Alona affinis*, *Alona* sp. and *Pleuroxus* sp. strongly correlated with pond age and did not tolerate high concentrations of heavy metals (Cu and Cd). This analysis indicated that the rate of water exchange by the river flow and the presence of aquatic plants, affect species composition more than pond age itself.

Keywords: Zn-Pb maine; subfossil; Cladocera; heavy metals; CCA analyses; anthropogenic impact

---

1. Introduction

Cladocera (Crustacea) are an important component of the small invertebrates living in freshwater. They occur in different environmental conditions; in shallow and deep waters; in alkaline, neutral and acidic conditions, and among environmental gradients. Their sensitivity to environmental conditions makes cladocerans good indicators for a wide range of environmental variables [1–3]. They respond rapidly to heavy metals and to other physical-chemical variables affected by a discharge of mine waters [1,3,4]. These environmental changes are recorded in bottom sediments and may be reconstructed from subfossil skeletal remains, preserved in the bottom layer of mud. For this reason, they are useful for the reconstruction of anthropogenic disturbances [4–7].

Extensive investigations conducted at mining sites throughout the world have recognized metal-contaminated wastes as the most important source of heavy metal contaminants. Lead and zinc (Pb, Zn) ores have been extracted for centuries in mines all over the world. In the second half of 20th century many mines have been closed, due to ore exhaustion, but leaving contaminated sediments as sources of metals leaking to the aquatic environment. The Chechło River valley in S. Poland is an example of an aquatic system affected by Pb-Zn mining, in which sediments of mining-related subsidence ponds are contaminated with heavy metals (Pb, Zn, Cd). For about 40 years, until 2009, the Chechło River has experienced pollution from a mine active since 1968. The lower reach of the Chechło River is affected by longwall coal mining. This resulted in subsidence of the river valley floor and emerging of two basins ponded with water. One of the basins started to subside at the beginning of the 1990s whereas subsidence of the other, 1 km upstream, started at about 2007. In both basins subsidence reached about 1.7–2.0 m resulting in shallow water bodies with plant succession much more advanced in the older than in the youngest basin [8]. Currently, the river is recovering, but many channels and floodplain locations still preserve the sediments that accumulated there during the mining time [8]. This allows the assessment of the impact of contaminated waters of the river and ponds on Cladocera communities based on the analysis of their remains in the sediments.

Our aim was to document changes in the species composition of Cladocera in response to mining cessation recorded in bottom sediments of subsidence ponds situated on the Chechło River floodplain (southern Poland). Our hypothesis assumes that regeneration of the cladoceran community was limited by high heavy metal concentrations (mainly heavy metals, such as: Cd, Pb, Zn and Cu) in the bottom sediments of ponds in the middle reach of the river valley. We analyzed the Cladocera assemblage and heavy metal concentrations in sediment cores from ponds which emerged (1) during the period of Zn and Pb ore extraction and (2) after the mine closed. The present study is the key to understanding the rate of ecosystem response and factors controlling ecosystem recovery from heavy metal contamination. We address this by analyzing Cladocera remains preserved in subsidence ponds sediment and correlate them with records of metal contamination from Pb-Zn ore mining.

2. Materials and Methods

2.1. Study Area

Our work was conducted in the middle course of the Chechło River, downstream of the point of mine waters discharge. The quality of the Chechło River was affected for about 50 years by the zinc and lead ore mining and by the other industrial and municipal sewage effluents from two towns Trzebinia and Chrzanów located in the middle reach of the river [8,9]. We distinguished two research areas about 1 km apart: Subsidence pond emerged after the closure of the mine (UP) and subsidence ponds ponded during the peak of the ore exploitation (DOWN) (Figure 1). All water bodies have been impacted by heavy metals contamination. They are small, with areas ranging from 0.5 to ca. 5 ha, whereas their average depth is about 1–2 meters. A part (ca. 20–50%) of the basins is overgrown with macrophytes (Figure 1).

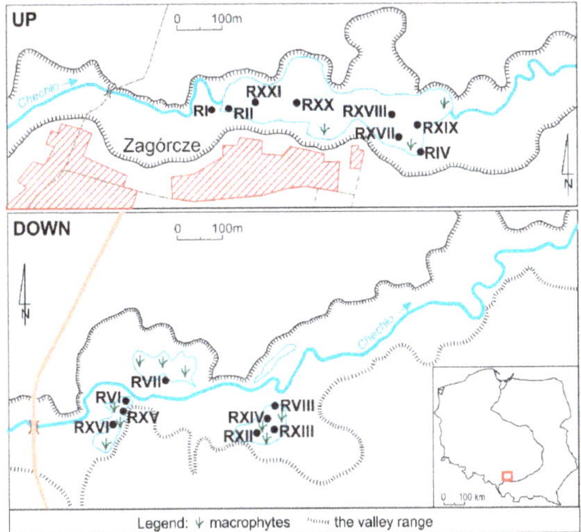

Figure 1. Sampling area (UP: Subsidence pond after the closure of the mine; cores: RI, RII, RIV, RXVII, RXVIII, RXIX, RXX, RXXI; DOWN: Subsidence ponds during peak exploitation; cores: RVI, RVII, RVIII, RXII, RXIII, RXIV, RXV, RXVI).

2.2. Sampling and Measurements

The concentrations of Cd, Pb, Zn and Cu and remains of aquatic organisms were analyzed in sediment cores from eight sites in pond formed after the closure of the mine (UP, cores: RI, RII, RIV, RXVII, RXVIII, RXIX, RXX, RXXI) and in eight sites of ponds existing during the peak of Zn and Pb ore exploitation (DOWN, cores: RVI, RVII, RVIII, RXII, RXIII, RXIV, RXV, RXVI) (in total 16 cores) (Figure 1). Cores were recovered using a Multisampler piston corer with diameter 4.5 cm (Eijkelkamp, Giesbeek, Netherlands). Cores were divided in the field depending on sediment lithology when macroscopic changes in color or grain-size of sediments were observed whereas profiles with no distinct strata were divided into 10 cm long subsamples and homogenized by mixing. Length of the retrieved profiles varied from 15 to 70 cm. Most of the sediments were composed of unstratified muds with a high content of organic matter which rested on the sandy sediments of the former floodplain. UP and DOWN pond cores were divided into sections with a thickness of 5–15 cm depending on the lithology of the sediment (see Figure 2).

Sediment samples for heavy metal analysis were dried at 105 °C and sieved through a 0.063 mm sieve. Then they (0.5 g) were digested with 10 cm^3 of 65% HNO_3 and 2 cm^3 of 30% H_2O_2 (both analytical grade) using a microwave digestion technique [6]. The Cd, Pb, Zn and Cu concentrations were determined using a flame atomic absorption spectrometer (F-AAS). Metal analyses were performed according to (standard certified) analytical quality control procedures.

Subfossil Cladocera preparation was conducted, according to Frey [5]. One centimeter cubed of fresh homogenized sediment was taken from the separation section from each core for cladoceran analysis. After elimination of carbonates using HCl, each sample was boiled for half an hour in a 10% KOH solution. A magnetic stirrer was used for dispersion. After boiling and washing, the remains were sieved through a 35 μm sieve. The residuum was stored in 10 ml of water with glycerine and safranine, in polycarbonate test-tubes in a fridge. Temporary slides were used for identification and to count the cladoceran remains. Taxa were identified and counted at 200 or 400× magnification under a Nikon 50i microscope. All skeletal parts were counted: headshields, shells, postabdomens, postabdominal claws, ephippia and filtering combs. The most abundant body part for each taxon was chosen to represent the number of individuals. The results of qualitative and quantitative analyses are presented in diagrams, in which an absolute number of specimens was calculated for 1 cm^3 sediment volume. Identification of the species was based on Frey [10] and Szeroczyńska and Sarmaja-Korjonen [11].

2.3. Statistical Analyses

Shannon's diversity index

The species diversity of Cladocera was evaluated using Shannon's diversity index. The analysis of species diversity was carried out using the MultiVariate Statistical Package (MVSP) 3.1 program [12].

Pearson's correlation

In order to investigate the relationship between heavy metals content in sediments and cladocerans, Pearson's correlation was used with the Statistica 13 program.

Cluster analysis

In order to compare qualitatively the Cladocera assemblages cluster analysis was made using the UPGMA method. The hierarchical classification was obtained using the MVSP 3.1 program [12].

Correspondence analysis

Indirect and direct ordination techniques were used to assess the impact of sediment environmental factors on the cladoceran community: Unimodal techniques—correspondence analysis (CA) and canonical correspondence analysis (CCA).

The analysis was carried out after a prior detrended correspondence analysis (DCA) to verify the nature of the data, based on the length of the gradient expressed in standard deviation (SD) units. During the DCA, CA and CCA analysis, the data was log-transformed [ln (x + 1)] and centered. During the CCA analysis, a forward selection was carried out to assess the role of environmental variables in shaping the structure of cladoceran communities. The used analyses were performed on Cladocera data and subsidence sediments samples to identify the changes in the ponds and to show the relationships between the environmental variables and the distribution of the Cladocera.

An evaluation of their statistical significance, as well as the statistical significance of canonical axes, was made using the Monte Carlo permutation test for 499 repetitions. The analyses were performed using CANOCO for Windows 4.5 program [13].

3. Results

3.1. Sediment—Heavy Metals Contamination Analysis

The heavy metal concentrations in sediment cores from studied UP and DOWN subsidence ponds ranged as follows: Zn, 0.5–23.1 mg g^{-1}; Cd, 6.1–612.3 µm g^{-1}; Pb, 0.3–10.2 mg g^{-1}; Cu, 21.4–397.0 µm g^{-1} (Figure 2). The sediment from the UP subsidence pond exhibited lower differences in heavy metal concentrations among particular strata of each individual core (with the exception of the core RXVIII). Only in core RXVIII (length 20 cm) collected from the lower part of the UP subsidence pond, they were considerable with maximum concentrations of the metals of all cores in UP subsidence pond were found in the bottom strata. The cores from DOWN subsidence ponds, usually of a larger length, exhibited higher variability in heavy metal concentrations which each core, comparing to UP subsidence pond. These differences were particularly large (up to 24 times for Zn, 19 times for Cd, 13 times for Pb, and 7 times for Cu) in 30 cm long cores RVIII and RXII. In the deepest section of core RXII the lowest metal concentrations for all DOWN subsidence ponds were found. Higher metal concentrations usually occurred in the middle or the surface strata of cores from DOWN subsidence ponds and reflected the period of most intensive mining of Zn and Pb ores at the end of the 20th century. The maximum concentrations of heavy metals in cores from DOWN subsidence ponds were several times higher (Zn, Cu and Pb 2–4, Cd eight times) than from the UP subsidence pond (with the exception of RXVIII core).

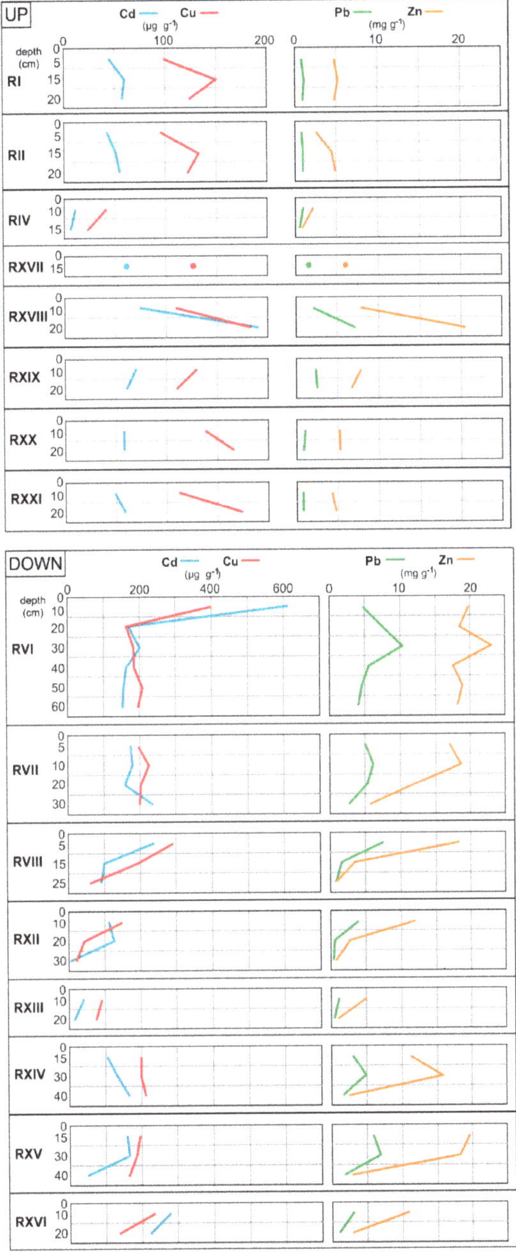

Figure 2. Heavy metal concentrations in the sediment cores in subsidence ponds (UP: Subsidence pond after the closure of the mine; cores: RI, RII, RIV, RXVII, RXVIII, RXIX, RXX, RXXI; DOWN: Subsidence ponds during peak exploitation; cores: RVI, RVII, RVIII, RXII, RXIII, RXIV, RXV, RXVI).

3.2. Subfossil Cladocera Analysis

Cladoceran remains in the sediments of the subsidence ponds represented 20 taxa and four families (Chydoridae, Eurycercidae, Bosminidae, Daphniidae). Chydoridae was the most diverse family (13 taxa), whereas the other families were represented by fewer taxa (Eurycercidae: 1; Bosminidae: 2; Daphniidae: 4). In the pond which emerged after the mine was closed (UP) 12 taxa were found while in ponds existing during the mine operation (DOWN) 17 taxa were found. The number of cladoceran taxa in each sediment core ranged from 1 to 13 (Table 1).

Table 1. Number of taxa and range (min.–max.) of total Cladocera specimens in cores of subsidence ponds.

Subsidence Ponds Area	Core Number	Number of Taxa in Core	Range (Min.–Max.) of Cladocera Individuals in 1 cm^3 of Sediment
UP (subsidence pond formed after the closure of the mine)	RI	5	1–6
	RII	3	0–2
	RIV	7	4–22
	RXVII	5	37
	RXVIII	3	1–12
	RXIX	4	0–15
	RXX	5	3–6
	RXXI	1	1
DOWN (subsidence ponds created during the peak of Zn and Pb ore exploitation)	RVI	3	3–8
	RVII	2	1–6
	RVIII	6	1–50
	RXII	13	23–109
	RXIII	3	22–101
	RXIV	6	38–86
	RXV	6	20–93
	RXVI	4	3–76

The pond existing during the mine operation had the highest diversity of Cladocera taxa in comparison to remaining subsidence ponds. Total density of Cladocera individuals in sediments (ind./1 cm^3) varied from 0 in UP subsidence pond formed after the closure of the mine (RII; RXIX) to over 100 individuals in DOWN ponds existed during mine exploitation (RXII; RXIII) (Table 1). In the UP pond Cladocera assemblage was rather poor and its density was not higher than 37 ind./1 cm^3. This result is probably related to the fact that in subsidence ponds where sediment cores were studied the river water flowed through the center of the pond (Figure 1). The DOWN subsidence ponds are supplied with river water by side channels, and characterized by stable water with well-developed macrophyte vegetation (Figure 1). *Chydorus sphaericus* was the species present in all studied ponds except one, with the presence of the only one species *Bosmina longirostris* (RXXI; Figure 3). *Ch. sphaericus* is species occurring in pelagic and littoral zones, and its high density is characteristic for eutrophic and polluted water. The highest densities of this species were observed mainly in the lowest layers of core sediments with more than 40 ind./1 cm^3 in subsidence ponds existed during the peak of Zn and Pb ore exploitation (DOWN). The smaller density of the discussed species, below 10 ind./ 1 cm^3 were observed in cores of the pond formed after the mine was closed (UP) (Figures 3 and 4).

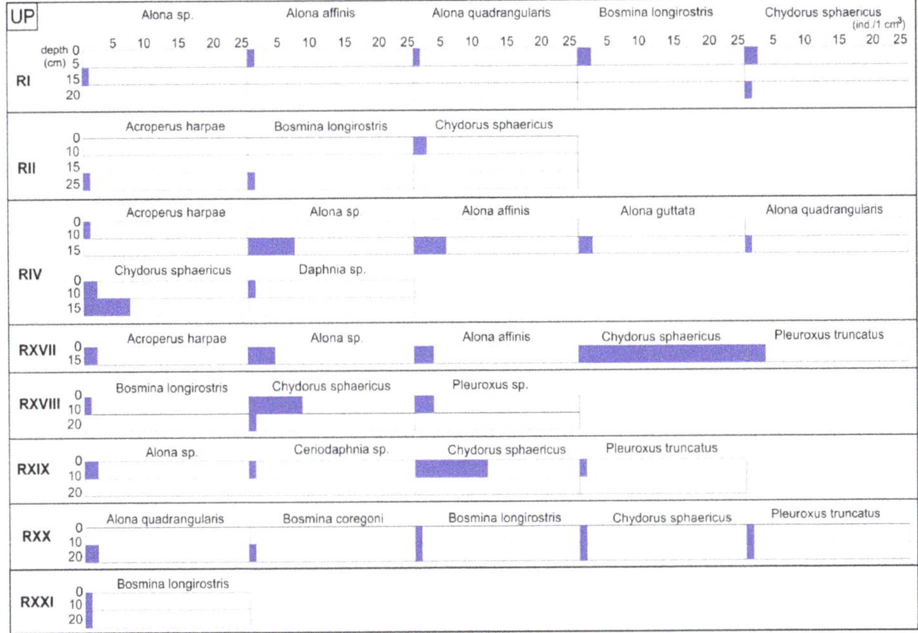

Figure 3. Diagram of the absolute number of Cladocera individuals in 1 cm^3 of sediment from subsidence pond formed after the closure of the mine (UP).

We found seven littoral taxa in the UP pond characterized by through-flow of river water and 12 littoral taxa in DOWN ponds characterized by stagnant water conditions and littoral zone overgrown with macrophytes. In all ponds were found littoral taxa belonging to six genera: *Alona, Alonella, Acroperus, Pleuroxus, Eurycercus* and *Graptoleberis.* Only in one subsidence pond was the highest number of littoral taxa (9 taxa—RXII: DOWN) with density reaching over 30 ind./1 cm^3 (Figure 3).

Figure 4. Diagram of the absolute number of Cladocera individuals in 1 cm^3 of sediment from subsidence ponds existed during mine exploitation (DOWN).

3.3. Relationship between Cladocera and Heavy Metal Concentration

The highest Shannon (H′) diversity rates (up to <2) occurred only in one core RXII in the subsidence pond formed during mine operation (DOWN). Moreover, the high H′ diversity rates were observed in five cores (up to <1) in the water ponds formed after of mine cessation (UP) (Table 2).

Table 2. Shannon (H′) diversity index value in investigated sediment cores. (UP—subsidence ponds formed after the closure of the mine; DOWN—subsidence ponds created during the peak of Zn and Pb ore exploitation).

Core (UP)	Index Value	Core (DOWN)	Index Value
RI	1.49	RVI	0.35
RII	1.04	RVII	0.58
RIV	1.61	RVIII	0.66
RXVII	1.07	RXII	2.12
RXVIII	0.79	RXIII	0.73
RXIX	0.85	RXIV	0.80
RXX	1.58	RXV	0.57
RXXI	0.00	RXVI	0.68

There were statistically significant correlations between the abundance of particular species and the heavy metal concentrations in sediment cores of UP and DOWN ponds. In the most contaminated sediments (DOWN ponds), a negative correllation was found between Cu concentration and occurrence of the genera *Alona*, *Alonella*, *Daphnia* and *Graptoleberis*; also, a negative correlation was revealed between the density of *Alona* sp. and Zn and Pb in sediments. Negative correlations with Cu concentration was found in less polluted sediments (UP pond), but only for *Alona* and *Daphnia* genera. The cladoceran *Alona* and *Daphnia* seem to be more sensitive to heavy metals contamination than the other investigated Cladocera (Table 3). Our results showed the negative impact of metal mining on cladocerans, as well as positive correlations between taxa.

Cluster analysis divided the studied cores into two groups. This was mainly caused by the different dominance structure of the identified species of Cladocera at sampling sites. The first group constitutes cores from ponds that existed during the mine exploitation activity; the second group included water bodies created after the mine was closed. In the first group, site RXII clearly stood out, due to the fact that it was characterised by the highest density and variety of Cladocera, as well as the highest values of Shannon's diversity index (H′ = 2.12). In the second group apart from cores from UP pond, also cores RVI and RVII from DOWN ponds were included (Figure 5). Such classification was likely to be related to the small diversity and density of Cladocera and low values of Shannon's diversity index (H′ = 0.35 and 0.58 respectively) in the cores RVI and RVII.

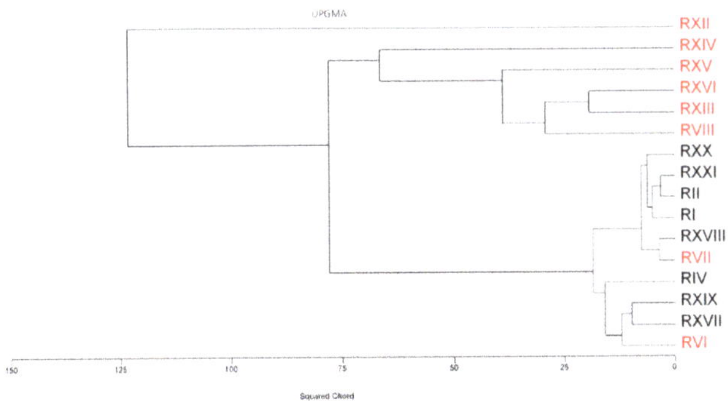

Figure 5. Dendrogram of faunistic similarities for Cladocera fauna in the studied sediment cores (black color: Cores from pond, created after mine was closed (UP); red color: Cores from ponds existed during mine exploitation (DOWN)).

To show the statistical importance and the associations of Cladocera assemblages with particular types of ponds correspondence analysis (CA) was used. The results showed that the first axis explains 25.4%, while the second 17% of the variability of the occurrence of the species of Cladocera at the sampling sites. The first ordination axis, with its eigenvalue $\lambda = 0.292$, differentiated sampling sites and clearly identified the RXII core. The core was characterized by the highest species diversity of Cladocera compared to other studied cores.

The gradient represented by the second ordination axis, with its eigenvalue $\lambda = 0.195$, clearly grouped the cores in two groups: cores from ponds existing during mine exploitation, and ponds created after its closing. In the upper part of the diagram was placed RXX and RXXI cores (created after mining stopped). In this part of the diagram there are also other cores of the same group: RI, RII, RXVIII and RXIX. In reservoirs mentioned above, a small number of species and at low density was found (Figures 3 and 4).

The distribution of the species in the ordination space of the diagram shows that they occur in particular cores. *Daphnia cucullata, Alonella nana, Graptoleberis testudinaria* and *Alonella excisa* were located in the right part of the diagram and were associated with the RXII core (Figure 6).

Table 3. Pearson's correlation—significant relationships between heavy metals concentrations and density of Cladocera (A: UP—subsidence ponds formed after the closure of the mine; B: DOWN—subsidence ponds created during the peak of Zn and Pb ore exploitation).

A Variables	Cu	Acroperus harpae	Alona sp.	Alona affinis	Alona quadrangularis	Chydorus sphaericus
Alona affinis	−0.53 p = 0.028		0.94 p = 0.000			
Alona guttata	−0.60 p = 0.01		0.83 p = 0.000	0.84 p = 0.000		
Bosmina coregoni					0.80 p = 0.000	
Chydorus sphaericus		0.65 p = 0.004	0.62 p = 0.008	0.57 p = 0.018		
Daphnia sp.	−0.48 p = 0.049					
Pleuroxus truncatus		0.65 p = 0.005				0.82 p = 0.000

B Variables	Zn	Pb	Cu	Alona sp.	Alona affinis	Alonella excisa	Alona quadrangularis	Alona rectangula	Alonella nana	Chydorus sphaericus	Eurycercus lamellatus	Pleuroxus truncatus
Alona sp.	−0.40 p = 0.042	−0.41 p = 0.034	−0.45 p = 0.021									
Alona affinis			−0.48 p = 0.012	0.66 p = 0.00								
Alonella excisa				0.71 p = 0.00	0.93 p = 0.00							
Alona guttata				0.64 p = 0.00								
Alona quadrangularis			−0.41 p = 0.037									
Alona rectangula				0.73 p = 0.00	0.90 p = 0.00	0.97 p = 0.00						
Alonella nana			−0.40 p = 0.041				0.99 p = 0.00					
Daphnia cucullata			−0.41 p = 0.037						0.99 p = 0.00			
Eurycercus lamellatus										0.65 p = 0.000		
Graptoleberis testudinaria			−0.41 p = 0.037	0.69 p = 0.00	0.96 p = 0.00	0.98 p = 0.00		0.95 p = 0.000				
Pleuroxus truncatus										0.44 p = 0.023	0.41 p = 0.039	
Pleuroxus trigonellus										0.52 p = 0.006	0.43 p = 0.029	0.72 p = 0.000

Canonical correspondence analyses (CCA) showed the importance of heavy metal concentration in sediment cores on the distribution of Cladocera taxa. The results showed that the first axis explains 50.0%, and the second 21.9% of the variability of the occurrence of species. In the CCA ordination diagrams, the location of species in relation to axis I and II, as well as the intensity of changes of environmental variables are presented. The model of CCA is highly statistically significant (The Monte Carlo permutation test: First canonical axis F = 6.148, p = 0.002; all canonical axes F = 2.757, p = 0.002). The location of the species in the right part of the ordination diagram is clearly visible, which indicates their intolerance to higher concentrations of heavy metals (Cu and Cd) (Figure 7).

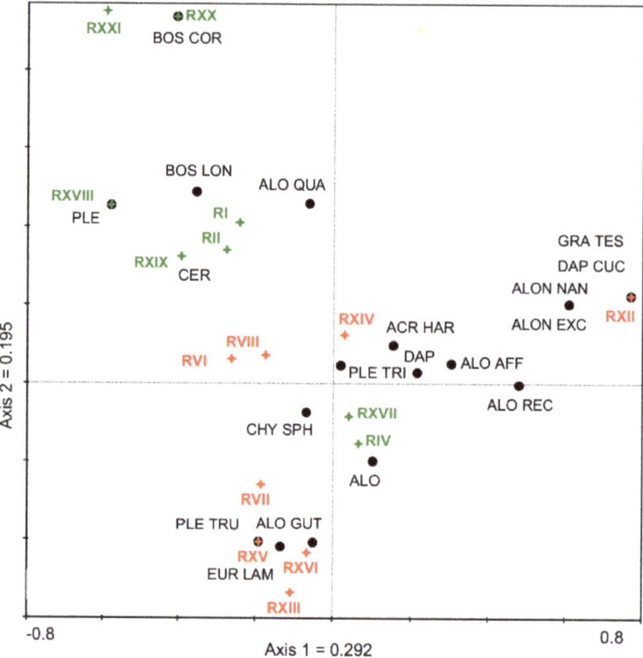

Figure 6. Correspondence analysis (CA) diagram for Cladocera fauna in sediment cores of subsidence ponds (ACR HAR: *Acroperus harpae*; ALO: *Alona* sp.; ALO AFF: *A. affinis*; ALO GUT: *A. guttata*; ALO QUA: *A. quadrangularis*; ALO REC: *A. rectangula*; ALON EXC: *Alonella excisa*; ALON NAN: *A. nana*; BOS COR: *Bosmina coregoni*; BOS LON: *B. longirostris*; CER: *Ceriodaphnia* sp.; CHY SPH: *Chydorus sphaericus*; DAP: *Daphnia* sp.; DAP CUC: *D. cucullata*; EUR LAM: *Eurycercus lamellatus*; GRA TES: *Graptoleberis testudinaria*; PLE: *Pleuroxus* sp.; PLE TRU: *P. truncatus*; PLE TRI: *P. trigonellus*). UP: Sediment cores are marked in green color; DOWN: Sediment cores are marked in a red color.

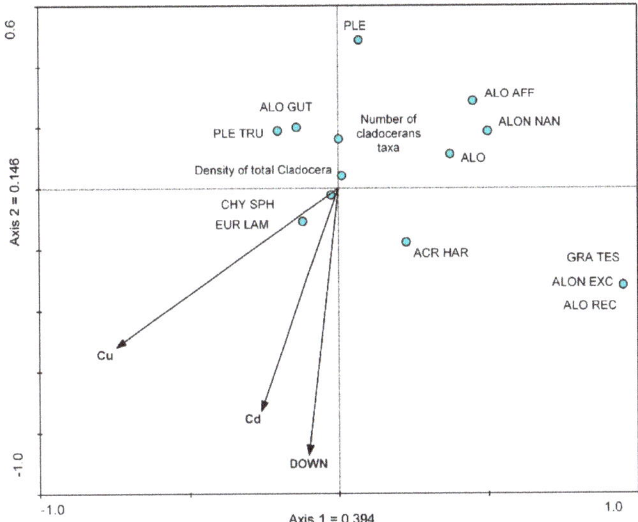

Figure 7. Canonical correspondence analysis (CCA) diagram for selected Cladocera species. The biplot illustrates the relationship between the heavy metal concentration in the sediment cores and Cladocera fauna. Taxa name code see Figure 6.

4. Discussion

The obtained results indicated strong contamination by Cd, Pb and Zn, and lower by Cu of sediment cores of UP and DOWN subsidence ponds of the Chechło River if compared to local geochemical background [8] and sediments of unpolluted areas [14]. The Cd, Pb and Zn concentrations were from tens to hundreds of times higher compared to those in the reference lake sediments (Figure 2) [14]. Such high heavy metals concentrations in the Chechło River sediments are typical for water ecosystems exposed to discharge from the Zn and Pb mines [6,15]. Metals discharged with mine waters and suspended sediments to the Chechło River during mine operation affected the DOWN subsidence ponds, what was reflected in maximal/very high Cd, Pb and Zn concentrations in the investigated sediment cores. The mine closure resulted in the drop of decrease in the river water contamination and lower metal concentrations [8] reflected in the large difference between contamination of the UP and DOWN subsidence ponds. Despite the progressive decrease of the most recent sediment contamination with Cd, Pb and Zn, their concentrations in all cores still exceed toxic levels for many aquatic organisms [16]. Moreover, Cu concentrations exceeded the threshold concentration (TEC, 86 μg g^{-1};16) for 80% of the samples of the UP and DOWN subsidence ponds sediments. The sediment-associated heavy metals might be released from the sediment to the overlying water under changing physical-chemical conditions particularly as a result of sediment resuspension or oxidation of surface sediment strata during floods after heavy rainfalls or during the melting period [17,18].

Anthropogenic pollution, such as heavy metal contamination, could be a detrimental and intense stressor of water ecosystems. The Cladocera are an important component of the zooplankton community in water bodies with both pelagic and littoral habitats. The Cladocera analysis raises a discussion about the influence of heavy metal pollution as a human activity. Cladocera assemblages from the subsidence ponds show significant differences in species composition, their densities and especially domination. In all subsidence ponds dominated *Ch. sphaericus* which is tolerant to water pollution and seems to be a 'specialist' in tolerance to a wide range of abiotic conditions [19]. However, some authors wrote that the long-term exposure of this species to Cu results in a reduction in the rate of population growth [7,20], but also Burton et al. [20] observed, that another species *Alona quadrangularis* is restricted to sites of low Cu contamination. Our results confirmed that *Ch. sphaericus* tolerated high concentration of Cu, but *A. quadrangularis* is not tolerant of this metal (Table 3; Figure 7). In both types of subsidence ponds (UP and DOWN) high concentration of Cu was not tolerated by species of *Alona, Alonella, Daphnia* and *Graptoleberis*. Sarma et al. [21] showed no single species of studied Cladocera (e.g., *Alona rectangula, Daphnia laevis, D. pulex, D. similis, Ceriodaphnia dubia*) was consistently sensitive for stress from heavy metal. The sensitivity of the Cladocera to the concentration of heavy metals in the environment is also dependent on other abiotic factors. Additionally, it is possible that the temporary presence of *B. longirostris*, dominant presence of *Ch. sphaericus* and littoral cladocerans, e.g., *Alona affinis, A. rectangula*, which prefer more fertile water, reflect a response of this group to changes in the local habitat (macrophyte abundance with more food for the cladocerans) [22]. In the investigated subsidence ponds, we also observed preferences of cladoceran assemblages to stagnant water overgrown by macrophytes.

Cluster analysis divided our sampling cores into two groups—one with sediment less polluted by the heavy metals that accumulated after mine cessation, and a second group of highly polluted sediments by heavy metal that existed during mine operation (Figure 2; Figure 5). The impact of metals on the Cladocera assemblage is reflected in the Shannon index (Table 2), where we observe a greater number of cores with the value of the index above 1 (in reservoir formed after the mine was closed), which may mean recovery an increase in biological diversity of organisms and improvement of the quality of the environment.

The total abundance of Cladocera remains decreased dramatically in the postindustrial sediments (UP subsidence pond) compared with the industrial sediments (DOWN subsidence ponds). The subfossil records for all Cladocera taxa investigated very clearly revealed that the onset of industrial activities dramatically altered the ecology of small subsidence ponds which emerged in the river valley. Based on existing knowledge of aquatic ecosystems and their responses to metal pollution, we attempt to interpret the observed industrial to postindustrial changes or trends in biota in the studied ponds. Despite more than 10 years since of heavy metal from Pb-Zn ore mining cessation, we observed little evidence of recovery in Cladocera assemblages, which we observed in the value of Shannon diversity index, as well as the occurrence of *Daphnia* remains in RXII, the more polluted subsidence pond existing during mine working. Doig et al. [4] observed similar results in Ross Lake sediment impacted by mining, metallurgical and municipal activities where Cladocera remains have low abundance in postindustrial times and shows minimal signs of recovering.

5. Conclusions

Sediment cores of ponds on the floodplain of River Chechło allowed reconstruction of human impact. Cores of UP and DOWN ponds show various contamination by Pb, Zn, Cd and Cu, the reaction of Cladocera assemblages by changing plankton species composition and their density observed in sediments.

The analysis of Cladocera remains allowed to reconstruct pre-mining condition in the studied subsidence ponds and also showed the environmental conditions of the studied ponds when the Zn-Pb main was operating and after when was closed. The occurrence of different ecological groups of cladocerans (diversity in taxa and density) in studied subsidence ponds may suggest differences in ponds types in past and present, and showed important changes in water quality during mine operation and after it was closed. In the sediment core layers with lower metal concentrations we observed a little recovery of Cladocera assemblages. In all subsidence ponds *Ch. sphaericus* was the dominant species, and most abundant in studied sediments. The dominance of less sensitive species confirmed communities adapted to chronic high metal contamination.

The sediment quality assessment will help to understand persistent adverse effects of sediment contamination on aquatic communities and to better recognize factors decisive on mechanisms of the succession of aquatic organisms in contaminated river systems.

Author Contributions: A.P. and D.C. were responsible for the research design. A.P., A.Z.W., E.S.-G. and D.C. analyzed the data, prepared drafted the text and figures. A.C. performed statistical analyses. All authors participated in discussions and editing.

Funding: This research was funded by National Science Center grant n. 2014/15/B/ST10/03862 and under Institute of Nature Conversation, Polish Academy of Sciences subvention.

Acknowledgments: We would like to thank Hanna Kuciel for their help with drawing works. We extend our thanks to Henri Dumont for the assistance with our text and in language editing. The researches are a part of a grant titled: Reconstruction and prognosis of response of a river system affected by Pb-Zn ore extraction to mining cessation.

Conflicts of Interest: The authors declare no conflict of interest.

References

1. Chen, G.; Dalton, C.; Taylor, D. Cladocera as indicators of trophic state in Irish lakes. *J. Paleolimnol.* **2010**, *44*, 465–481. [CrossRef]
2. Rumes, B.; Eggermont, H.; Verschuren, D. Distribution and faunal richness of cladocera in western Uganda crater lakes. *Hydrobiologia* **2011**, *676*, 39–56. [CrossRef]
3. Zawisza, E.; Zawiska, I.; Correa-Metrio, A. Cladocera community composition as a function of physicochemical and morphological parameters of dystrophic lakes in NE Poland. *Wetlands* **2016**, *36*, 1131–1142. [CrossRef]
4. Doig, L.E.; Schiffer, S.T.; Liber, K. Reconstructing the ecological impacts of eight decades of mining, metallurgical, and municipal activities on a small boreal lake in northern Canada. *Integr. Environ. Assess. Manag.* **2015**, *11*, 490–501. [CrossRef] [PubMed]
5. Frey, D.G. Cladocera analysis. In *Handbook of Holocene Palaeoecology and Palaeohydrology*; Berglund, B.E., Ed.; Wiley: New York, NY, USA, 1986; pp. 667–692.
6. Ciszewski, D.; Aleksander-Kwaterczak, U.; Pociecha, A.; Szarek-Gwiazda, E.; Waloszek, A.; Wilk-Woźniak, E. Small effects of a large sediments contamination with heavy metals on aquatic organisms in the vicinity of an abandoned lead and zinc mine. *Environ. Monit. Assess.* **2013**, *185*, 9825–9842. [CrossRef] [PubMed]
7. Koivisto, S.; Ketola, M.; Walls, M. Comparison of five cladoceran species in short-and long-term copper exposure. *Hydrobiologia* **1992**, *248*, 125–136. [CrossRef]
8. Ciszewski, D. The past and prognosis of mining cessation impact on river sediment pollution. *J. Soils Sediments* **2019**, *19*, 393–402. [CrossRef]
9. Michailova, P.; Ilkova, J.; Szarek-Gwiazda, E.; Kownacki, A.; Ciszewski, D. Genome instability in Chironomus annularius sensu Strenzke (Diptera, Chironomidae): A biomarker for assessment of the heavy metal contaminants in Poland. *J. Limnol.* **2018**, *7*, 15–24. [CrossRef]
10. Frey, D.G. The taxonomy and biogeography of the Cladocera. *Hydrobiologia* **1987**, *145*, 5–17. [CrossRef]
11. Szeroczyńska, K.; Sarmaja-Korjonen, K. *Atlas of Subfossil Cladocera from Central and Northern Europe*; Towarzystwo Przyjaciół Dolnej Wisły: Świecie, Polska, 2007; pp. 1–84.
12. Kovach, W.L. *A multivariate statistical package for Windows, version 3.1*; Kovach Computing Services: Pentraeth, Wales, UK, 1999; p. 133.
13. Ter Braak, C.J.F.; Šmilauer, P. *CANOCO v. 4.5*; Microcomputer Power-Ithaca: New York, NY, USA, 2002.
14. Forstner, U.; Salomons, W. Trace metal analysis in polluted sediments. *Environ. Technol. Lett.* **1980**, *1*, 1–494. [CrossRef]
15. Byrne, P.; Reid, I.; Wood, P.J. Sediment geochemistry of streams draining abandoned lead/zinc mines in central Wales: The Afon Twymyn. *J. Soils Sediments* **2010**, *10*, 683–697. [CrossRef]
16. EC and MENVIQ (Environment Canada and Ministère de l'Environnement du Québec). *Interim Criteria for Quality Assessment of St. Lawrence River Sediment*; Environment Canada: Ottawa, ON, Canada, 1992; ISBN 622-19849-2.
17. Calmano, W.; Von Der Kammer, F.; Schwartz, R. Characterization of redox conditions in soils and sediments: Heavy metals. In *Soil and Sediment Remediation*; Lens, P., Grotenhuis, T., Malina, G., Tabak, H., Eds.; IWA Publ.: London, UK, 2005; pp. 102–120.
18. Szarek-Gwiazda, E.; Ciszewski, D. Variability of heavy metal concentrations in waters of fishponds affected by the former lead and zinc mine in southern Poland. *Environ. Prot. Eng.* **2017**, *43*, 121–137. [CrossRef]

19. Belyaeva, M.; Deneke, R. Colonization of acidic mining lakes: Chydorus sphaericus and other Cladocera within a dynamic horizontal pH gradient (pH 3–7) in Lake Senftenberger See (Germany). *Hydrobiologia* **2007**, *594*, 97–108. [CrossRef]
20. Burton, S.M.; Rundle, S.D.; Jones, M.B. The relationship between trace metal contamination and stream meiofauna. *Environ. Pollut.* **2001**, *111*, 159–167. [CrossRef]
21. Sarma, S.S.S.; Peredo-Alvarez, V.M.; Nandini, S. Comparative study of the sensitivities of neonates and adults of selected cladoceran (Cladocera: Crustacea) species to acute toxicity stress. *J. Environ. Sci. Health Part A* **2007**, *42*, 1449–1452. [CrossRef] [PubMed]
22. Petera-Zganiacz, J.; Dzieduszyńska, D.A.; Twardy, J.; Pawłowski, D.; Płóciennik, M.; Lutyńska, M.; Kittel, P. Younger Dryas flood events: A case study from the middle Warta River valley (Central Poland). *Quat. Int.* **2015**, *386*, 55–69. [CrossRef]

MDPI
St. Alban-Anlage 66
4052 Basel
Switzerland
Tel. +41 61 683 77 34
Fax +41 61 302 89 18
www.mdpi.com

Water Editorial Office
E-mail: water@mdpi.com
www.mdpi.com/journal/water